Hair Analysis in Clinical and Forensic Toxicology

Hair Analysis in Clinical and Forensic Toxicology

Edited by

Pascal Kintz
X-Pertise Consulting, Oberhausbergen, France and Institute of Legal Medicine, University of Strasbourg, Strasbourg, France

Alberto Salomone
Centro Regionale Antidoping e di Tossicologia "A. Bertinaria", Orbassano (TO), Italy

Marco Vincenti
Dipartimento di Chimica, Università di Torino, Turin, Italy and Centro Regionale Antidoping e di Tossicologia "A. Bertinaria", Orbassano (TO), Italy

AMSTERDAM • BOSTON • HEIDELBERG • LONDON
NEW YORK • OXFORD • PARIS • SAN DIEGO
SAN FRANCISCO • SINGAPORE • SYDNEY • TOKYO

Academic Press is an imprint of Elsevier

Academic Press is an imprint of Elsevier
32 Jamestown Road, London NW1 7BY, UK
525 B Street, Suite 1800, San Diego, CA 92101-4495, USA
225 Wyman Street, Waltham, MA 02451, USA
The Boulevard, Langford Lane, Kidlington, Oxford OX5 1GB, UK

ISBN: 978-0-12-801700-5

British Library Cataloguing-in-Publication Data
A catalogue record for this book is available from the British Library.

Library of Congress Cataloging-in-Publication Data
A catalog record for this book is available from the Library of Congress.

For Information on all Academic Press publications
visit our website at http://store.elsevier.com/

Typeset by MPS Limited, Chennai, India
www.adi-mps.com

Printed and bound in the United States of America

Contents

Foreword

A Google search for "Society of Hair Testing" brings nearly 40 million hits, showing the impressive activities of SoHT; for example, it has passed statements concerning the examination of drugs in human hair in 1997, 2011, and 2014, published consensus reports on hair analysis in forensic cases, on the use of alcohol markers in hair for the assessment of both abstinence and chronic excessive alcohol consumption, and of hair testing for doping agents, and organized proficiency tests of drugs in hair.

This was beyond our imagination when we joined the "International Meeting and Workshop on Hair Analysis in Forensic Toxicology" organized by the General Directorate of the Abu Dhabi Police in cooperation with The International Association of Forensic Toxicologists in November 1995. We "brainstormed" at that meeting about the future of hair analysis and decided to found a society. Three weeks later, the foundation of the Society of Hair Testing (SoHT) was initiated in Munich and judicially completed in Strasbourg in late December 1995. It started with 12 members, growing to 354 members worldwide in 2014.

The workshop and the proceedings brought attention to the basic scientific knowledge of the incorporation of xenobiotics into hair, and also brought the development of analytical procedures for the different drugs into focus, as well as the interpretation of the results and acceptance in court.

It is to the credit of Pascal Kintz to continue this series as editor of two books with similar and extended topics *Drug Testing in Hair* (1996) and *Analytical and Practical Aspects of Drug Testing in Hair* (2006). However, the main focuses of all monographs concerning drug testing in hair are basically identical: anatomy and physiology of hair, drug incorporation into hair, pharmacokinetics, hair collection, grinding, washing, extraction and screening procedures, analytical identification methods, interpretation of results, cut-offs, and proficiency tests.

At the time of the twentieth birthday of SoHT, this new book in the series will appear with an extended title, compared to the publication of 1995: *Hair Analysis in Clinical and Forensic Toxicology*. An impressive list of distinguished authors of international reputation will indicate the scientific progress in the last decade, discuss the actual knowledge and pinpoint new problems, and, hopefully, also solutions on interesting new topics.

Congratulations to the co-editors Pascal Kintz, Alberto Salomone, and Marco Vincenti, and Happy Birthday to SoHT: *ad multos annos*.

Manfred R. Moeller, Ph.D.

List of Contributors

Patricia Anielski, Institute of Doping Analysis, Kreischa, Germany

Brice M.R. Appenzeller, Laboratory of Analytical Human Biomonitoring – Luxembourg Institute of Health, Luxembourg

Marta Baber, Department of Pharmacology and Toxicology, Faculty of Medicine, University of Toronto, Ontario, Canada

Craig Chatterton, Office of The Chief Medical Examiner, in Edmonton, AB, Canada

Gail Audrey Ann Cooper, Cooper Gold Forensic Consultancy Ltd, Fife, Scotland, UK

Malin Forsman, National Board of Forensic Medicine, Department of Forensic Genetics and Forensic Toxicology, Linköping, Sweden

Carmen Jurado, National Institute of Toxicology and Forensic Sciences, Seville, Spain

Pascal Kintz, X-Pertise Consulting, Oberhausbergen, France; Institute of Legal Medicine, Strasbourg, France

Robert Kronstrand, National Board of Forensic Medicine, Department of Forensic Genetics and Forensic Toxicology, Linköping, Sweden and Division of Drug Research, Linköping University, Linköping, Sweden

Simona Pichini, Department of Therapeutic Research and Medicine Evaluation, Istituto Superiore di Sanità, Rome, Italy

Fritz Pragst, Charité—University Medicine Berlin, Berlin, Germany

Evan Russell, Department of Physiology and Pharmacology, University of Western Ontario, London, Ontario, Canada

Alberto Salomone, Centro Regionale Antidoping e di Tossicologia "A. Bertinaria", Orbassano (TO), Italy

Tor Seldén, National Board of Forensic Medicine, Department of Forensic Genetics and Forensic Toxicology, Linköping, Sweden and Division of Drug Research, Linköping University, Linköping, Sweden

Detlef Thieme, Institute of Doping Analysis, Kreischa, Germany

Lolita Tsanaclis, Cansford Laboratories, The Cardiff Medicentre; Heath Park, Cardiff, UK and Laboratório ChromaTox Ltda, São Paulo-SP, Brazil

Marco Vincenti, Dipartimento di Chimica, Università di Torino, Turin, Italy and Centro Regionale Antidoping e di Tossicologia "A. Bertinaria", Orbassano (TO), Italy

John Wicks, Cansford Laboratories, The Cardiff Medicentre; Heath Park, Cardiff, UK and Laboratório ChromaTox Ltda, São Paulo-SP, Brazil

Chapter 1

Anatomy and Physiology of Hair, and Principles for its Collection

Gail Audrey Ann Cooper
Cooper Gold Forensic Consultancy Ltd, Fife, Scotland, UK

1.1 INTRODUCTION

The main advantage of hair as a testing matrix is the ability to provide historical detail of an individual's exposure to drugs following chronic use, but it is also possible to detect drugs in hair following a single exposure [1]. Unlike many traditional biological samples collected for toxicological investigations such as blood, urine, or oral fluid, drugs incorporated into the hair matrix remain relatively stable for many months or even years.

As a consequence of these advantages, hair samples are routinely collected and analyzed as part of criminal investigations (drug-related deaths, drug-facilitated crimes (DFCs), child protection) and for monitoring drug misuse (drug rehabilitation programs, workplace drug testing). The number of laboratories offering hair testing continues to increase and although there are currently no recognized standardized methodologies, the importance of adherence to industry best practice and recognition of the implementation of international quality standards (e.g., ISO/IEC 17025:2005) for testing laboratories who provide this service has been reported [2].

With continual improvement in instrument sensitivity and recognition of the importance of method validity with acceptable detection limits, the main challenges to expert evidence relating to hair testing are not as a result of unsuitable testing methodologies but are most commonly a consequence of over interpretation of the analytical findings. This is invariably due to a lack of understanding of the many factors affecting the presence of drugs in hair and no consideration given to the role of alternative routes of incorporation or indeed the degradation of drugs in hair over time.

Although researchers have reported finding arsenic in hair collected from the Emperor Napoleon Bonaparte [3] and cocaine in the hair of Peruvian mummies [4] centuries after their deaths, drug concentrations in hair decrease over time due to natural wash out [5]. The stability of drugs in hair is dependent on both the morphology and physicochemical properties of hair [6]. On a daily basis, hair is subjected to exposure to sunlight and weathering that has been shown to reduce concentrations of cannabinoids [7], methadone, cocaine and their metabolites, and heroin metabolites in hair [8]. External contamination and loss of drugs from hair is facilitated by diffusion into and out of the hair in the presence of water [9]. Although daily shampooing does not significantly affect drug concentrations [10,11], drying your hair, curling or straightening your hair with heat can damage or destroy the cuticle providing routes for contamination and loss of incorporated drugs.

Further damage is evident when hair is subjected to harsher cosmetic treatments [10,12–16]. Studies reported in the literature are varied with respect to the drugs and cosmetic treatments investigated but are all in agreement that dyeing, perming, relaxing, and bleaching have a deleterious effect on the drug concentration detected in hair. Decreases in drug concentrations were more prominent for bleached hair compared with dyed hair and damaged hair showed significantly lower concentrations.

In order to fully understand and interpret the results of hair tests, it is essential that the forensic toxicologist has a clear understanding of what factors influence the incorporation of drugs into the growing hair shaft including contamination but must also consider the factors affecting loss of the incorporated drug. The focus of this chapter will be on the routes of incorporation and the factors influencing incorporation. This chapter will also address the importance of the sample collection process and will highlight case-type-specific collection protocols to ensure the collection of a representative sample for all toxicological investigations.

1.2 HAIR ANATOMY AND PHYSIOLOGY

Hair covers almost the entire surface of the human body with the exception of the outer surface of the lips, the palms of the hands, soles of the feet, and some parts of the external genitalia [17]. The main purpose of body hair is to protect the surface of the skin from injury and to help regulate body temperature. The appearance of body hair varies from fine almost colorless hair found on most parts of the body surface to thicker, longer hair on the scalp.

The hair shaft, visible above the surface of the skin, consists of tightly packed fully keratinized cells with the cuticle forming an outer protective layer [18]. The cuticle is susceptible to damage from a number of sources including, exposure to light or heat, chemical treatments including bleaching and perming, and also physical damage. As a consequence, over time the surface of the hair may be compromised exposing the inner layers and is particularly evident in the distal ends of the hair.

The interior structure of the hair shaft contains cortical cells forming the cortex which encompasses the bulk of the hair shaft and also contains cuticle cells where melanin, the principle pigment in hair, is located. The innermost region of hair within the cortex is the medulla and may be continuous along the length of the shaft, discontinuous, or completely absent.

The hair follicle is embedded 3–4 mm below the surface of the skin and is the key structure responsible for hair growth. The structure of the hair follicle is illustrated in Figure 1.1.

The main structures of the hair follicle include the outer root sheath (ORS), the inner root sheath (IRS), and the root bulb. The ORS forms a bulge area at the base of the erector pili muscle, close to the sebaceous gland and is believed to be the source of stem cells critical for hair follicle development and pigmentation [19]. The IRS provides support for the growing hair producing intracellular binding material and directing the growing hair upward [20].

The cells located within the lower region of the root bulb, the matrix, are mitotically active while the upper region of the root bulb contains the keratogenous region. At the base of the root bulb is the dermal papilla which contains the blood supply. The IRS degrades as the growing hair dehydrates and keratinization takes place [21].

Close to the hair follicle are the sebaceous and apocrine glands both of which secrete directly into the follicle. The apocrine glands, unlike the sebaceous glands, are not present over the entire surface of the body but are localized in the axilla and pubic regions. The eccrine sweat glands are located close to the follicle but secrete near the exit of the hair follicle not directly into the follicle [22].

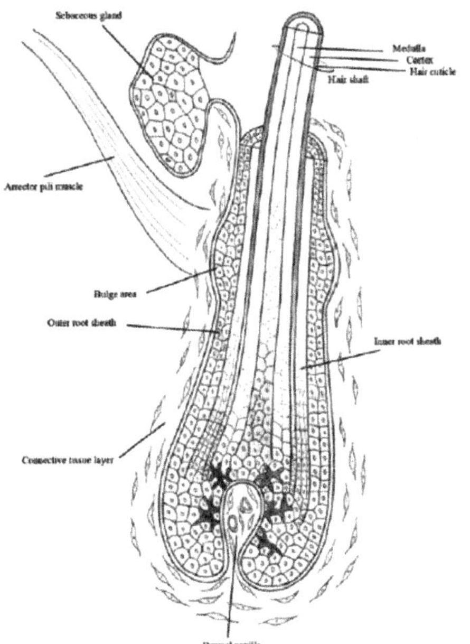

FIGURE 1.1 Structure of the hair follicle.

1.3 CLASSIFICATION OF HAIR TYPES

The three basic hair types on the human body are classified as vellus, intermediate, and terminal. Vellus hair covers the majority of the body surface of both children and adults, is fine, short, and non-pigmented, and is also found on the eyelids and forehead. Where vellus hair is produced by non-sexual hair follicles and unaffected by hormones, intermediate and terminal hairs are influenced by hormones and changes during puberty. Terminal hair is found on the scalp, beard, eyebrow, eyelash, armpit, and pubic areas and in contrast to vellus hair is coarse, long, and pigmented with a large cross-sectional area. Intermediate hair has characteristics of both vellus and terminal hairs and is found on the arms and legs of adults.

Human hair is also commonly classified using the ethnic subgroups of African, Asian, or European. An alternative classification system proposes eight different subgroups based on whether the hair is straight or curly and provides a more objective approach than a subjective ethnicity-based approach [23]. The classification system involves the measurement of three parameters: the curve diameter, the curl index, and the number of waves.

1.4 HAIR GROWTH RATES

A number of factors affect hair growth including age, stage of development, sex, pregnancy, metabolic and genetic disorders, nutrition, and seasonal changes [24,25]. Hair growth is a cyclical process driven by changes in the activity of cytokines (hormones) resulting in individual hairs on the body at various stages of the growth cycle [26]. The three recognized stages in hair growth are the anagen, catagen, and telogen phases. The anagen or growth phase can last for several years and is characterized by the formation of the hair shaft protruding from the surface of the skin. It is estimated that 85% of the hairs on the human scalp are in the anagen phase. The catagen or transitional phase follows this period of active growth. During this stage, cell division stops, the hair shaft becomes fully keratinized, and the bulb begins to degenerate, the length of which varies depending on the type of hair. The telogen or resting phase follows the catagen phase and is a period when there is no hair growth but the dermal papilla remains in the resting phase. It is estimated that 10–15% of all hairs are in the telogen phase at any given time with the length of time increasing with age and is also dependent on the hair type. The hair is easily removed during this stage and is often referred to as "shedding." A short time later the growth phase recommences by stimulating stem cells from the bulge area of the ORS [27].

The Society of Hair Testing (SoHT) recommends utilizing an average growth rate of 1 cm/month for head hair [28]. The variation in the reported growth rates for scalp hair is considerable and is further compounded by the

TABLE 1.1 Published Growth Rates for Different Human Hair Types (Centimeters/Month)

Hair Type	Mean/Range	Reference
Head	0.60–3.36	[18]
	0.75–1.35	[29]
	1.05	[30–32]
	0.6–1.5	[33]
	0.84–1.41	[34]
	1.1 (0.6–1.5)	[35–38]
Pubic	0.60–0.90	[33]
	0.75 (0.6–0.9)	[35–38]
Axillary	0.87–1.00	[33,34]
	0.9	[30–32]
	0.9 (0.87–1.0)	[35–38]
Beard	0.75–0.87	[33,37]
	1.2	[39]
	0.75–0.81	[35–38]
Body	0.30	[39]
Chest	0.66–0.96	[33,37]
Arm hair	1.05	[30,31]
Leg hair	0.81	[40]
	1.0 (0.81–1.05)	[35–38]
	0.9	[30,31,35–38]
	0.63	[30,31]
	0.6 (0.39–0.75)	[35–38,40]
	0.6	[32]

variability of growth rates for different hair types as summarized in Table 1.1 [18,29–40].

Scalp hair is preferred because it has the fastest growth rate with the highest percentage of follicles in the anagen phase. In comparison, pubic hair has a slower growth rate with a longer resting phase while beard hair is thicker with the slowest growth rate. When analyzing non-head hair samples, it is important to consider not only the variation in growth rates but also the variation in the proportion of hair actively growing or in the resting phase [18]. Where head hair is not available, pubic hair is considered a good alternative but additional consideration should be given to the potential for contamination from urine.

The majority of the published growth rates for scalp hair range from 0.6 to 1.5 cm each month supporting the average growth rate of 1 cm/month. However, this is an oversimplification as a growth rate of as high as 3.36 cm/month has been reported and these rates are for adults hair only [18]. There is a significant lack of published information relating to the variation in growth rates for younger children or teenagers. Hair growth rates in children were reported in a small study published in 1964 [41] with 13 females and 7 males whose ages ranged from 3 to 9 years. Hair growth rates were faster for males across all regions of the scalp ranging from 0.300 to 0.355 mm/day for males and 0.273 to 0.331 mm/day for females.

It is recognized that in early life many infants are born with a full head of hair while others have little or no hair for many months and those born with hair may lose some or all of their hair in the first year of life. Barth [42] described the emergence of hair on the body of the fetus from 9 weeks to actively growing hair follicles with roots covering the surface of the scalp at 20 weeks' gestation. The hair follicles change from the anagen phase to the catogen phase during the 26th to 28th week gestation period and then on to the telogen phase in what is described as a progressive wave from the frontal to parietal regions of the scalp. Many of the telogen hairs are shed *in utero* while other hairs continue to grow until near birth before also entering the telogen phase.

Consideration should always be given when interpreting timescales represented by the length of a hair sample not only as a consequence of the variation in recognized growth rates but also to the age of the subject. In addition, as it has been estimated to take approximately 7–10 days for the growing hair to reach the surface of the scalp, hair cut from the scalp does not represent the most recent period of hair growth.

1.5 HAIR COLOR

Pigmentation within the hair shaft accounts for as little as 0.1–5% of the hair mass with proteins and lipids accounting for 65–95% and 1–9%, respectively [18]. Differences in hair color are a consequence of the variation in the type and quantity of melanin present. Follicular melanogenesis or pigment formation takes place within organelles named melanosomes which are present within specialized cells named melanocytes [43]. This process takes place exclusively within the hair follicle and is regulated by enzymes, receptors, and proteins during the anagen phase of active hair growth [44].

Four types of melanin are thought to determine the color of hair, namely eumelanin, pheomelanin and their oxidative products, oxyeumelanin, and oxypheomelanin [45]. In general darker colors of black and brown hair are associated with predominantly eumelanin, with lighter shades associated with increasing amounts of oxyeumelanin. Pheomelanin results in red shades of hair color with lighter shades being associated with increasing amounts of oxypheomelanin.

1.6 MECHANISMS OF DRUG INCORPORATION

The exact mechanisms by which drugs or analytes of interest are incorporated into hair and the factors affecting their stability are not fully understood. It is believed that drugs and trace elements circulating in the body are incorporated into hair during periods of increased metabolic activity and cell division synonymous with the anagen growing phase [46]. There are however three recognized routes of incorporation of drugs into hair including directly from the blood supply, from sebum and sweat bathing the hair, and from external contamination. Figure 1.2 illustrates the three recognized routes of incorporation of drugs into hair [18,47]. The extent to which each of these routes contributes to drug incorporation is unclear or indeed to what extent they each exert their influence, but what is clear is that it does vary from drug to drug [21].

There are different models of incorporation proposed that attempt to explain the drug profiles observed in hair. The most simplistic model involves passive diffusion of drugs directly from the blood supplying the hair follicle. With passive diffusion it would be expected that drug

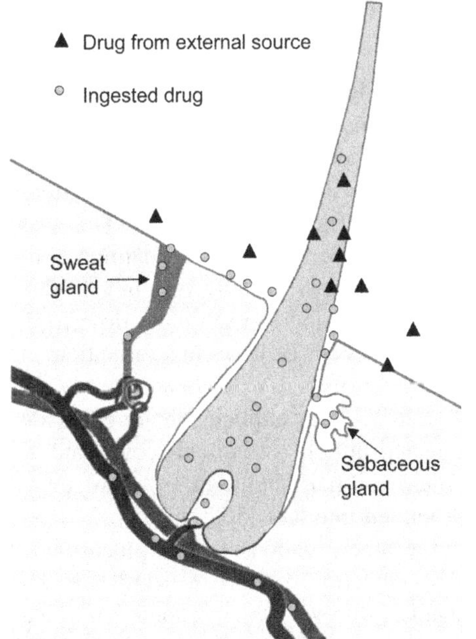

FIGURE 1.2 Incorporation routes into the hair follicle.

concentrations in hair would correlate with the drug concentration in the blood at the time of hair synthesis [48,49]. This model does not however explain the different metabolic profiles seen in hair and blood. In blood, parent drugs are less commonly detected in comparison to the corresponding primary metabolites, while in hair the presence of the parent drug is more common, for example, cocaine and 6-monoacetylmorphine (6-MAM), the primary metabolite of heroin, are generally found in higher concentration in hair than their metabolites, benzoylecgonine and morphine, respectively.

The "biochemical concept" was proposed to explain the endogenous incorporation of drug molecules into growing hair [50]. This concept tentatively explains the high parent drug to metabolite ratio in hair, the dependence of incorporation on the physicochemical properties of the drug, the incorporation of drugs into non-pigmented hair, and the dependence of drug content on hair pigmentation.

It is clear that hair has different affinities and binding capacities for different drugs with unique binding mechanisms for each drug [51,52]. Drug pKa, structure, size, lipophilicity, protein binding capacity, and melanin affinity are the factors known to affect the incorporation and binding of drugs [29,53,54]. Basic drugs, such as amphetamines and cocaine, incorporate into hair to a greater extent than neutral or acidic drugs and are consequently present in higher concentrations in hair compared with benzodiazepines and cannabinoids.

1.7 INCORPORATION FROM THE BLOODSTREAM, SEBUM AND SWEAT

Lipid solubility of a drug is a critical factor in determining the rate of transport from the blood stream across the cell membrane into the growing root bulb. The pH gradient from plasma, pH 7.3, to more acidic conditions within the melanocytes/keratinocytes (pH 3–6) provides advantageous conditions for basic drugs to incorporate preferentially to more acidic drugs. This was clearly demonstrated using structurally related compounds, rhodamine (cation) and fluorescein (anion), administered intraperitoneally resulting in distinct fluorescent bands of rhodamine corresponding to each daily dose [55]. Although fluorescein was detected in the matrix cells during formation, it was not detected in the keratinized hair.

Other routes must also be considered to help explain the drug profile observed in hair. Drugs have been detected in both sweat and sebum and as both of these secretions bathe the follicle and hair shaft as they grow, there will be contribution to the drugs incorporated into hair [56,57].

Evidence to support the role of sweat incorporation of ethyl glucuronide (EtG) into hair was reported following detection of EtG in beard hair 9 h after a single dose of alcohol (ethanol), however, the predominant route was incorporation from the blood stream directly into the root bulb [58].

EtG concentrations increased to a maximum of 72–242 pg/mg within 2–4 days post dose and gradually decreased to below the limit of quantitation by day 8–10 as the root bulb grew and emerged from the surface of the skin.

1.8 INCORPORATION FROM EXTERNAL CONTAMINATION

Drug contamination from the environment must also be considered including passive exposure from being close to individuals smoking drugs (e.g., heroin, crack cocaine, cannabis), through directly handling the drugs, touching surfaces contaminated with drugs and then touching your own or someone else's hair. In particular, children, and infants to a greater extent, are at high risk of exposure if living within a household where drug misuse is prevalent [59–61]. Consideration should also be given to the unborn fetus who will be exposed to drugs if their mother continues to misuse during her pregnancy [62–65].

External contamination and the role it plays in drug incorporation has been extensively investigated by researchers, with the greatest challenge in attempting to recreate realistic contamination scenarios. Consideration should also be given to the condition of the hair sample which may be damaged through natural wear and tear or chemical treatment and also to the porosity of the hair, all of which will affect the incorporation of drugs from external contamination.

Studies have demonstrated significant contamination of drug-free hair exposed to smoke containing cannabis [66], blood containing cocaine [67] and heroin metabolites [68], and from the direct application of cocaine to the hair [69].

Site-specific contamination should also be taken into account, including the potential for urine and gland secretions contaminating pubic hair and pieces of epidermis contaminating beard hair when collected by shaving. Concentrations of EtG measured in different hair sample types varied significantly and raise concerns about the use of pubic or axillary hair samples as alternative specimens to head hair in the assessment of chronic use of alcohol due to positive and negative biases, respectively [70].

The SoHT provides guidelines for washing to remove gross contamination from the surface of the hair, however, the selective removal of drugs on the surface while preserving drugs incorporated into the hair matrix is an unlikely outcome when using both organic solvents and aqueous solutions. In one study, complete decontamination was not achieved by participating laboratories who were sent drug-free hair samples contaminated with cocaine. The laboratories all used different wash protocols but none of them were able to decontaminate the hair [71].

Proposals for discriminating between contamination and incorporated drugs have been proposed [72,73] and may provide sufficient confidence in the future to be accepted as a standardized approach but until such time, the potential role of external contamination must always be considered when interpreting positive hair tests.

External contamination is discussed in greater detail in Chapter 3.

1.9 DOSE–RESPONSE RELATIONSHIP

Although controlled dosage studies with animals and humans have supported a linear relationship between dose and drug concentrations found in hair for certain drugs [48,74–77], the vast majority of studies have demonstrated that a linear relationship does not exist [78–82]. The most likely explanation is due to inter-subject variation with a number of additional variables including hair color and cosmetic treatments being of particular significance.

Following controlled low and high dosage regimes of methamphetamine, good intra-subject correlation was observed with methamphetamine and amphetamine concentrations measured in hair [83]. A large inter-subject variation was also reported in this study, however, when corrected for melanin content demonstrated a good correlation, supporting a dose-related relationship for the incorporation of methamphetamine and its metabolite amphetamine and resulting concentrations measured in hair.

Evidence of a dose-response relationship for methamphetamine was reported under controlled conditions and correcting for the melanin content. However, controlled conditions are not representative of real-life exposure and the role of contamination must also be considered. Children exposed to methamphetamine were reported to have methamphetamine/amphetamine concentrations in hair of similar magnitude to those measured in adults misusing methamphetamine [84,85].

1.10 MELANIN BINDING

A number of studies have investigated the incorporation of different drugs into hair involving animal models which have provided useful information on the role of melanin [86–91]. Darker hair contains more pigmentation (melanin) than light colored hair and so animals with both pigmented and non-pigmented hair were administered drugs of different basicity. The findings supported the greater binding affinity of basic drugs to melanin as they were present in higher concentrations in pigmented hair. Neutral and acidic drugs demonstrated no greater affinity to pigmented or non-pigmented hair.

It is important to remember when relating these findings to human hair that animal and human hair differs with respect to growth cycles and structure which may result in different drug distribution.

Differing incorporation profiles of fatty acid ethyl esters in rat hair compared with human hair were reported [92] and was also observed for erectile dysfunction drugs and their metabolites [93]. Sildenafil and its primary metabolite, desmethyl sildenafil, were measured in pigmented and non-pigmented rat hair and in two human hair samples collected from individuals suspected of using illegal supplies of sildenafil. The concentrations in pigmented hair were higher than those measured in non-pigmented hair supporting the role of melanin in drug incorporation. However, the metabolite concentration was significantly higher than the parent drug in rat hair with a mean ratio of drug to metabolite of 0.1 (no range reported) and was in direct contrast to the ratio observed in the two human hair samples of 0.83 and 3.6. The authors acknowledged previously published research that also observed higher metabolite concentrations in rat hair for sildenafil but they also reported a similar profile for human hair samples analyzed, not consistent with their findings [94]. In both studies the number of human hair samples tested was small and therefore a much larger study is required to investigate the drug to metabolite profiles in human hair. The effect of the hair sample preparation conditions used to extract the drugs from the matrix was also reported as influential on the apparent ratios measured. In addition, both the concentrations measured and the parent drug to metabolite ratios varied for the other erectile dysfunction drugs investigated (mirodenafil and vardenafil). Although structurally related compounds, other factors affect their incorporation in additional to melanin.

Kim et al. [95] investigated the role of melanin with respect to the incorporation of synthetic cannabinoids in hair and found melanin was not a significant factor in their incorporation into hair.

Additional factors affect the incorporation of drugs into hair including the amount and type of melanin present and binding to keratin [96].

1.11 SAMPLE COLLECTION PROTOCOLS

Guidance is available on recommended best practice for the collection of hair samples, including specific recommendations for postmortem hair sampling, workplace drug testing, and DFCs [97–100]. The majority of the steps involved with the collection process are the same irrespective of the case type, but there are case-type-specific protocols that must also be considered.

The SoHT recommends cutting hair as close as possible to the scalp from the vertex region of the head as illustrated in Figure 1.3 [101]. Collection of a sample from the vertex region is preferred as this is the site where there is least variation in growth rates compared with other regions of the scalp or when compared to other body hair types. In addition, hair growing on the scalp has the highest percentage of follicles actively growing and not in the resting phase.

The volume of hair required for analysis is a "lock of hair" proportionate to the thickness of a pencil. The collection of a sufficient volume of hair is critical to ensure the completion of all necessary tests. Not surprisingly, the most common concern from both collectors and the donor is that the amount of hair sampled should not leave a visible "bald patch" and is of particular concern for parents of small children or individuals suffering from baldness or thinning hair. To avoid distress to individuals, the

FIGURE 1.3 Vertex region of the scalp.

collection of smaller hair samples from multiple sites within the vertex region is also an acceptable protocol. The smaller individual locks should then be combined into one larger lock of hair and aligned with the root end identified.

In addition to collecting sufficient volume to ensure all tests required can be carried out, it is particularly important when carrying out segmental analysis of hair. Cutting hair into smaller segments of most commonly 0.5–3 cm in length provides a more detailed historical profile of an individual's exposure and may also provide a means of determining external contamination in postmortem cases [102]. However, smaller segments will further compromise the volume available for testing.

Although specialist facilities are not required for the collection of scalp hair, when head hair is not available, alternative collection sites should be considered including intimate samples (e.g., pubic hair). In these circumstances, the privacy of the donor must be prioritized while ensuring the integrity of the collection process.

Collection of hair samples should only be carried out by a competent collector who must recognize and adhere to good practice guidelines for sample collection to eliminate the potential for contamination of the sample(s) and will ensure chain of custody is maintained throughout the process. Le Beau et al. [103] also highlighted the importance of ensuring the sample is collected as close to the scalp as possible following an evaluation of the hair samples taken by collectors with varying levels of experience. The length of hair remaining on the scalp after cutting was 0.8 ± 0.1 cm leading to significant impact on the timescale of the corresponding segments from the collected samples and the subsequent interpretation of the findings.

Hair samples are routinely collected by medical personnel (general practitioner, forensic medical examiner, nurse) but the collector does not require medical qualifications. The collector may however be required to give evidence in a court of law and should therefore be prepared to demonstrate their competence through experience and appropriate training records.

1.12 COLLECTION PROCEDURE

Clear instructions must be provided with the hair collection kit to provide guidance and to allow the collector to check the contents are all present and correct. The collection kit should be sealed so that the collector can check that the seals have not been tampered with prior to opening the kit to use for the first time and to check the contents. Scissors used to cut the hair sample should be cleaned with an alcohol-free wipe or sterilized prior to use. The standard contents of a hair collection kit should include the following:

- Chain of custody form
- Foil and collection envelope
- Security seals
- Evidence bag (optional)
- Transportation envelope (optional)
- Instructions for collection of a hair sample

The main purpose of the chain of custody form is to document details relating to the collection and handling of the sample and to facilitate identification of the sample. Duplicate or carbon copies of the form may be required if there are multiple agencies involved. In the case of workplace drug testing, the anonymity of the donor must be maintained therefore anonymized copies of the form are required for dispatch with the sample to the laboratory for analysis.

A secure mail service is recommended to ensure chain-of-custody is maintained, and specialist courier services are available with tracking systems and secure transportation to allow complete traceability of the samples from the collector to the laboratory.

The following step-by-step guide summarizes the collection process:

Step One: Cut a lock of hair close to the scalp from the vertex region and align the hair identifying the root end (Figure 1.4).
Step Two: Fold the foil lengthways and avoid folding in the middle as this will kink the hair making it difficult to handle (Figure 1.5).
Step Three: Place the foil-wrapped hair sample within the collection envelope, seal, initial, and date.
Step Four: Place the completed chain of custody form and sealed collection envelope into the evidence bag and transportation envelope and send to the laboratory for analysis.

For long-term storage, hair samples should be stored at room temperature, dry, and away from direct sunlight.

1.12.1 Workplace Drug Testing

The European Workplace Drug Testing Society published guidelines for European workplace drug and alcohol testing in hair in 2010 [98]. Within

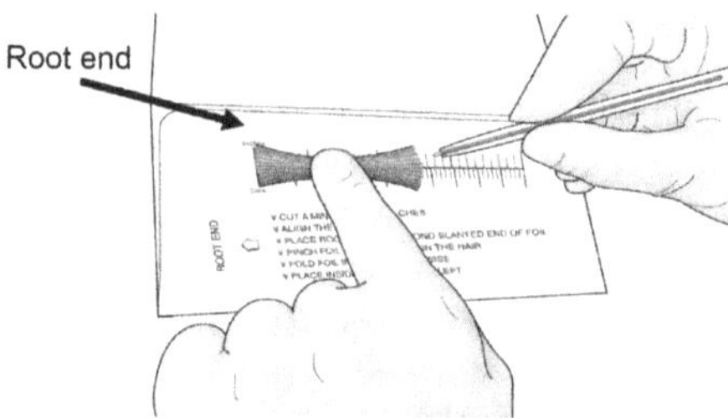

FIGURE 1.4 Cut hair aligned with the root end identified.

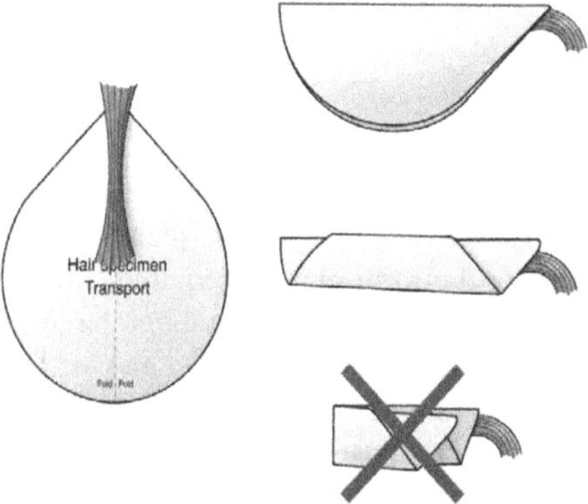

FIGURE 1.5 Fold the foil lengthways to secure the aligned hair.

the guidelines the collection of hair was covered in detail as a critical part of the process and in particular the role of the collector. An immediate supervisor of the donor or coworker, relative or close friend may not perform the collection. Additionally, an employee of the drug testing laboratory may not collect a sample from the donor unless it can be demonstrated that the sample could not be linked to the donor.

Access to the collection facility should be restricted and entry to the room prohibited during the collection process. The chain of custody form should have a minimum of three parts or copies: one for the testing laboratory, one for the donor, and one retained by the collector. A list of information to be recorded on the form was also detailed within the guidelines including a declaration from the donor on their use of prescribed medication, sample authenticity, correctness of sample labeling and packaging, and permission for the sample to be analyzed at the testing laboratory.

Importantly the collection of a lock of hair as described above is then split in two to ensure the donor has the option to have a representative sample retested by a different laboratory if they want to contest the results from the primary testing laboratory. The collection kit must therefore contain duplicate components for storage and transport of two hair samples.

1.12.2 Drug-Facilitated Crime

The majority of cases involving DFC involve administration of a single dose of a drug, and as a consequence, the collection of the hair sample(s) is time critical and dependent on the period of time that has elapsed since the incident took place [104]. The SoHT recommends that a hair sample is collected at least 4–6 weeks after the incident and a second sample collected 1 month later but only if the first sample is positive [101]. The second sample is collected to act as a control to corroborate the findings in the first sample. To avoid compromising the investigation, the complainant must not have chemical treatments on their hair, for example, bleach, dyes, perming lotions, and must not cut their hair until the investigation is over. This is particularly important if the complainant's hair is short.

Alternative protocols have been described regarding the collection timescale. Baciu recommends waiting 8 weeks after the incident before collecting a sample to ensure the hair sample collected does represent the exposure period [105]. Scott [99] described the process of collecting the first hair sample within 1 week of the incident for early reporting cases and then collecting a second sample 1 month later. The Faculty of Forensic and Legal Medicine reviewed this practice following concerns raised by collectors that many complainants do not report early and those who do were concerned about providing the first hair sample and did not return for the second collection. Their recommendations for the collection of forensic specimens from complainants and suspects were amended to collect hair 4–6 weeks after the incident [106]. They also only recommend collecting hair if the incident has taken place in the previous 6-month period.

Sample volume is especially critical when dealing with DFC cases due to the low concentrations in hair following a single dose and the range of possible drugs that could have been administered. One recommendation was to collect four locks of approximately 100 hairs with one lock of hair to be used for screening for drugs most commonly associated with DFC, one lock for drugs of abuse, one for specifically testing for gamma-hydroxybutyrate, and the fourth to be retained for independent analysis if required [107].

Although the ideal is for large volumes of hair to be submitted, the reality is that a single lock of hair is submitted for analysis and the tests must be prioritized based on the symptomology described by complainant.

1.12.3 Postmortem Investigations

The collection of postmortem hair samples can pose a challenge depending on the condition of the deceased's hair. Ideally, hair samples should be dry and collected with the root end of the hair identified, but this is not always possible. Hair heavily soiled with blood is commonplace following road traffic accidents or violent assaults and heavy soiling from body fluids is also routinely encountered in cases of decomposition. In these circumstances, the standard hair collection kit foil and envelope may be inappropriate and an alternative more robust collection vessel required. Hair samples heavy soiled with biological material will require extensive cleanup procedures compared with standard washing protocols.

Although the collection of cut hair will provide information relating to historical drug use in the months prior to death, the collection and analysis of hair roots can provide additional information relating to recent drug use prior to death. In circumstances where information relating to recent use is required, the collection of both a cut hair sample and a pulled hair sample with roots intact is also recommended [101].

Handling hair roots requires considerable skill and patience but as the information provided from their analysis may be crucial to the outcome of a case, for example, in cases of advanced decomposition, the additional effort required would be justified.

1.13 DISCUSSION

Hair as a testing matrix has the advantage of providing information relating to historical drug exposure lending itself to many applications within forensic and clinical toxicology. When interpreting findings in a particular case, it is important to consider all of the factors that may have contributed to that result. It is understood that drugs incorporate into the growing hair directly from the blood stream into the root of the developing hair shaft and that the physicochemical properties of the drug will determine how well the drug will incorporate. Acidic drugs incorporate less well than basic drugs and although melanin binding affinity is an important factor, it is also known that many other binding mechanisms also contribute, due to the presence of drugs in hair samples with little or no melanin.

Drugs not only incorporate via the bloodstream as the drug/metabolite profile in hair does not reflect that observed in blood. Drugs are also incorporated from sweat and sebum bathing the hair and do not simply remain on the surface of the hair. This is also the case with external contamination which can result in a positive hair test even though the individual has not actively used the drug. By understanding all of the possible reasons for a positive hair test, we can better provide a reasoned interpretation and avoid over interpretation which undermines the efficacy of hair testing.

Of course, fundamentally the collection of hair samples must be carried out in accordance with best practice guidelines to ensure the sample is fit-for-purpose and chain of custody has been maintained. Special consideration must be given to case-type-specific collection protocols as differences do exist between samples collected for workplace drug testing, postmortem, or DFCs.

Significant progress has been made in our understanding of drug incorporation into hair and the factors that affect its stability. Ultimately, the key aim would be to develop a standardized approach to discriminating between drugs incorporated through direct use and those present from sources of external contamination. The presence of metabolites and their relative concentrations in comparison to the parent drug may provide an answer but until that time, the role of contamination must always be considered for all positive hair tests.

1.14 CONCLUSION

Drugs are incorporated from three main routes, namely, directly from the bloodstream, from sebum and sweat, and from external contamination. At present, it is not possible to selectively remove all trace of external contamination and although we better understand the mechanisms of drug incorporation and the factors influencing the stability of drugs in hair, the role of contamination must always be considered. Further research is required to gain a better understanding of the mechanisms of incorporation of drugs and their metabolites, and how their relative concentrations can be utilized to discriminate drugs use from external contamination.

REFERENCES

[1] Kintz P. Value of the concept of minimal detectable dosage in human hair. Ther Forensic Sci Int 2012;218:28–30.

[2] Cooper GAA. Current status of accreditation for drug testing in hair. Forensic Sci Int 2008;176:9–12.

[3] Smith H, Forshufvud S, Wassen A. Distribution of arsenic in Napoleon's hair. Nature 1962;194:725–6.

[4] Springfield AC, Cartmell LW, Aufderheide AC, Buikstra J, Ho J. Cocaine and metabolites in the hair of ancient Peruvian coca leaf chewers. Forensic Sci Int 1993;63:269–75.

[5] Chatterton C, Kintz P. Hair analysis to demonstrate administration of amitriptyline, temazepam, tramadol and dihydrocodeine to a child in a case of kidnap and false imprisonment. J Forensic Legal Med 2014;23:26–31.

[6] Pötsch L, Moeller MR. On pathways for small molecules into and out of human hair fibres. J Forensic Sci 1996;41:121–5.

[7] Skopp G, Pötsch L, Mauden M. Stability of cannabinoids in hair exposed to sunlight. Clin Chem 2000;46:1846–8.

[8] Favretto D, Tucci M, Monaldi A, Ferrara SD, Miolo G. A study on photodegradation of methadone, EDDP, and other drugs of abuse in hair exposed to controlled UVB radiation. Drug Test Anal 2014;6:78–84.

[9] Skopp G, Pötsch L, Aderjan R. Experimental investigations on hair fibres as diffusion bridges and opiates as solutes in solution. J Forensic Sci 1996;41:117–20.

[10] Donahue TD, Rhode KA, Berkable DR. Effects of various hair stripping treatments on Analytical Results. Presented at The SOFT/TIAFT Meeting in Albuquerque, NM (USA); 1998.

[11] Röhrich J, Zorntlein S, Pötsch L, Skopp G, Becker J. Effect of the shampoo Ultra Clean on drug concentrations in human hair. Int J Legal Med 2000;113:102–6.

[12] Skopp G, Pötsch L, Moeller MR. On cosmetically treated hair—aspects and pitfalls of interpretation. Forensic Sci Int 1997;84:43–52.

[13] Dawber R. Hair: Its structure and response to cosmetic preparations. Clin Dermatol 1996;14:105–12.

[14] Jurado C, Kintz P, Menedez M, Repetto M. Influence of cosmetic treatment of hair on drug testing. Int J Legal Med 1997;110:159–63.

[15] Cajkovac M, Oremovic L, Cajkovac V. The effect of hair colours and perm products on the state of hair. Acta Pharmeutica 1996;46:39–49.

[16] Pötsch L, Skopp G. Stability of opiates in hair fibers after exposure to cosmetic treatment. Forensic Sci Int 1996;81:95–102.

[17] Buffoli B, Rinaldi F, Labanca M, Sorbellini E, Trink A, Guanziroli E, et al. The human hair: from anatomy to physiology. Int J Dermatol 2014;53:331–41.

[18] Harkey MR. Anatomy and physiology of hair. Forensic Sci Int 1993;63:9–18.

[19] Paus R, Arck P, Tiede S. (Neuro-)endocrinology of epithelial hair follicle stem cells. Mol Cell Endocrinol 2008;288:38–51.

[20] Randall VA, Botchkareva NV. The biology of hair growth. In: Ahluwalia GS, editor. Cosmetic application of laser and light-based system. Norwich, NY: William Andrew Inc.; 2009. p. 3–35.

[21] Kronstrand R, Scott K. Drug Incorporation into hair. In: Kintz P, editor. Analytical and practical aspects of drugs testing in hair. Boca Raton, FL: CRC Press; 2007. p. 1–23.

[22] Montagna W, Van Scott EJ. The anatomy of the hair follicle. In: Montagna W, Ellis RA, editors. The biology of hair growth. New York, NY: Academic Press; 1958. p. 39–64.

[23] De la Mettrie R, Saint-Léger D, Loussouarn G, Garcel A, Porter C, Langaney A. Shape variability and classification of human hair: a worldwide approach. Hum Biol 2007;79:265–81.

[24] Randall VA, Ebling FJ. Seasonal changes in human hair growth. Br J Dermatol 1991;124:146–51.

[25] Wolfram LJ. Human hair: a unique physicochemical composite. J Am Acad Dermatol 2003;48:s106–14.

[26] Paus R. Principles of hair cycle control. J Dermatol 1998;25:793–802.

[27] Wosicka H, Cal K. Targeting to the hair follicles: current status and potential. J Dermatol Sci 2010;57:83–9.

[28] Society of Hair Testing. Recommendations for hair testing in forensic cases. Forensic Sci Int 2004;145:83–4.

[29] Pötsch L. A discourse on human hair fibres and reflections on the conservation of drug molecules. Int J Legal Med 1996;108:285–93.

[30] Richards RN, Uy M, Meharg GE. Temporary hair removal in patients with hirsutism: a clinical study. Cutis 1990;45:199–202.

[31] Richards RN, Meharg GE. Cosmetic and medical electrolysis and temporary hair removal: a practice manual and reference guide. 2nd ed. Toronto: Medric Ltd; 1991/1997.

[32] Vogt A, McElwee KJ, Blume-Peytavi U. Biology of the hair follicle. In: Blume-Peytavi U, Tosti A, Whiting DA, Trüeb RM, editors. Hair growth and disorders. Berlin, Heidelberg: Springer-Verlag; 2008. p. 13.

[33] Pragst F, Rothe M, Spiegel K, Sporkert F. Illegal and therapeutic drug concentrations in hair segments—a timetable of drug exposure? Forensic Sci Rev 1998;10:81–112.

[34] Pragst F. Pitfalls in hair analysis. TIAFT Bull 2005;35:10–17.

[35] Baumgartner M. (2014). Personal communication.

[36] Pianta A, Liniger B, Baumgartner MR. Ethyl glucuronide in scalp and non-head hair: an intra-individual comparison. Alcohol Alcohol 2013;48:295–302.

[37] Hartwig S, Auwärter V, Pragst F. Fatty acid ethyl esters in scalp, pubic, axillary, beard and body hair as marker for alcohol misuse. Alcohol Alcohol 2003;38:163–7.

[38] Kerekes I, Yegles M, Grimm U, Wennig R. Ethyl glucuronide determination: head hair versus non-head hair. Alcohol Alcohol 2009;44:62–6.

[39] Myers RJ, Hamilton JB. Regeneration and rate of growth of hairs in man. Ann N Y Acad Sci 1951;53:562–8.

[40] Pragst F, Sachs H. Die haarprobe als untersuchungsmatrix zur toxikologischen fahreignungsdiagnostik. Satelliten symposium "Toxikologische aspekte der fahreignung". Tagungsband zum XV. GTFCh-Symposium. Mosbach/Baden; 2007. pp. 84–99.

[41] Pecoraro V, Astore I, Barman J, Araujo CI. The normal trichogram in the child before the age of puberty. J Invest Dermatol 1964;42:427–30.

[42] Barth JH. Normal hair growth in children. Pediatr Dermatol 1987;4:173–84.

[43] Prota G. The role of peroxidase in melanogenesis revisted. Pigment Cell Res 1992;(Suppl. 2):25–31.

[44] Slominski A, Wortsman J, Plonka PM, Schallreuter KU, Paus R, Tobin DJ. Hair follicle pigmentation. J Invest Dermatol 2005;124:13–21.

[45] Prota G. Melanins, melanogenesis and melanocytes: looking at their functional significance from the chemist's viewpoint. Pigment Cell Res 2000;13:283–93.

[46] Ryder ML. Nutritional factors influencing hair and wool growth. In: Montagna W, Ellis RA, editors. The biology of hair growth. New York, NY: Academic Press; 1958. p. 305–34.

[47] Chittleborough G, Steel BJ. Is human hair a dosimeter for endogenous zinc and other trace elements? Sci Total Environ 1980;15:25–35.

[48] Baumgartner WA, Hill VA, Bland WH. Hair analysis for drugs of abuse. J Forensic Sci 1993;34:1433–53.

[49] Harkey MR. Technical issues concerning hair analysis for drugs of abuse. NIDA Res Monogr 1995;154:218–34.

[50] Pötsch L, Skopp G, Moeller MR. Biochemical approach on the conservation of drug molecules during hair formation. Forensic Sci Int 1997;84:25–35.

[51] Kikura R, Nakahara Y, Mieczkowski T, Tagliaro F. Hair analysis for drug abuse XV. Disposition of 3,4-methylenedioxy-methamphetamine (MDMA) and its related compounds into rat hair and application to hair analysis for MDMA abuse. Forensic Sci Int 1997;84:165–77.

[52] Nakahara Y, Kikura R. Hair analysis for drugs of abuse XIII. Effect of structural factors on incorporation of drugs into hair: the incorporation rates of amphetamine analogs. Arch Toxicol 1996;70:841–9.

[53] Joseph ER, Su TP, Cone EJ. *In vitro* binding studies of drugs to hair: influence of melanin and lipids on cocaine binding to caucasoid and africoid hair. J Anal Toxicol 1996;20:338–44.

[54] Knorle R, Schniz E, Feuerstein TJ. Drug accumulation in melanin: an affinity chromatographic study. J Chromatogr B Biomed Sci Appl 1998;714:171–9.

[55] Stout P, Ruth JA. Comparison of *in vivo* and *in vitro* deposition of rhodamine and fluorescein in hair. Drug Metab Dispos 1998;26:943–8.

[56] Huestis MA, Oyler JM, Cone EJ, Wstadik AT, Schoendorfer D, Joseph Jr. RE. Sweat testing for cocaine, codeine and metabolites by gas chromatography-mass spectrometry. J Chromatogr B Biomed Sci Appl 1999;733:247–64.

[57] Joseph Jr. RE, Oyler JM, Wstadik AT, Ohuoha C, Cone EJ. Drug testing with alternative matrices. I. Pharmacological effects and disposition of cocaine and codeine in plasma, sebum and stratum corneum. J Anal Toxicol 1998;22:6–17.

[58] Schräder J, Rothe M, Pragst F. Ethyl glucuronide concentrations in beard hair after a single alcohol dose: evidence for incorporation in hair root. Int J Legal Med 2012;126:791–9.

[59] Joya X, Papaseit E, Civit E, Pellegrini M, Vall O, Garcia-Agar O, et al. Unsuspected exposure to cocaine in preschool children from a Mediterranean city detected by hair analysis. Ther Drug Monit 2009;31:391–5.

[60] Garcia-Bournissen F, Nesterenko M, Karaskov T, Koren G. Passive environmental exposure to cocaine in Canadian children. Pediatr Drugs 2009;11:30–2.

[61] Pichini S, Garcia-Algar O, Avarez AT, Mercadal M, Mortali C, Gottardi M, et al. Pediatric exposure to drugs of abuse by hair testing: monitoring 15 years of evolution in Spain. Int J Environ Res Public Health 2014;11:8267–75.

[62] Iwersen S, Schmoldt A, Schulz F, Puschel K. Evidence of gestational heroin exposure by comparative analysis of fetal and maternal body fluids, tissues, and hair in a heroin-related death. J Anal Toxicol 1998;22:296–8.

[63] Gray T, Huestis M. Bioanalytical procedures for monitoring *in utero* drug exposure. Anal Bioanal Chem 2007;388:1455–65.

[64] Gareri J, Koren G. Prenatal hair development: implications for drug exposure determination. Forensic Sci Int 2010;196:27–31.

[65] Kintz P. Contribution of *in utero* drug exposure when interpreting hair result in young children. Forensic Sci Int 2015; <http://dx.doi.org/10.1016/j.forsciint.2014.09.014> (In press).

[66] Thorspecken J, Skopp G, Pötsch L. *In vitro* contamination of hair by marijuana smoke. Clin Chem 2004;50:596–602.

[67] Paterson S, Lee S, Cordero R. Analysis of hair after contamination with blood containing cocaine and blood containing benzoylecgonine. Forensic Sci Int 2010;194:94–6.

[68] Paterson S, Lee S, Cordero R. Analysis of hair after contamination with blood containing 6-acetylmorphine and blood containing morphine. Forensic Sci Int 2011;210:129–32.

[69] Romano G, Barbera N, Lombardo I. Hair testing for drugs of abuse: evaluation of external cocaine contamination and risk of false positives. Forensic Sci Int 2001;123:119–29.

[70] Pirro V, Di Corci D, Pellegrino S, Vincenti M, Sciutteri B, Salomone A. A study of distribution of ethyl glucuronide in different keratin matrices. Forensic Sci Int 2011;210:271–7.

[71] Stout PR, Ropero-Miller JD, Baylor MR, Mitchell JM. External contamination of hair with cocaine: evaluation of external cocaine contamination and development of performance-testing materials. J Anal Toxicol 2006;30:490–500.

[72] Baumgartner WA, Hill VA. Sample preparation techniques. Forensic Sci Int 1993;63:121–35.

[73] Morris-Kukoski CL, Montgomery MA, Hammer RL. Analysis of extensively washed hair from cocaine users and drug chemists to establish new reporting criteria. J Anal Toxicol 2014;38:628–36.

[74] Ferko AP, Barbieri EJ, DiGregorio GJ, Ruch EK. The accumulation and disappearance of cocaine and benzoylecgonine in rat hair following prolonged administration of cocaine. Life Sci 1992;51:1823–32.

[75] Forman R, Scheiderman J, Klein J, Graham K, Greenwald M, Koren G. Accumulation of cocaine in maternal and fetal hair: the dose response curve. Life Sci 1992;50:1333–41.

[76] Scheidweiler KB, Cone EJ, Moolchan ET, Huestis MA. Dose-related distribution of codeine, cocaine, and metabolites into human hair following controlled oral codeine and subcutaneous cocaine administration. J Pharmacol Exp Ther 2005;313:909–15.

[77] Rollins DE, Wilkins DG, Kreuger G. Models for studying the cellular processes and barriers to the incorporation of drugs into hair. NIDA Res Monogr 1995;154:235–44.

[78] Henderson GL. Mechanisms of drug incorporation into hair. Forensic Sci Int 1993;63:19–29.

[79] Pragst F, Rothe M, Hunger J, Thor S. Structural and concentration effects on the deposition of tricyclic antidepressants in human hair. Forensic Sci Int 1997;84:225–36.

[80] Girod C, Staub C. Methadone and EDDP in hair from human subjects following a maintenance program: results from a pilot study. Forensic Sci Int 2001;117:175–84.

[81] Paterson S, Cordero R, McPhillips M, Carman S. Interindividual dose/concentration relationship for methadone in hair. J Anal Toxicol 2003;27:20–3.

[82] Ditton J, Cooper GAA, Scott KS, Allen DL, Oliver JS, Smith ID. Hair testing for "ecstacy" (MDMA) in volunteer Scottish drug users. Addict Biol 2000;5:207–13.

[83] Polettini A, Cone EJ, Gorelick DA, Huestis MA. Incorporation of methamphetamine and amphetamine in human hair following controlled oral methamphetamine administration. Anal Chim Acta 2012;726:35–43.

[84] Farst K, Meyer JAR, Bird TM, James L, Robbins JM. Hair drug testing of children suspected of exposure to the manufacture of methamphetamine. J Forensic Legal Med 2011;18:110–14.

[85] Bassindale T. Quantitative analysis of methamphetamine in hair of children removed from clandestine laboratories—evidence of passive exposure? Forensic Sci Int 2012;219:179–82.

[86] Pötsch L, Skopp G, Moeller MR. Influence of pigmentation on the codeine content of hair fibers in guinea pigs. J Forensic Sci 1997;42:1095–8.

[87] Slawson MH, Wilkins DG, Rollins DE. The incorporation of drugs into hair: relationship of hair color and melanin concentration to phencyclidine incorporation. J Anal Toxicol 1998;22:406–13.

[88] Borges CR, Wilkins DG, Rollins DE. Amphetamine and *N*-acetylamphetamine incorporation into hair: an investigation of the potential role of drug basicity in hair color bias. J Anal Toxicol 2001;25:221–7.

[89] Scott KS, Nakahara Y. A study into the rate of incorporation of eight benzodiazepines into rat hair. Forensic Sci Int 2003;133:47–56.

[90] Kikura-Hanajiri R, Kawamura M, Saisho K, Kodama Y, Goda Y. The disposition into hair of new designer drugs; methylone, MBDB and methcathinone. J Chromatogr B 2007;855:121–6.

[91] Lee S, Han E, Kim E, Choi H, Chung H, Oh SM, et al. Simultaneous quantification of opiates and effect of pigmentation on its deposition in hair. Arch Pharm Res 2010;33:1805–11.

[92] Kulaga V, Velazquez-Armenta Y, Aleksa K, Vergee Z, Koren G. The effect of hair pigment on the incorporation of fatty acid ethyl esters (FAEE). Alcohol Alcohol 2009;44:287–92.

[93] Lee S, Choi B, Kim J, In S, Baeck S, Oh SM, et al. An LC-MS/MS method for the determination of five erectile dysfunction drugs and their selected metabolites in hair. J Chromatogr B 2015;978-979:1–10.

[94] Saisho K, Scott KS, Morimoto S, Nakahara Y. Hair analysis for pharmaceutical drugs. II. Effective extraction and determination of sildenafil (Viagra) and its *N*-desmethyl metabolite in rat and human hair. Biol Pharm Bull 2001;24:1384–8.

[95] Kim J, Park Y, Park M, Kim E, Yang W, Baeck S, et al. Simultaneous determination of five naphthoylindole-base synthetic cannabinoids and metabolites and their deposition in human and rat hair. J Pharm Biomed Anal 2015;102:162–75.

[96] Kronstrand R, Ahlner J, Dizdar N, Larson G. Quantitative analysis of desmethylselegiline, methamphetamine, and amphetamine I hair and plasma from Parkinson patients on long-term selegiline medication. J Anal Toxicol 2003;27:135–41.

[97] Cooper GAA. Hair testing is taking root. Ann Clin Biochem 2011;48:516–30.

[98] Agius R, Kintz P. Guidelines for European workplace drug and alcohol testing in hair. Drug Test Anal 2010;2:367–76.

[99] Scott KS. The use of hair as a toxicological tool in DFC casework. Sci Justice 2009;49:229–308.

[100] Villain M. Applications of hair in drug facilitated crime evidence. In: Kintz P, editor. Analytical and practical aspects of drug testing in hair. Boca Raton, FL: CRC Press; 2007. p. 255–71.

[101] Cooper GAA, Kronstrand R, Kintz P. Society of Hair Testing guidelines for drug testing. Forensic Sci Int 2012;218:20–4.

[102] Kintz P. Segmental hair analysis can demonstrate external contamination in post-mortem cases. Forensic Sci Int 2012;215:73–6.

[103] Le Beau MA, Montgomery MA, Brewer JD. The role of variations in growth rate and sample collection on interpreting results of segmental analyses of hair. Forensic Sci Int 2011;210:110–16.

[104] Cheze M, Gaulier JM. Drugs involved in drug-facilitated crimes (DFC), analytical aspects: 2—Hair. In: Kintz P, editor. Toxicological aspects of drug-facilitated crimes. London, UK: Academic Press; 2014. p. 181–222.

[105] Baciu T, Borrull F, Aguilar C, Calull M. Recent trends in analytical methods and separation techniques for drugs of abuse in hair. Anal Chim Acta 2015;856:1–26.

[106] Faculty of Forensic and Legal Medicine. Recommendations for the collection of forensic specimens from complainants and suspects. 2014 <www.fflm.ac.uk>.

[107] Kintz P. Bioanalytical procedures for detection of chemical agents in hair in the case of drug-facilitated crimes. Anal Bioanal Chem 2007;388:1467–74.

Chapter 2

Hair Sample Preparation, Extraction, and Screening Procedures for Drugs of Abuse and Pharmaceuticals

Robert Kronstrand[1,2], Malin Forsman[1] and Tor Seldén[1,2]
[1]*National Board of Forensic Medicine, Department of Forensic Genetics and Forensic Toxicology, Linköping, Sweden,* [2]*Division of Drug Research, Linköping University, Linköping, Sweden*

2.1 INTRODUCTION

The aims of this chapter are to give the reader a good insight into the practical preanalytical aspects of hair testing and also to inform about the different screening strategies that can be used. The chain of events leading up to the accurate interpretation of results starts already at the sampling site but this chapter will focus on the preparation of hair samples, performed at the laboratory, to ensure a valid result. This includes segmentation, homogenization, and the extraction of drugs from the hair matrix. After extraction, screening of drugs and medications is usually performed and if presumptive positive a confirmatory analysis is performed. Depending on the problem trying to be solved, the strategies for screening will vary. Common strategies include immunoassays that are quick, sensitive, and cost-effective, especially when the proportion of positive samples is low. However, chromatographic techniques are used more and more particularly by forensic laboratories where the positivity rate is much higher. Sometimes, the chromatographic screening methods are used as the end point even though, in contrast to forensic best practice, only one analysis has been performed. The reason is the high power of identification of specific analytes as well as accurate determination of their concentration that can be retrieved from such methods in combination with an increased security of sample integrity using barcodes and scanners throughout the laboratory. Also, the use of time-of-flight mass spectrometric methodology enables the analytical toxicologist to further explore many analytes in the same run even though true untargeted screening methods are rare. However, the analytical work begins with obtaining an appropriate and representative aliquot of the hair to be used for extraction and analysis.

2.2 SAMPLE PREPARATION

2.2.1 Segmentation

Even though analysis may be performed on the whole hair, it is very common to use one or more segments of the sample. A commonly used segment length is the 3-cm proximal portion representing hair grown during the last 3 months. However, depending on the case, different strategies for segmentation may apply. Several short segments can provide a much more detailed profile of an individual's drug exposure than a single segment. The accuracy of such segmental analysis depends on both the sampling and the segmentation procedure at the laboratory. Poor sampling procedures with unparallel hair strands may severely lower the segmental resolution.

When given a single dose of a drug or medication one would assume that only one segment of hair, grown at the time after administration, would be positive. However, even under controlled study conditions the drug may be found in more than one segment [1]. This is illustrated in Figure 2.1 depicting data from a study with single intake of flunitrazepam. The subject received a single 2 mg dose of flunitrazepam and hair samples were obtained after 2 and 4 months. The samples were segmented in 5 mm segments and analyzed with liquid chromatography tandem mass spectrometry (LC-MS/MS). In the 2-month sample the concentration of the metabolite, 7-aminoflunitrazepam, is high in S3 corresponding to the time of intake whereas the surrounding segments have much less drug. In the 4-month sample, segment S7 should represent the time of intake but both segments S7 and S8 have similar concentrations, significantly lower than those in the previous sample. This illustrates how the results are obscured over time when hair from different growth cycles is included in the measurement or if hair has shifted during sampling, transport, and segmentation. It is important to understand that when investigating a single intake, the use of large segments will dilute the incorporated drug and cause false negative results. Figure 2.2 illustrates in a schematic way, how the resolution in time is increasing with shorter segments and also that the measured concentrations increases because of less dilution from drug-free hair. The Society of Hair Testing (SoHT) recommends that only hair cut from the scalp with the root end identified should be subjected to segmental analysis. One should also be aware of the possibility that the hair is not cut right at the scalp, as recommended. Indeed, LeBeau et al. showed this in a study where they instructed experienced and unexperienced samplers to obtain hair samples [2].

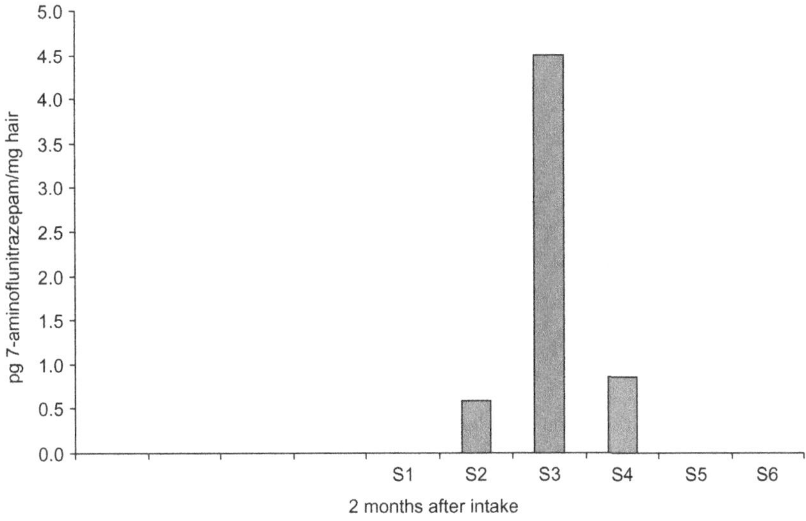

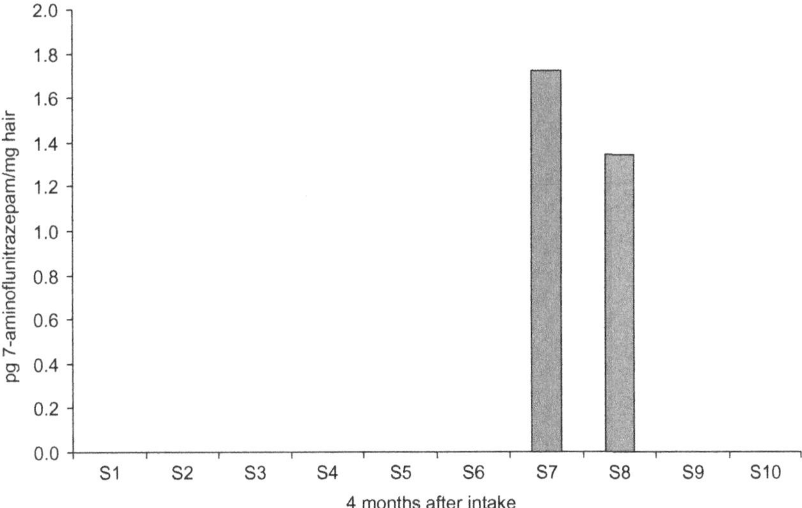

FIGURE 2.1 Segmental outcome of hair samples obtained from a subject approximately 2 and 4 months after a single intake of 2 mg flunitrazepam using 5 mm segments for analysis. Segments S3 and S7 correspond to the time of intake.

The mean length of hair remaining on the scalp was 0.8 cm. Therefore, one should be careful when extrapolating the hair length used to a particular time period using the sampling date as baseline.

Segmental hair analysis can also provide detailed temporal information about medication intake or drug abuse pattern. If the purpose is to evaluate the possible

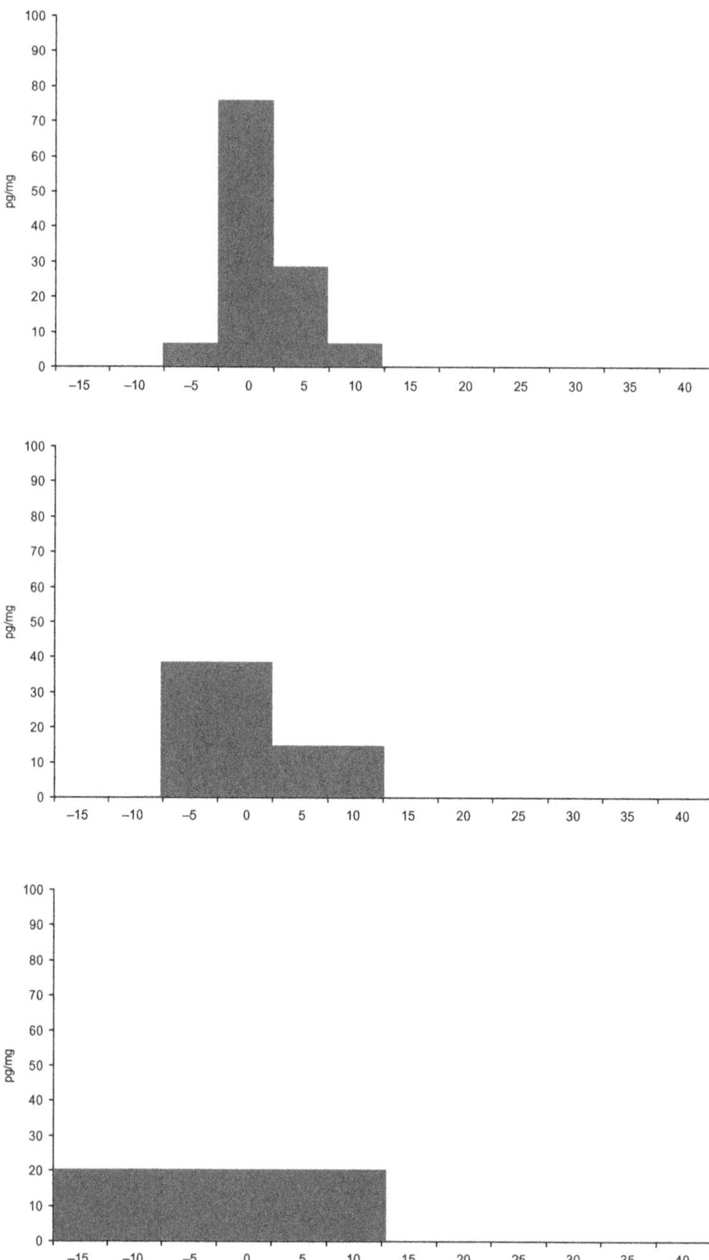

FIGURE 2.2 Schematic results from analysis of different segment lengths of a given hair sample; upper panel 0.5-cm segment, middle panel 1-cm segment, and lower panel 3-cm segment. Shorter segments provide not only higher resolution but also higher concentrations in the segment corresponding to the time of intake because the dilution with negative hair is less than in a longer segment.

tolerance or nontolerance to certain drugs or groups of drugs, the more recent hair segments are most relevant to analyze. For opioid drugs, tolerance develops gradually, as does the loss of tolerance. The time frame for these processes are poorly understood, and complicated by the fact that tolerance to the euphoric effects may be different from that to respiratory depression. A period of abstinence of 7–14 days is however likely to result in a marked reduction in tolerance to opioid drugs, implying that hair segments should be rather short to capture this time period accurately. Positive detections in hair segments of 1 cm or longer may not discriminate between intakes that occurred during a few days before the demise from exposure about 4 weeks back. Segmentation into 5 mm long segments seems to be relevant in opioid overdose cases to separate cases with a recent abstinence from cases with continuous use [3]. To accurately follow the distribution of drugs in the hair over time washing is important. The washing steps should be performed after the segmentation to avoid contamination along the hair shaft during the washing process.

2.2.2 Washing

Since hair is growing on the body surface there is a risk of contamination from outer sources. (This is described in Chapter 3 of this book.) Washing of the hair before analysis is therefore an important step. The washing has two main purposes: first, to remove hair care products, sweat, sebum, or surface material that may interfere with the analysis or that may reduce extraction recovery and second, to remove potential external contamination of drugs from the environment. However, there are probably as many washing procedures as there are laboratories and the debate about the correct way to perform and interpret washing results is still ongoing. The SoHT recommends that laboratories should have a procedure for washing hair samples prior to analysis, that the procedure should include washing steps with both organic solvent and aqueous solutions, and that the laboratory should investigate to what extent their wash procedure removes surface contamination [4].

The use of a first step with an organic solvent that does not swell the hair or penetrate the inner structure of hair will remove hair care products and surface contamination only whereas aqueous solutions also will extract drugs from the hair matrix. Today, it is accepted that drugs deposited into hair are from several sources including the blood stream, sweat, sebum, and other secretions. This means that a drug user or patient, in addition to incorporating drugs through the bloodstream also exposes the hair to an "internal external" contamination that can be very difficult to differentiate from solely external contamination. However, washing is not necessary in screening procedures unless to remove grease or debris that might affect the extraction recovery or sample homogeneity. A negative result can be reported with confidence and any positive screening results should be confirmed using another method that includes washing of the sample.

2.2.3 Obtaining a Representative Sample Aliquot

In contrast to body fluids, which are easily aliquoted prior to analysis, hair is a solid matrix that needs some sort of homogenization prior to analysis. Since hair as a matrix many times is limited in quantity, there is a tendency to use less and less hair for the analysis when more sensitive techniques and instrumentation become available. Still, it must be stressed that the fewer hairs that are used for the analysis, the less representative they are for the different hair growth cycles on the scalp. The SoHT recommends that hair samples are cut into smaller pieces and mixed or milled to a powder to make sure that a representative sample aliquot is used for analysis. The mixing/powdering as well as the amount of hair accurately weighed should be sufficient to ensure reproducibility. An aliquot of 10–50 mg is recommended for analysis. In reality that means that 50–100 hairs should be used for the analysis. The impact of powdering time on the analytical variation is shown in Figure 2.3. An authentic hair sample positive for several analytes were milled 6, 15, or 25 min before mixing and analysis of 10 mg aliquots in five replicates (see also Figure 2.4) by incubation in a solvent: buffer mixture. The results show that a short milling time increases the variation. The reason for the greater variation in less powdered hair might be the variable active surface in a given 10 mg aliquot as well as the less likelihood to aliquot of five equal samples. To ensure a representative and homogenous sample the original sample amount should be large enough and the powdering time should be long enough to obtain a fine powder as depicted in Figure 2.4D. The powdering of hair before aliquoting is not only a way to increase the surface area but also a homogenization that improves the analysis even if the hair sample is subject to an extraction that involves dissolution of the hair matrix.

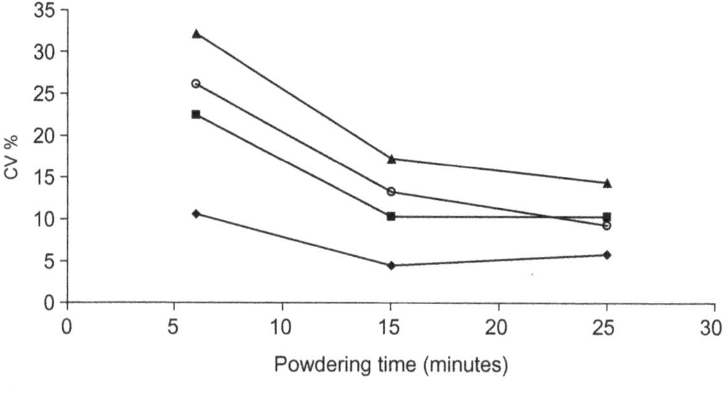

FIGURE 2.3 Impact of powdering time on the analytical variation using an authentic hair sample milled 6, 15, and 25 min before aliquoted into 10 mg portions ($N = 5$). A short milling time increases the variation due to inhomogeneous sample aliquots.

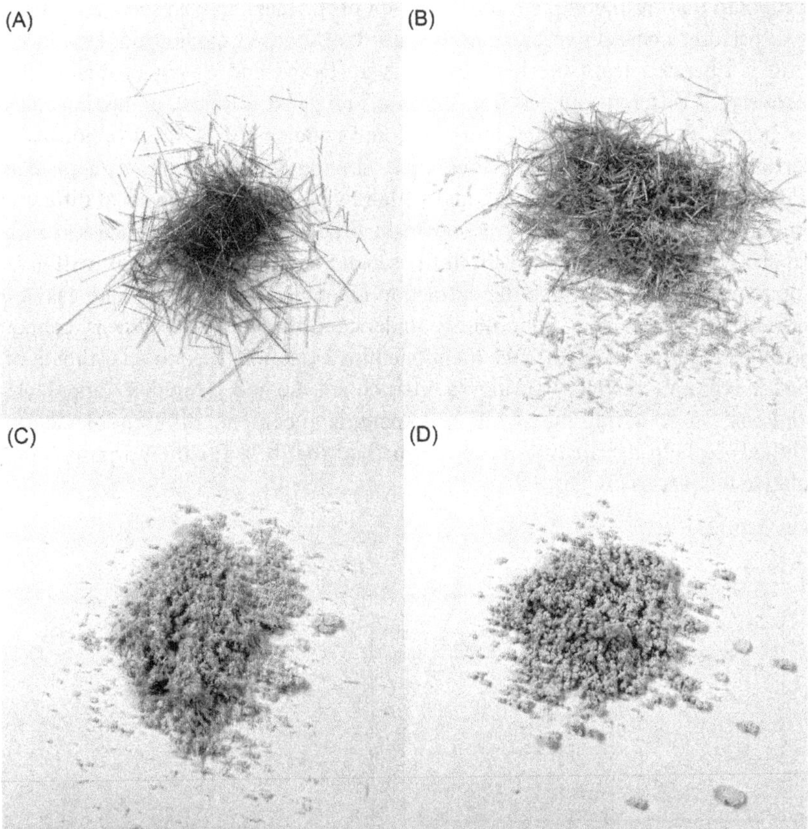

FIGURE 2.4 Photographs of hair samples powdered for different times. The intact hair before milling is shown in panel (A); (B) 6 min, (C) 15 min, or (D) 25 min. To ensure a representative and homogenous sample a fine powder should be obtained as shown in panel (D).

2.3 ANALYTE EXTRACTION

As mentioned above, the recovery of analytes from the hair matrix depends on several factors such as surface area, incubation time, the degree of hair disintegration, physical and chemical properties of solvents and analytes, and extraction conditions in general. Extraction recovery and reproducibility are important parameters for any analytical method and concerning hair, it must be evaluated using authentic hair from users of drugs or medications [4]. Several ways to evaluate extraction recovery has been used and published. One method is to compare with a reference method that is considered to have 100% recovery of the analytes from hair. Usually, such a method includes the disintegration of the hair matrix by chemical or enzymatic hydrolysis. This strategy was used by Kronstrand et al. to investigate the recovery of amphetamine and methylenedioxymethamphetamine (MDMA) from a solvent:buffer incubation and found it equal to that from complete solubilization of the hair [5]. The second strategy is to perform consecutive extractions with the proposed method until no more drug is liberated from the hair. In this way Rothe and Pragst evaluated the recovery of different opiates from hair and revealed different extraction rates for heroin, 6-acetylmorphine, morphine, and codeine [6]. A third option is to perform the extraction over a long time, driving the extraction to a plateau where no more drug is extracted, and withdraw aliquots for analysis at different times. Some examples of this is shown in Figure 2.5 and also discussed in a paper by Kronstrand et al. [5]. Both the second and the third strategy will give important information about the extraction kinetics and what drugs are easy or difficult to liberate. It is important to understand that the true recovery cannot easily be determined. Rather the mentioned investigations lead to an estimate of the maximum recovery using a particular method. Another approach, commonly used during method development, is to compare different extraction methods to help the analyst chose the method that best fits the purpose using relative recoveries [7–9].

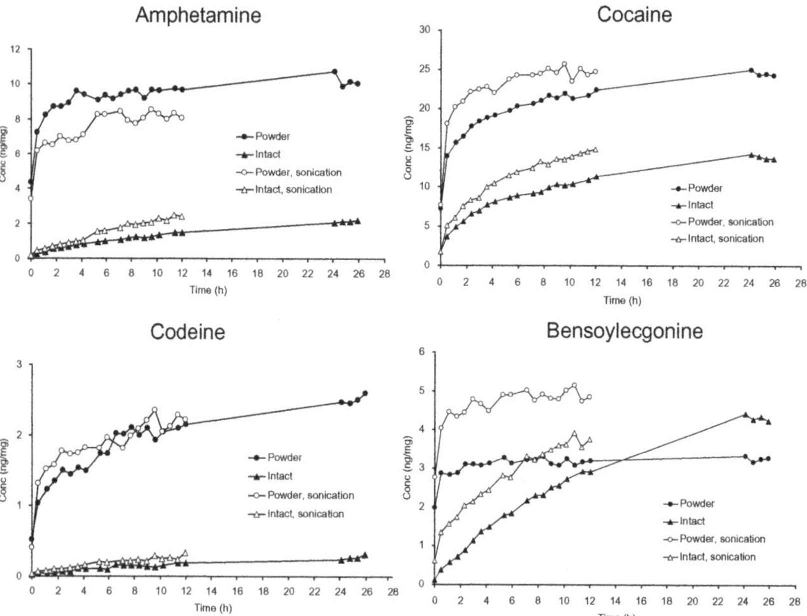

FIGURE 2.5 Comparison of two incubation procedures, using a shaking water bath or sonication. Five replicates of each experiment were performed using a pool of extensively washed authentic hair to ensure only extracting drug from within the hair. To 10 mg of either cut or powdered hair was added 2.0 mL of methanol and internal standard and the sample was incubated in a water bath with orbital shaking at 37°C for 26 h or in an ultrasonic bath at 50°C for 12 h. At 30 min intervals during the first 12 h a 20 μL aliquot was removed from the incubation. The investigated drugs were amphetamine, codeine, cocaine, and bensoylecgonine.

2.3.1 Hydrolysis of the Hair Matrix

In hair analysis, hydrolysis, or solubilization, means the disintegration of the hair matrix to liberate the incorporated substances as solutes in a liquid media enabling further extraction or direct analysis. Depending on the chemical properties of the analyte care is needed to keep the analyte intact during hydrolysis. Cocaine and 6-acetylmorphine are unstable at high pH and several benzodiazepines and their metabolites are unstable at both high and low pH [10–13]. During alkaline hydrolysis using sodium hydroxide, the hair matrix is completely destroyed and the analytes are liberated into the aqueous medium. Thus, alkaline hydrolysis ensures a high recovery for stable analytes such as amphetamines but the harsh condition makes it less suitable for general screening procedures where drugs from different groups are investigated. Incubation in diluted hydrochloric acid has been suggested as a method that is suitable also for unstable compounds [14,15]. Enzymatic hydrolysis has also been used and proven successful but is more expensive than chemical hydrolysis [13]. To be useful as a screening method, the extraction conditions need to be mild so that unstable compounds are protected from breakdown. Therefore, incubation in an organic solvent or a buffer has been extensively used.

2.3.2 Solvent Incubation

Solvent incubation requires special attention since the active surface of the sample is an important factor in recovery and reproducibility. In a series of experiments at the authors' laboratory this was investigated for amphetamine, codeine, cocaine, and bensoylecgonine.

Two incubation procedures were compared, one using a shaking water bath and a procedure using sonication. The experiments were performed on a pool of authentic hair. Prior to the extraction, hair samples were washed thoroughly following a standard protocol including an initial wash with isopropanol, and then three washes with phosphate buffer, and a final isopropanol wash. Thus, we aimed at only extracting drug from within the hair when comparing the extraction methods. Five replicates of each experiment were performed using either intact or powdered hair. To 10 mg of hair was added 2.0 mL of methanol and 50 μL of internal standard and the sample was incubated in a water bath with orbital shaking at 37°C for 26 h or in an ultrasonic bath at 50°C for 12 h. At 30 min intervals during the first 12 h a 20 μL aliquot was removed from the incubation mixture, diluted 1:2 with water and 1 μL was injected into the LC-MS/MS system. The LC-MS/MS analysis was performed on a SCIEX API 4000 MS/MS instrument equipped with an electrospray interface.

Comparing the 12 h fractions, sonication showed higher recoveries than incubation in a normal water bath but to a varying degree. In addition, powdered hair generally showed much more rapid kinetics initially, followed by a slower increase or a plateau for the remaining incubation. For all analytes except bensoylecgonine, the powdering of hair markedly increased the recovery; however the impact was different depending on the drug as depicted in the Figure 2.5. For amphetamine and codeine, the recovery was approximately five times higher using powdered hair whereas for cocaine it was twice as high. The slow extraction kinetics of codeine from intact hair was also demonstrated by Rothe et al. [6]. Different extraction recoveries for cocaine and bensoylecgonine from powdered and intact hair resulted in different metabolite/parent compound ratios. Using powdered hair, the ratio became 2–3 times lower because of the higher recovery of cocaine. In light of this, one should be careful to use general guidelines for parent/metabolite ratios since they might depend on the hair pretreatment or extraction procedure. Some authors have proposed ultrasonication as a good means to shorten the incubation time but in our experience it is rather the increased surface area from pulverization than the ultrasonication that shortens the time to maximum recovery. This has been verified by Broecker et al. [16] and Nielsen et al. [17].

Assuming that the average hair strand is cylindrical and has a diameter of 100 μM, the increase in surface area can be calculated based on the additional cross-section area. Theoretically, the total surface area is doubled at a length of about 40 μM, which corresponds to dividing a 1 cm strand into 250 segments. Subsequent cuts will exponentially increase the surface area. This means that cutting the hair in small segments with scissors will not significantly increase the ability for drugs to be extracted.

Miyaguchi et al. showed, using a scanning electron microscope, that micropulverization for 1 min highly damaged the cuticle layer and formed deep cracks [18]. Pulverization for more than 3 min resulted in the disintegration of cortical fibers, which contain melanin granules that are the main site of interaction for basic drugs. The destruction of the cuticle layer promotes permeation of a solvent into the inner parts of hair, which is necessary for extraction of drugs located in cortex and medulla. Disintegration of cortical fibers would strongly promote and facilitate the extraction of basic drugs that are believed to strongly bind to melanin.

In several recent studies, micropulverized extraction (MPE) of hair samples has been used [7–9,18–22]. The MPE methodology allows a rapid simultaneous pulverization and extraction of hair samples into appropriate solvent using metal beads in a small plastic tube. In some reports, the hair is micropulverized first and then solvent is added to the tube. In contrast to the SoHT guidelines regarding the amount of hair used for analysis, some of the reported MPE methods use as little as 2 mg of hair. However, we believe that this procedure will become a very important method for general screening of hair samples in the future. Extraction efficiencies have been determined for several drugs in comparison to conventional extraction methods.

2.4 SCREENING STRATEGIES

Screening refers to the sieving out the presumptive positive samples for further analysis. Historically immunoassays have been the primary means to do this in both clinical and forensic toxicology using several techniques over the years [23–33]. Immunoassays are based on the cross-reactivity of several substances to an antibody and the response is the sum of all binding substances. This can sometimes complicate both qualitative and quantitative results from a corresponding confirmatory analysis. More recently, chromatographic techniques have been proposed as possible screening tools [5,16,17,34–45]. An advantage with chromatography is that one can design the screening method to cover all the analytes that are included in the confirmation methods and customize the reporting threshold for each analyte, overriding the drawbacks with immunoassays. Chromatographic methods, however, are more expensive and labor-intensive than immunoassays and inconvenient when screening large numbers of samples, especially if those are mainly negative.

2.4.1 Immunoassays

The first report about the incorporation of drugs into hair came in the mid-1950s by Goldblum [46] who used spectrophotometry to detect barbiturates in the hair from dosed guinea pigs. However, it took until the groundbreaking efforts of Baumgartner's group in the late 1970s before hair analysis for drugs was introduced on a broader scale using radioimmunoassay [23,25]. For years radioimmunoassay was the only technique sensitive enough to detect the minute traces of drugs in human hair. Today, the dominating immunoassay for screening is enzyme-linked immunosorbent assay (ELISA). ELISA has been successfully used for the preliminary analysis of amphetamines [22,47–49], cocaine, opiates [47,49–52], cannabinoids [47,49], and benzodiazepines [47,49,53]. Typically 10–20 mg of hair is incubated in organic solvent or in buffer to liberate the drugs from the hair matrix.

A small volume of the incubate is then added to antibody-coated wells followed by the addition of an enzyme-conjugated drug that competes with the free drug in the sample for binding sites. After incubation and washing of the wells to remove unbound drug an enzyme substrate is added. After a second incubation the reaction is stopped by the addition of acid and the absorbance is read [51]. The absorbance is inversely proportional to the concentrations of drug in the sample. ELISA can be used in both qualitative mode with a threshold calibrator run in each batch of samples or quantitatively using multiple calibrators. Most of the published methods using ELISA are quantitative. Table 2.1 gives some characteristics of methods recently published.

TABLE 2.1 Characteristics of Recently Published ELISA Methods

Year	Reference	Substance Groups	Hair Amount (mg)	ELISA Kit	Quantitation	Cut-Off (ng/mg)	Confirmation in New Extract	Validation of Specificity and Sensitivity
2013	[47]	Amphetamines, cannabinoids, cocaine, opiates, methadone, and benzodiazepines	10, powdered	Immunalysis	Yes	Amphetamines 0.1, cannabinoids 0.02, cocaine 0.1, opiates 0.1, methadone 0.1, benzodiazepines 0.05	Yes, GC-MS, LC-MS/MS	ROC
				LUCIO-Direct				
2010	[27]	Opiates, methamphetamine, amphetamine, cocaine, and THC	10	Immunalysis	Yes	Opiates 0.2, cocaine 0.5, methamphetamine, and amphetamine 0.2, THC 0.1	No	No
2009	[21]	Methamphetamine	1	microELISA	Yes	0.2	Yes, LC-MS/MS	No
2007	[49]	THC, cocaine, opiates, amphetamines, methamphetamine, MDMA	50	One-Step	Yes	THC 0.1	No, GC-MS	Sensitivity 100%
						Cocaine 0.5 opiates/amphetamine 0.2		Specificity 70–90%
2006	[48]	Methamphetamine	10	Immunalysis	Yes	0.5	Yes, GC-MS	Sensitivity 97%
								Specificity 100%

(Continued)

TABLE 2.1 (Continued)

Year	Reference	Substance Groups	Hair Amount (mg)	ELISA Kit	Quantitation	Cut-Off (ng/mg)	Confirmation in New Extract	Validation of Specificity and Sensitivity
2006	[53]	Benzodiazepines	10	Immunalysis	Yes	0.1	Yes, LC-MS/MS	Sensitivity 100%
								Specificity 81%
2005	[52]	Methadone	20–30	Cozart Microplate	Yes	0.2	Yes, GC-MS	ROC
								Sensitivity 95 ± 2%
								Specificity of 100 ± 3.5%
2004	[50]	Buprenorphine	30, powdered	One-Step	Yes	0.01	Yes, LC-MS	Yes
2003	[51]	Opiates	20–30	Cozart Microplate	Yes	0.2–0.3	Yes, GC-MS	ROC
								Sensitivity 98% ± 2%
								Specificity of 92.7% ± 3.5%

THC, tetrahydrocannabinol

Svazier and coworkers [33] published a procedure using a commercial reagent kit called VMA-T, a basic aqueous buffer, used to liberate buprenorphine from hair prior to immunoassay screening. More recently, the procedure was reported for several drugs of abuse by another Italian research group, Pichini and coworkers, as well as from groups in Switzerland and Spain [24,28,31]. The procedure specifies that 33 mg of finely cut hair is washed with 1 mL of SLV-VMA-T for 10–20 s, and that the latter is discarded. After incubation with 0.4 mL VMA-T at 100°C for 1 h, the samples are cooled to room temperature and screened with immunometric methods. This procedure eliminates time-consuming extraction procedures commonly used, making it suitable for laboratories with high workloads. The extracts are also readily available for immunological screening assays on already existing equipment, providing an alternative to urine testing. According to the manufacturer of VMA-T, cocaine and 6-monoacetylmorphine are hydrolyzed to bensoylecgonine and morphine, respectively, during the procedure. Baumgartner et al. [24] reported that cocaine was indeed completely hydrolyzed, thus making it impossible to discriminate between cocaine consumption and external contamination. However, other authors have suggested the measurement of additional metabolites to rule out contamination [54,55]. The heroin metabolite 6-monoacetylmorphine, on the other hand, was only slightly degraded, enabling discrimination between heroin, codeine, and morphine consumption.

Regardless of the screening technique, they operate with thresholds for a presumptive positive result. Recommended thresholds for the most common drug groups have been published by the SoHT [4]. The sensitivity and specificity of a screening method should be determined in-house using the screening-confirmation method pair. In order to do this one needs to analyze both positive and negative samples with both methods and calculate the true positives, true negatives, false positives, and false negatives for each analyte or analyte group. The proportion of positive samples in the population of interest should be taken into account when choosing samples for the sensitivity/specificity study and the number of samples included should not be less than 50. An example of an evaluation of a screening method and a confirmation method with fixed thresholds was published by Lopez et al. in 2010 [56] where they showed a sensitivity of 89.5% and a specificity of 100% for their ELISA cocaine assay compared to their LC-MS/MS confirmation method. Another approach is to use receiver operating curves (ROCs) to establish the screening threshold that provides the best sensitivity and specificity [47,52].

2.4.2 Chromatographic Procedures

The development of chromatographic techniques for the determination of drugs and medications in hair has followed the same path as that for the common matrices of blood and urine with a progression from liquid chromatography ultra violet (LC-UV) [36] to gas chromatography mass spectrometry (GC-MS) [37,43], LC-MS/MS [5,38,41], and more recently liquid chromatography time of flight (LC-TOF) [16,17,34,40,44,57]. Chromatographic techniques have historically been used for confirmation and quantitation but after the introduction of bench-top LC-MS/MS in the early 2000s [5] Kronstrand et al. presented a preliminary screening method using a simple incubation of the hair in a buffer:organic solvent mixture and subsequent direct analysis with LC-MS/MS that included 14 analytes from the most common groups of drugs of abuse. The method's intended use was to investigate drug-related deaths and all results were confirmed in a new sample cut in shorter segments and analyzed with a different technique. Another paper by Hegstad et al. was published in 2008 using a similar incubation methodology but with 22 analytes also including some medications with abuse potential [38]. An important difference between the methods was that Hegstad et al. used the method as end point and no further analysis was carried out. A similar approach was reported by Lendoiro et al. in 2012 [41] but with a reinjection of the extract acquiring several transitions and using ratios to confirm the identity of an analyte. Using mass spectrometric methods indeed increases the selectivity and it is not uncommon to report positive findings from a single determination or a single sample aliquot. However, the well-established forensic praxis of two agreeing determinations to ensure a true positive is eliminated.

LC-MS/MS methods are almost exclusively targeted and operated in multiple reaction mode measuring a set of predefined transitions for the analytes. Systematic toxicological analysis (STA) has been reported on matrices such as blood or urine but for hair they are scarcer. High-resolution mass spectrometry and time of flight technology do enable the analyst to acquire and analyze much more data than an LC-MS/MS method. In 2008, Pelander et al. reported a screening method for basic drugs in hair [44]. After basic hydrolysis and solid phase extraction of 150–250 mg of hair, the samples were analyzed with a 29-min chromatography and the data matched to an in-house database comprised of 815 compounds. The authors predicted that LC-TOF will become a more prominent methodology to comprehensively screen for drugs also in hair samples. The next effort again turned to a more targeted screening with a publication from Nielsen et al. in 2010 [17]. Now, instrumentation was more sensitive requiring only 10 mg hair for the analysis and the use of ultrahigh performance liquid chromatography separated 52 substances within 17 min. limit of detection (LOD) and limit of quantitation (LOQ) were determined for all analytes, in contrast to the previous work by Pelander where only two analytes were validated. Nielsen used the method as an end point quantitative method reporting concentrations of all drugs. In 2011, Dominguez-Romero et al. [34] reported another combined screening and quantitative method targeting 30 substances from different classes of abused drugs. Identification of substances was achieved by matching the retention time and accurate mass even though additional information from in source fragmentation was used as complementary information. An STA method was published by Broecker et al. in 2012 where they used liquid chromatography hybrid quadrupole time-of-flight mass spectrometry to comprehensively screen hair for more than 2,500 substances and metabolites [16]. Validation was performed on 24 representative substances from different drug of abuse groups. Results were not confirmed with another technique but reported using identification by accurate mass and spectra comparison. A recent report by Kronstrand et al. confirmed the application of ultrahigh performance liquid chromatography with time-of-flight mass spectrometry for the targeted screening of drugs and medications in unwashed intact hair [40]. Using a simple incubation of hair in a solvent:buffer mixture and direct analysis of the extract they were able to screen for 30 drugs at the recommended thresholds provided by the SoHT. The method was used as a screening with a daily threshold calibrator and all presumptive positive results were confirmed and quantified with another technique. A comparison between the screening and confirmation in 267 positive results revealed 15% false positives. These were typically low concentrations of codeine and buprenorphine. In summary, the chromatographic techniques commonly used to screen for drugs in hair today are either LC-MS/MS or LC-TOF methods, and the most common strategy is to report the identification of a drug either qualitatively or quantitatively directly from the screening method, sometimes after a confirmatory reinjection acquiring additional transitions or spectra. For LC-MS/MS, the identification criteria have been adapted from those recommended for blood or urine analysis. That is to confirm the identity using recommended ranges for relative retention time and transition ratios compared to injected reference materials. For TOF methods, there are still no recommendations for identification parameters, let alone ranges that they should fall within. Each instrument supplier has their own algorithms to produce scores or hit lists. Therefore, the laboratory should carefully consider this before launching end point TOF methods. Table 2.2 summarizes the existing LC-TOF methods for hair together with their identification criteria.

An important advantage of chromatographic screening methods, especially those that use TOF, is the relative ease by which one can add new substances into the method. Data acquisition using TOF is comprehensive, with all m/z recorded over the chromatographic run. This means that additional analytes can be validated into a method without changing the acquisition parameters. An increasingly important advantage is that many new drugs of abuse are introduced on the illegal market every year in addition to an increased internet sale of medications regardless of a country's pharmaceutical specialties.

TABLE 2.2 Characteristics of Recently Published LC-(Q)-TOF Methods

Year	Reference	Substances	Hair Amount (mg)	Extraction from Hair	Purification/ Concentration	Chromatography	Detection/ Acquisition	Identification Parameters	ID Criteria	Quantitation
2014	[57]	(8 compounds)	100	Intact hair	LLE	Zorbax eclipse XDB (2.1 × 150 mm, 5 μm)	ES + QTOF	Accurate mass	Not stated	Yes
		Synthetic cannabinoids		Incubation overnight 45°C in 0.5M NaOH		Formic acid/methanol		Isotopic pattern recognition		
2013	[40]	(30 compounds)	20	Intact hair	None	Acquity UPLC T3 (2.1 × 150 mm, 1.8 μm)	ES + TOF	Retention time	±0.15 min	No
		Amphetamines		Incubation overnight ACN: MeOH:buffer (10:10:80)		Formiate:ACN		Accurate mass	±10 ppm	
		Opiates						Isotopic pattern		
		Opioids						Combined score	>80 (max100)	
		Cocaine								
		Bensodiazepines								
		Z-drugs								

(Continued)

TABLE 2.2 (Continued)

Year	Reference	Substances	Hair Amount (mg)	Extraction from Hair	Purification/ Concentration	Chromatography	Detection/ Acquisition	Identification Parameters	ID Criteria	Quantitation
2012	[16]	2,500 (7,500)	20	Intact hair	Evaporation	Zorbax eclipse C18 (2.1 × 100 mm, 3.5 μm)	ES + QTOF	Accurate mass	Not stated	Yes
				Incubation 2 × 18 h at 37°C ACN: MeOH:buffer (25:25:50)		NH4Ac:MeOH		Spectra match		Selected compounds
2011	[34]	(30 compounds)	20	Intact hair	Evaporation	XDB-C18 (4.6 × 50 mm, 1.8 μm)	ES + TOF	Retention time	±0.2 min	Yes
		Amphetamines		Incubation with ultrasonication 8 h at 50°C MeOH		Formic acid:ACN		Accurate mass (screening)	±10 ppm	
		Opiates								
		Opioids								
		Cocaine								
		THC						Accurate mass (confirmation) + additional CID fragmentation	±5 ppm	
		Bensodiazepines								
		Steroids								
		Selected pharmaceuticals								

(*Continued*)

TABLE 2.2 (Continued)

Year	Reference	Substances	Hair Amount (mg)	Extraction from Hair	Purification/ Concentration	Chromatography	Detection/ Acquisition	Identification Parameters	ID Criteria	Quantitation
2010	[17]	(52 compounds)	10	Intact hair	None	Acquity HSS T3 (2.1 × 100 mm, 1.8 μm)	ES + TOF	Retention time	±0.2 min	
		Amphetamines		Incubation 18 h at 37°C ACN: MeOH:buffer (25:25:50)		Formic acid:MeOH		Accurate mass	15 mDa	
		Analgesics								
		Antidepressants								
		Antipsychotics						Isotopic pattern		
		Bensodiazepines								
		Cocaine								
		Opioids								
		Ketamine								
2008	[44]	(815 compounds	250	Intact hair	SPE	Luna C-18 (2 × 100 mm, 3 μm)	ES + TOF	Accurate mass	<5 ppm	No
		Basic drugs and medications		Dissolution in 1M NaOH at 100°C for 1 h		NH4Ac:ACN		Isotopic pattern	<0.03 (SigmaFit)	
								Retention time (when available)	±0.1 min	

CID, collision induced dissociation; SPE, soild phase extraction; LC-(Q)-TOF, liquid chromatography quadrupole time of flight

2.5 CONCLUDING REMARKS

In a chain of events each step is important and has impact on the final result. In this chapter the first part concerns the pretreatments of hair. Segmentation, which is the foundation for temporal interpretations and should be undertaken with the at most care. Failing to accurately segment the hair sample will lead to obscure results and to false conclusions. Extraction of the drugs from the hair matrix is the foundation for quantitative interpretations. Whenever the toxicologist compares the quantitative results with a previous sample, a dataset from the own laboratory or even with the literature one needs confidence in that each time an extraction is performed, it is quantitative or that there is knowledge about the method recovery as well as the variation in recovery for the analytes measured. Without this, a quantitative value is of little use. From our review of the literature, we have found that few papers actually address this and few have validated their methods with authentic hair samples trying to estimate the true recovery. Another important factor is how to obtain a representative aliquot of the material sent for analysis. It seems common today to weigh very small sample amounts and use those directly for the analysis. Doing that decreases the probability to obtain a representative aliquot. For each method, sample aliquoting should be investigated and only the use of authentic samples can shed light over the variation in quantitative result.

The second part of this chapter concerns the simultaneous screening of many substances in hair as a means to exclude or verify use of drugs. For large-scale screening, immunoassays are by far the most cost-effective, especially if the majority of samples are negative. In our opinion, ELISA seems to be the state-of-the-art methodology used today. Results are always confirmed by another technique, however not necessarily on a new portion of hair. Another observation is that many of the chromatographic screening methods are used as end point methods from which results are reported. The lack of criteria for a positive result in high resolution mass spectrometry must be addressed in the clinical and forensic community in general and guidelines adopted by the professional organizations. Finally, the scope of screening methods should be continuously updated to include the drugs and medications used at the time of analysis.

REFERENCES

[1] Jakobsson G, Kronstrand R. Segmental analysis of amphetamines in hair using a sensitive UHPLC-MS/MS method. Drug Test Anal 2014;6(Suppl. 1):22–9.

[2] LeBeau MA, Montgomery MA, Brewer JD. The role of variations in growth rate and sample collection on interpreting results of segmental analyses of hair. Forensic Sci Int 2011;210:110–16.

[3] Druid H, Strandberg JJ, Alkass K, Nystrom I, Kugelberg FC, Kronstrand R. Evaluation of the role of abstinence in heroin overdose deaths using segmental hair analysis. Forensic Sci Int 2007;168:223–6.

[4] Cooper GA, Kronstrand R, Kintz P, Society of Hair Testing. Society of Hair Testing guidelines for drug testing in hair. Forensic Sci Int 2012;218:20–4.

[5] Kronstrand R, Nystrom I, Strandberg J, Druid H. Screening for drugs of abuse in hair with ion spray LC-MS-MS. Forensic Sci Int 2004;145:183–90.

[6] Rothe M, Pragst F. Solvent optimization for the direct extraction of opiates from hair samples. J Anal Toxicol 1995;19:236–40.

[7] Miyaguchi H. Determination of zolpidem in human hair by micropulverized extraction based on the evaluation of relative extraction efficiency of seven psychoactive drugs from an incurred human hair specimen. J Chromatogr A 2013;1293:28–35.

[8] Favretto D, Vogliardi S, Stocchero G, Nalesso A, Tucci M, Ferrara SD. High performance liquid chromatography-high resolution mass spectrometry and micropulverized extraction for the quantification of amphetamines, cocaine, opioids, benzodiazepines, antidepressants and hallucinogens in 2.5 mg hair samples. J Chromatogr A 2011;1218:6583–95.

[9] Kim JY, Shin SH, Lee JI, In MK. Rapid and simple determination of psychotropic phenylalkylamine derivatives in human hair by gas chromatography-mass spectrometry using micro-pulverized extraction. Forensic Sci Int 2010;196:43–50.

[10] Miller EI, Wylie FM, Oliver JS. Simultaneous detection and quantification of amphetamines, diazepam and its metabolites, cocaine and its metabolites, and opiates in hair by LC-ESI-MS-MS using a single extraction method. J Anal Toxicol 2008;32:457–69.

[11] Polettini A, Stramesi C, Vignali C, Montagna M. Determination of opiates in hair. Effects of extraction methods on recovery and on stability of analytes. Forensic Sci Int 1997; 84:259–69.

[12] Romolo FS, Rotolo MC, Palmi I, Pacifici R, Lopez A. Optimized conditions for simultaneous determination of opiates, cocaine and bensoylecgonine in hair samples by GC-MS. Forensic Sci Int 2003;138:17–26.

[13] Kronstrand R, Nystrom I, Josefsson M, Hodgins S. Segmental ion spray LC-MS-MS analysis of benzodiazepines in hair of psychiatric patients. J Anal Toxicol 2002;26:479–84.

[14] Girod C, Staub C. Analysis of drugs of abuse in hair by automated solid-phase extraction, GC/EI/MS and GC ion trap/CI/MS. Forensic Sci Int 2000;107:261–71.

[15] Cordero R, Paterson S. Simultaneous quantification of opiates, amphetamines, cocaine and metabolites and diazepam and metabolite in a single hair sample using GC-MS. J Chromatogr B Analyt Technol Biomed Life Sci 2007;850:423–31.

[16] Broecker S, Herre S, Pragst F. General unknown screening in hair by liquid chromatography-hybrid quadrupole time-of-flight mass spectrometry (LC-QTOF-MS). Forensic Sci Int 2012;218:68–81.

[17] Nielsen MK, Johansen SS, Dalsgaard PW, Linnet K. Simultaneous screening and quantification of 52 common pharmaceuticals and drugs of abuse in hair using UPLC-TOF-MS. Forensic Sci Int 2010;196:85–92.

[18] Miyaguchi H, Kakuta M, Iwata YT, Matsuda H, Tazawa H, Kimura H, et al. Development of a micropulverized extraction method for rapid toxicological analysis of methamphetamine in hair. J Chromatogr A 2007;1163:43–8.

[19] Favretto D, Vogliardi S, Stocchero G, Nalesso A, Tucci M, Terranova C, et al. Determination of ketamine and norketamine in hair by micropulverized extraction and liquid chromatography-high resolution mass spectrometry. Forensic Sci Int 2013;226:88–93.

[20] Inagaki S, Makino H, Fukushima T, Min JZ, Toyo'oka T. Rapid detection of ketamine and norketamine in rat hair using micropulverized extraction and ultra-performance liquid chromatography-electrospray ionization mass spectrometry. Biomed Chromatogr 2009;23: 1245–50.

[21] Miyaguchi H, Iwata YT, Kanamori T, Tsujikawa K, Kuwayama K, Inoue H. Rapid identification and quantification of methamphetamine and amphetamine in hair by gas chromatography/mass spectrometry coupled with micropulverized extraction, aqueous acetylation and microextraction by packed sorbent. J Chromatogr A 2009;1216:4063–70.

[22] Miyaguchi H, Takahashi H, Ohashi T, Mawatari K, Iwata YT, Inoue H, et al. Rapid analysis of methamphetamine in hair by micropulverized extraction and microchip-based competitive ELISA. Forensic Sci Int 2009;184:1–5.

[23] Baumgartner AM, Jones PF, Baumgartner WA, Black CT. Radioimmunoassay of hair for determining opiate-abuse histories. J Nucl Med 1979;20:748–52.

[24] Baumgartner MR, Guglielmello R, Fanger M, Kraemer T. Analysis of drugs of abuse in hair: evaluation of the immunochemical method VMA-T vs. LC-MS/MS or GC-MS. Forensic Sci Int 2012;215:56–9.

[25] Baumgartner WA, Black CT, Jones PF, Blahd WH. Radioimmunoassay of cocaine in hair: concise communication. J Nucl Med 1982;23:790–2.

[26] Cheong JC, Suh S, Ko BJ, Lee JI, Kim JY, Suh YJ, et al. Screening method for the detection of methamphetamine in hair using fluorescence polarization immunoassay. J Anal Toxicol 2013;37:217–21.

[27] Coulter C, Tuyay J, Taruc M, Moore C. Semi-quantitative analysis of drugs of abuse, including tetrahydrocannabinol in hair using aqueous extraction and immunoassay. Forensic Sci Int 2010;196:70–3.

[28] de la Torre R, Civit E, Svaizer F, Lotti A, Gottardi M, Miozzo M. High throughput analysis of drugs of abuse in hair by combining purposely designed sample extraction compatible with immunometric methods used for drug testing in urine. Forensic Sci Int 2010;196:18–21.

[29] Haasnoot W, Stouten P, Schilt R, Hooijerink D. A fast immunoassay for the screening of beta-agonists in hair. Analyst (Lond) 1998;123:2707–10.

[30] Lachenmeier K, Musshoff F, Madea B. Determination of opiates and cocaine in hair using automated enzyme immunoassay screening methodologies followed by gas chromatographic-mass spectrometric (GC-MS) confirmation. Forensic Sci Int 2006;159:189–99.

[31] Pichini S, Gottardi M, Marchei E, Svaizer F, Pellegrini M, Rotolo MC, et al. Rapid extraction, identification and quantification of drugs of abuse in hair by immunoassay and ultra-performance liquid chromatography tandem mass spectrometry. Clin Chem Lab Med 2014;52:679–86.

[32] Segura J, Stramesi C, Redon A, Ventura M, Sanchez CJ, Gonzalez G, et al. Immunological screening of drugs of abuse and gas chromatographic-mass spectrometric confirmation of opiates and cocaine in hair. J Chromatogr B Biomed Sci Appl 1999;724:9–21.

[33] Svaizer F, Lotti A, Gottardi M, Miozzo MP. Buprenorphine detection in hair samples by immunometric screening test: preliminary experience. Forensic Sci Int 2010;196:118–20.

[34] Dominguez-Romero JC, Garcia-Reyes JF, Molina-Diaz A. Screening and quantitation of multiclass drugs of abuse and pharmaceuticals in hair by fast liquid chromatography electrospray time-of-flight mass spectrometry. J Chromatogr B Analyt Technol Biomed Life Sci 2011;879:2034–42.

[35] Fisichella M, Morini L, Sempio C, Groppi A. Validation of a multi-analyte LC-MS/MS method for screening and quantification of 87 psychoactive drugs and their metabolites in hair. Anal Bioanal Chem 2014;406:3497–506.

[36] Gaillard Y, Pepin G. Screening and identification of drugs in human hair by high-performance liquid chromatography-photodiode-array UV detection and gas chromatography-mass spectrometry after solid-phase extraction. A powerful tool in forensic medicine. J Chromatogr A 1997;762:251–67.

[37] Gentili S, Cornetta M, Macchia T. Rapid screening procedure based on headspace solid-phase microextraction and gas chromatography-mass spectrometry for the detection of many recreational drugs in hair. J Chromatogr B Analyt Technol Biomed Life Sci 2004;801:289–96.

[38] Hegstad S, Khiabani HZ, Kristoffersen L, Kunoe N, Lobmaier PP, Christophersen AS. Drug screening of hair by liquid chromatography-tandem mass spectrometry. J Anal Toxicol 2008;32:364–72.

[39] Koster RA, Alffenaar JW, Greijdanus B, VanDernagel JE, Uges DR. Fast and highly selective LC-MS/MS screening for THC and 16 other abused drugs and metabolites in human hair to monitor patients for drug abuse. Ther Drug Monit 2014;36:234–43.

[40] Kronstrand R, Forsman M, Roman M. A screening method for 30 drugs in hair using ultrahigh-performance liquid chromatography time-of-flight mass spectrometry. Ther Drug Monit 2013;35:288–95.

[41] Lendoiro E, Quintela O, de Castro A, Cruz A, Lopez-Rivadulla M, Concheiro M. Target screening and confirmation of 35 licit and illicit drugs and metabolites in hair by LC-MSMS. Forensic Sci Int 2012;217:207–15.

[42] Montenarh D, Hopf M, Warth S, Maurer HH, Schmidt P, Ewald AH. A simple extraction and LC-MS/MS approach for the screening and identification of over 100 analytes in eight different matrices. Drug Test Anal 2014.

[43] Paterson S, McLachlan-Troup N, Cordero R, Dohnal M, Carman S. Qualitative screening for drugs of abuse in hair using GC-MS. J Anal Toxicol 2001;25:203–8.

[44] Pelander A, Ristimaa J, Rasanen I, Vuori E, Ojanpera I. Screening for basic drugs in hair of drug addicts by liquid chromatography/time-of-flight mass spectrometry. Ther Drug Monit 2008;30:717–24.

[45] Villain M, Concheiro M, Cirimele V, Kintz P. Screening method for benzodiazepines and hypnotics in hair at pg/mg level by liquid chromatography-mass spectrometry/mass spectrometry. J Chromatogr B Analyt Technol Biomed Life Sci 2005;825:72–8.

[46] Goldblum RW, Goldbaum LR, Piper WN. Barbiturate concentrations in the skin and hair of guinea pigs. J Invest Dermatol 1954;22:121–8.

[47] Agius R, Nadulski T. Utility of ELISA screening for the monitoring of abstinence from illegal and legal drugs in hair and urine. Drug Test Anal 2014;6(Suppl. 1):101–9.

[48] Han E, Miller E, Lee J, Park Y, Lim M, Chung H, et al. Validation of the immunalysis microplate ELISA for the detection of methamphetamine in hair. J Anal Toxicol 2006;30:380–5.

[49] Pujol ML, Cirimele V, Tritsch PJ, Villain M, Kintz P. Evaluation of the IDS One-Step ELISA kits for the detection of illicit drugs in hair. Forensic Sci Int 2007;170:189–92.

[50] Cirimele V, Etienne S, Villain M, Ludes B, Kintz P. Evaluation of the One-Step ELISA kit for the detection of buprenorphine in urine, blood, and hair specimens. Forensic Sci Int 2004;143:153–6.

[51] Cooper G, Wilson L, Reid C, Baldwin D, Hand C, Spiehler V. Validation of the Cozart microplate ELISA for detection of opiates in hair. J Anal Toxicol 2003;27:581–6.

[52] Cooper G, Wilson L, Reid C, Baldwin D, Hand C, Spiehler V. Comparison of Cozart microplate ELISA and GC-MS detection of methadone and metabolites in human hair. J Anal Toxicol 2005;29:678–81.

[53] Miller EI, Wylie FM, Oliver JS. Detection of benzodiazepines in hair using ELISA and LC-ESI-MS-MS. J Anal Toxicol 2006;30:441–8.

[54] Hoelzle C, Scheufler F, Uhl M, Sachs H, Thieme D. Application of discriminant analysis to differentiate between incorporation of cocaine and its congeners into hair and contamination. Forensic Sci Int 2008;176:13–18.

[55] Morris-Kukoski CL, Montgomery MA, Hammer RL. Analysis of extensively washed hair from cocaine users and drug chemists to establish new reporting criteria. J Anal Toxicol 2014.

[56] Lopez P, Martello S, Bermejo AM, De Vincenzi E, Tabernero MJ, Chiarotti M. Validation of ELISA screening and LC-MS/MS confirmation methods for cocaine in hair after simple extraction. Anal Bioanal Chem 2010;397:1539–48.

[57] Gottardo R, Sorio D, Musile G, Trapani E, Seri C, Serpelloni G, et al. Screening for synthetic cannabinoids in hair by using LC-QTOF MS: a new and powerful approach to study the penetration of these new psychoactive substances in the population. Med Sci Law 2014;54:22–7.

Chapter 3

External Contamination: Still a Debate?

Craig Chatterton
Office of The Chief Medical Examiner, in Edmonton, AB, Canada

3.1 INTRODUCTION

Drug in hair analysis is a critical and increasingly common component of modern forensic science. It is a powerful evidential tool that has been relied upon in many criminal cases during the last decade.

The most crucial issue regarding hair analysis and subsequent interpretation of analytical results is the avoidance of technical or evidentiary false positives [1]. This chapter provides a brief overview on how drugs can be incorporated into hair, for example, via body fluids as a result of ingestion; via body fluids as a result of passive inhalation; and via direct external contamination of the hair, that is, by exposure to smoke, fumes, and/or poor housekeeping. This chapter focuses on the various forms of external contaminations that are a potential source of evidentiary false positives. Technical false positives, caused for example by errors in the collection, processing, and analysis of specimens, are discussed in depth in other parts of this book.

3.1.1 Incorporation from the Bloodstream

The unique physiology and structure of hair allows multiple pathways for drug incorporation and loss. From an interpretive viewpoint, incorporation from the bloodstream is the most important route when we are interested in answering questions about the time of intake or even the dose taken. Several studies have been carried out to explain the factors that influence the incorporation of drugs from the bloodstream [2–14]. A schematic view of pathways for incorporation of drugs into hair is shown in Figure 3.1.

Baumgartner et al. [15] postulated that drugs enter hair via the bloodstream in direct proportion to their concentration. Nakahara et al. [8,16]

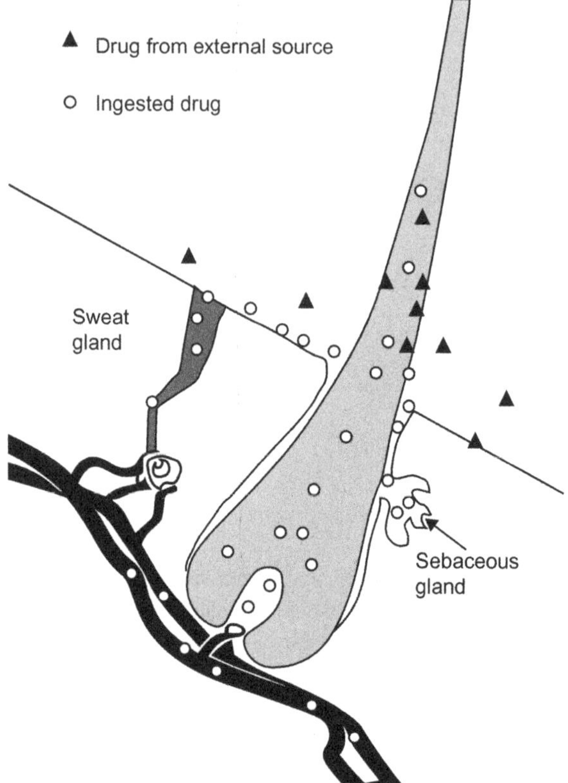

FIGURE 3.1 Three models of drug incorporation. Ingested drugs can enter the hair from the bloodstream feeding the dermal papilla as well as by sweat and sebum bathing the mature hair fiber. External drug from vapors or powders may also incorporate into the mature hair fiber.

studied the incorporation of cocaine and its metabolite benzoylecgonine (BE) into rat hair after administration of cocaine, and found that although the plasma concentration of BE was approximately four times higher than that of cocaine, its concentration in hair was ten times lower. The physiochemical properties of the drugs seem to be more important than their plasma concentrations. Other controlled dosing studies in animals support the concept of a linear relationship between dose and concentration of cocaine [17] and/or cocaine metabolite [15,17,18] and methamphetamine/amphetamine [19].

Drugs and their metabolites are distributed throughout the body primarily by passive diffusion from blood. Due to rapid cell division in the cells forming hair, the hair follicle is provided with a good blood supply. Drugs circulating in the blood will thus be delivered to the hair follicle and passive diffusion from arterial capillaries to matrix cells in the base of the follicle is therefore considered a primary means of drug deposition. The rate at which the drugs diffuse across the cell membrane is related to the lipid solubility of the drug; it can also be affected by the pH gradient between the plasma and the cell. Only drug molecules not bound to protein may participate in this diffusion.

Many drugs are either weak bases or weak acids that can be ionized by protonation or deprotonation. The pH of plasma is 7.3, whereas the pH of the keratinocytes and melanocytes is lower, varying between 3 and 6 [20]. The interaction between melanin and basic compounds is therefore essential for incorporation. Basic drugs, in contrast to acidic drugs, may accumulate in keratinocytes and melanocytes as diffusion into the cell is favored by the pH gradient, and once in the cell cytosol, the molecule will be protonated and hence unable to diffuse back into the plasma. The binding of drugs to the cell proteins may also enhance this effect, as the drug concentration in the cytosol decreases when the molecules are associating with structures within the cell.

Kronstrand et al. noted a strong correlation between codeine [21], selegiline [22], and melanin concentrations in human hair; however, the nature of the interaction is still unclear [23]. Cationic drugs are thought to be ionically bound to the polyanionic melanin molecule [24]. However, indications of covalent bonding of drugs [25] and of amphetamine, in particular [26], have been noted, and hydrophobic and electrostatic interactions have also been claimed [27]. The incorporation of neutral and acidic compounds in hair does not appear to be melanin-correlated as demonstrated by Borges et al. in studies of the incorporation of amphetamine and its nonbasic analogue *N*-acetylamphetamine in pigmented and nonpigmented rats [3] and melanocytes [28] and by recent studies on ethyl glucuronide (EtG) [29–31] and fatty acid ethyl esters (FAEE) [32].

3.1.2 Incorporation from Sweat and Other Secretions

Drug incorporation into hair can occur through exposure to sweat and sebaceous secretions; it is well known that drugs and their metabolites are excreted in sweat, and several papers have addressed this issue in the context of drug incorporation into hair [33–40]. Cone [41] studied the time frame of drugs appearing in hair. He administered codeine to several subjects and found that some drug appeared in hair after 24 h. This time is too short for initial formation of the hair in the root and for that hair to emerge above the skin. A later but larger bolus of codeine did appear in the correct time frame, which would be more supportive of the binding during hair formation. While administering the drug, sweat would contain the highest concentrations. Henderson et al. [42] reported that deuterated cocaine was found in multiple segments after a single dose, supporting sweat or other secretions as a route for drug deposition in hair. Raul et al. [43] suggested that cortisol and cortisone incorporates into hair not through the bloodstream, but mainly through diffusion from sweat.

In 1999, Stout and Ruth [14] evaluated the incorporation of cocaine, flunitrazepam, and nicotine and demonstrated insignificant deposition of the drugs into hair from sebum. More recently, Stout et al. [44] have evaluated the dynamics of incorporation concerning cocaine. Following contamination with cocaine HCl, hair locks were treated with a synthetic sweat solution and hygienic treatments to model real-life conditions. The results obtained generally confirmed the results of previous work and the group concluded that the addition of moisture to the hair as artificial sweat markedly increased the concentrations of drug in the hair. Once the analytes were absorbed into the hair, they were resistant to removal by shampooing the hair and/or laboratory decontamination wash procedures [45].

3.1.3 Incorporation from External Contamination

Efforts to differentiate between deposition from internal or external sources have been made both *in vitro* and *in vivo*, for example, by using the fluorescent compounds rhodamine and fluorescein [10,46,47]. The deposition of fluorescein was highly pH dependent and less compared with rhodamine, which showed no pH dependence. Stout and Ruth also performed *in vivo* studies on mice and found that the deposition of both fluorescein and rhodamine was markedly different from the *in vitro* results. The *in vivo* deposition was mainly in the cortex and the medulla as compared to the cuticle junctions observed when soaking the hair. Independent of the route of deposition the dyes could not be removed by extensive washing. This suggests that even though the endogenous and exogenous deposition of these model compounds can be distinguished, the analytical results after extraction remain difficult to interpret.

Monitoring unique metabolites (if present) together with carefully chosen cutoff levels can be helpful in reducing false positives, though it will not eliminate them entirely.

3.2 EXTERNAL CONTAMINATION: COMMONLY ENCOUNTERED DRUGS

External contamination constitutes a recurrent problem in forensic interpretation of hair analysis results. Passive exposure to a drug, resulting in an evidentiary false positive, can critically affect interpretation and can compromise employment and recruitment credentials. Hair analysis is only valuable if the drugs that are measured in hair arise from ingestion rather than from other sources. Therefore, it is imperative that drugs arising from the external environment be removed prior to analysis. The Society of Hair Testing (SoHT) has recommended a number of approaches to differentiate actual ingestion of a substance from external contamination of hair [48]. There are many published papers reporting external contamination of hair with drugs of abuse [44,45,49–55]. Exposure to illicit drugs in the workplace has been demonstrated during investigations involving narcotic police

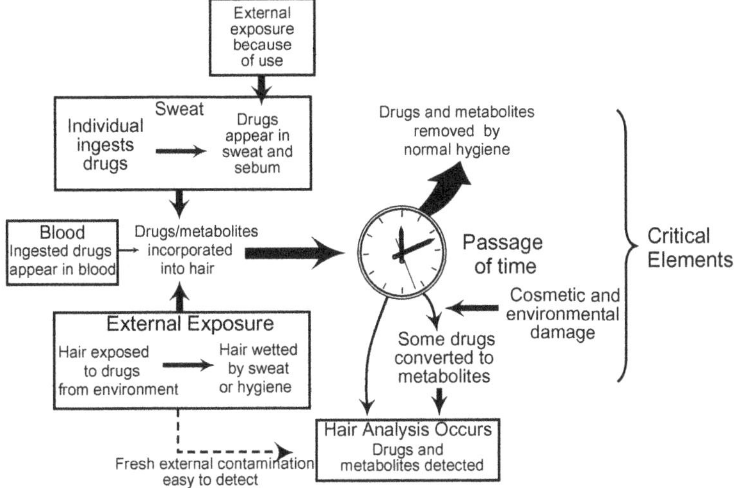

FIGURE 3.2 Framework for the incorporation and removal of drugs into and out of hair. The passage of time and hygiene are important components for contamination. Freshly contaminated hair is more readily detected as contaminated, than is hair that has been rinsed because of the high concentration of drugs on the outside. Over time, the loosely bound drug on the surface can migrate into the interior and become more tightly bound. Additionally, hygiene removes the surface-bound drugs. *Taken from Blank and Kidwell [59].*

officers [56,57] and more recently, external contamination from vapor exposure to chemical warfare agents has been reported [58].

To evaluate whether or not passive exposure can contribute to drugs in hair, the mechanisms of appearance and binding of drugs to hair both from external (passive) sources and ingestion must be understood. A schematic model for sources of drugs in hair is shown in Figure 3.2[59].

As Figure 3.2 further illustrates, there is usually some passage of time between ingestion or exposure and hair analysis. During that time, drugs loosely bound to the surface of the hair could be washed away by normal hygienic hair care. The removal of drugs will depend upon several variables, not the least of which are the characteristics of the solutions used to wash or treat the hair.

3.2.1 Smoke and Passive Inhalation

It is possible that a nondrug user may test positive for the presence of a drug/drug metabolite as a result of indirect inhalation of the smoke produced during drug use (by a third party).

Testing for drug use by employees and applicants for employment is a common practice in the United States, where corporations often model their drug programs after the Federal Drug Free Workplace programs.

The Department of Health and Human Services (DHSS) of the Mandatory Guidelines for Drug Testing of Federal Employees, which became known as the Substance Abuse and Mental Health Services Administration (SAMHSA), is considering expanding the guidelines to include hair specimens. Today, hair specimens are used outside the federally regulated workplaces for workplace and/or criminal justice testing.

Table 3.1 details the SAMHSA confirmation cutoffs for hair specimens [60,61].

In situations where drugs are known to be present in the environment, it is easy to demonstrate that passive exposure can produce positive hair analysis results. Cannabis, cocaine, and heroin are the most commonly abused (smoke inhalation) drugs. Cannabinoids, cocaine, plus its primary metabolite BE, and opiates (6-acetylmorphine and morphine) are all detectable in hair as a result of passive exposure.

Nicotine can be included in the aforementioned list though it is not an illicit substance. Nonsmokers, as demonstrated by Haley and Hoffmann in

TABLE 3.1 SAMHSA Confirmation Cutoffs for Hair Specimens

Drug/Drug Class	Drug Tested	Confirmation Cutoff (pg/mg)
Amphetamines	Amphetamine	300
	Methamphetamine	300
	MDMA	300
	MDA	300
	MDE	300
Cannabis	THC	0.05
	THC-COOH	
Cocaine	BE	50
	Cocaine	500
Opiates	Morphine	200
	Codeine	200
	6-Acetylmorphine	200
Phencyclidine	PCP	300

Note:
For a specimen to be positive for cocaine, both cocaine and BE must confirm positive and the ratio of parent cocaine to metabolite must be equal to or greater than 20.
If the specimen confirms positive for 6-acetylmorphine, it must also contain morphine at a concentration equal to or greater than 200 pg/mg.
For a specimen to be positive for methamphetamine it must contain amphetamine at a concentration equal to or greater than 50 pg/mg.

1985 [62], had appreciable levels (average 2.42 ng/mg) of nicotine that overlapped those of smokers. Kintz et al. in 1992 [63] proposed a cutoff level of 2 ng of nicotine per milligram of hair to identify smoking individuals, but even at this level, not all nonsmokers in their study would be negative.

Although one may assume that the hair of a nonsmoker would have considerably lower levels of nicotine and cotinine to that of an active smoker (based on the assumption that the nonsmoker inhales less smoke than the active smoker), a recent review by Okoli et al. indicates this may not be a valid assumption [64].

In 2007, Okoli et al. reviewed published studies examining nicotine levels related to exposure to secondhand tobacco smoke. This group reported that in two studies, active smokers had only slightly higher hair nicotine levels than highly exposed nonsmokers [65,66], indicating that secondhand smoke exposure can engender quantities of nicotine among nonsmokers similar to the amounts obtained during active smoking. If the same is true for cannabis, heroin, and cocaine, then the pharmacological and physical effects of long-term, indirect exposure to these illicit drugs must be considered when offering a toxicological opinion. The likelihood of acquired tolerance to the effects of the drug, albeit tolerance that was acquired indirectly, could affect the toxicologists' opinion as to whether the presence of a drug or drugs was a contributory factor in the death of a person, or a primary cause. Similar considerations must be given to cases of alleged drug-induced impairment.

The potential contamination of hair by external sources of drugs, generating evidentiary false positives, is a major limitation for the interpretation of forensic results. As mentioned previously, Δ-9-tetrahydrocannabinol (THC) cannabidiol (CBD), and cannabinol (CBN) are present in marijuana smoke [67], therefore positive screening test results must be confirmed by the identification of THC-COOH (11-nor-Δ-9-tetrahydrocannabinol-9-carboxylic acid) if the toxicologist wishes to offer an interpretation confirming active drug use. For example, Kauert and Röhrich [68] attempted to classify self-reported cannabis use and THC levels found in hair to user groups: THC concentrations in the range 0.1 ng/mg to 1 ng/mg as suggestive of weekly up to daily consumption and THC concentrations above 1 ng/mg as being associated with multiple daily cannabis use. A recent study which involved participants being exposed in a small, confined room to the smoke of one marijuana joint in the morning of every weekday for a 3-week period (leading to a total exposure to the smoke of 15 joints), returned THC concentrations ranging from 0 to 0.99 ng/mg depending on the time of collection relative to the time of exposure [69]. It is evident from this recent study that the earlier "classification" of cannabis use based solely on THC concentration is inherently dangerous and could lead to erroneous interpretation.

Interpretation should be able to differentiate active use from passive exposure, though even after confirmed consumption, THC-COOH is not detected in all cases even when extremely sensitive methods are applied

[67,70–74]. Recent studies confirmed that the biogenetic precursor of THC, THCA-A (Δ-9-tetrahydrocannabinolic acid A) can also be detected, albeit at low concentration, in the hair of noncannabis users after passive exposure to smoke. It is suggested that the majority of the relatively low concentration of THCA-A in the hair samples in these cases was not caused by sidestream marijuana smoke, but by other sources. The authors indicate that the THCA-A is present in marijuana smoke though it is not incorporated into the hair through the bloodstream [69].

In an attempt to distinguish passive exposure to cocaine from active drug use, Baumgartner and Hill proposed at least three criteria based upon the kinetics of drug removal from the hair under very specific conditions. These were designated the Extended Wash Ratio (R_{EW}), the Curvature Ratio (R_C), and the Safety Zone Ratio (R_{SZ}). The authors were able to specify cutoff values for these criteria that were reported to discriminate between passive exposure and active ingestion for cocaine based on their results from a large number of hair specimens [75]. Blank and Kidwell [49] investigated a number of decontamination scenarios employing detergents and solvents: none of which decontaminated the hair. This led them to suggest exercising caution when offering an interpretation of forensic results, as kinetic wash ratios were inadequate to identify contaminated samples.

Wang and Cone reported that substantial amounts of cocaine could be deposited in hair as a result of exposure to cocaine vapor: the degree of contamination being dependent upon the amount and length of exposure. A large portion of the cocaine contamination could be removed easily by bathing and shampooing and, if freshly contaminated, the contaminant would be found in the wash fraction in high ratio when compared to the hair extract. Analysis of the hair samples revealed the presence of anhydroecgonine methyl ester, thus excluding this molecule as a valid biological marker for active use. However, neither cocaethylene nor norcocaine were identified, indicating that their presence in hair would appear to be valid markers for active cocaine use [76].

3.2.2 Surface/Environmental Contamination: Poor Housekeeping

The presence of drugs in hair is the result of one or more mechanisms of incorporation. Drugs may be incorporated into hair from (a) blood, (b) sweat and sebaceous gland secretions, and (c) external environment [35]. The forensic scientist aims to determine whether the donor of the hair sample actively used drugs, or whether the donor came into contact with drugs. Interpretation of drug concentrations in hair is only valid if their presence is due to active use (ingestion) rather than from other sources; the scientist must find proof of use, such as the detection of unique compounds derived only from *in vivo* metabolism.

As previously discussed, the SoHT has recommended four approaches to differentiate actual ingestion of a substance from external contamination of hair: use of cutoff levels, analysis of metabolites, analysis of washings, and calculation of metabolite to parent drug concentration ratios [77,78].

In two separate studies, Smith and Kidwell [79] examined the children and spouses of cocaine users. In these studies, the children lived in a family where cocaine was used and thereby was present in the environment. The quantity present in the environment was not known. This study assumed that children 1–13 years of age are unlikely to be self-administering cocaine, so that any cocaine in the hair of the children must have come from passive exposure. Skin wipes were obtained from the children by swiping their foreheads with a cotton swab to assess external exposure. All of the skin wipes in the environmental population ($n = 29$) were positive for cocaine, indicating extensive surface contact with cocaine. In the active adult using population, 80% were positive for cocaine in their hair, and 85% of the children in this study were positive for cocaine. The distribution of the concentrations of cocaine in the hair of these two groups was also similar. Analysis of the data showed that several children of cocaine users had both cocaine and BE in their hair in varying quantities compared with the adult users. These amounts were less than, greater than, or equal to that of the drug-using parent (Figure 3.3). The concentrations of cocaine within a family group varied greatly, which is

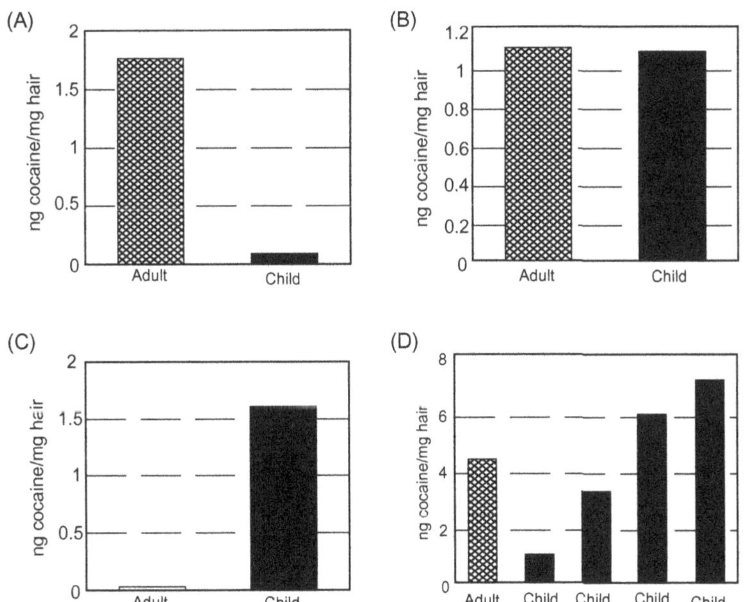

FIGURE 3.3 Concentrations of cocaine in the hair of family groups. Cocaine was used by all adults and presumably not by the children. *From Smith and Kidwell [79]. With permission.*

consistent with passive exposure being a random event (Figure 3.3D). Therefore, no simple cutoff level could distinguish between the adult users and the young children living in that environment. Furthermore, this study demonstrates that certain metabolites, such as BE, may not be an indicator of drug use. The presence of BE in the environment and in hair has been noted before [80]. Alternatively, it may be produced by degradation of the cocaine once the cocaine is incorporated into the hair [45].

Romano et al. [50] proposed a pattern of drug contamination corresponding to a realistic scenario, with the aim to verify the validity of hair testing for heroin in order to discriminate active users from false positives due to external contamination. In this study, each subject (six drug-free volunteers; three males and three females) was asked to apply 5 mg of a powdered mixture of heroin hydrochloride and acetylcodeine hydrochloride (10:1 w/w) to their hands for 5 min so that it was uniformly distributed. Each subject was then instructed on how to contaminate their hair with their hands, as uniformly as possible, from the roots to the ends for 10 min. This contamination procedure was carried out daily for 5 consecutive days. Hair samples were washed before collection and each sample (100 mg; 5 cm long) was cut as close as possible to the scalp from the posterior vertex. Samples were collected:

before contamination;
24 h after first contamination, and before second contamination (2nd day);
24 h after the last contamination (6th day);
each week for 14 weeks.

Each sample was decontaminated and analyzed on the day it was collected.

All hair samples that were tested were positive for opiates; data shown in Table 3.2. Heroin concentrations were high during the first 2 weeks and decreased during the next 2 weeks; it was no longer detectable between the 29th and 36th day. The 6-acetylmorphine concentrations decreased during the 1st week. Three months after contamination, every subject showed detectable levels of morphine (0.05–0.18 ng/mg) and 6-acetylmorphine (0.15–1.18 ng/mg).

If evaluation criteria, as suggested in literature, were applied to these test results, the contaminated subjects would be defined as active users. If one compares the 6-acetlymorphine concentrations reported here to the concordance between self-reported drug use and findings in hair, as published by Pepin and Gaillard [81], many of these six subjects would be categorized as having "low" or "medium" consumption habits, that is, the consumption of up to 3 g of heroin per week.

Romano et al. [45,50] reported the same fundamental conclusion following their separate investigations of external contamination with heroin and also with cocaine. Neither normal hygienic and cosmetic care nor

TABLE 3.2 Concentrations of Opiates (ng/mg) Found in Hair Samples after Decontamination

Subject	Analytes	2	6	8	15	22	29	36	43	50	57	64	71	78	85	92	99
A	H	0.42	1.72	0.83	0.28	0.09	–	–	–	–	–	–	–	–	–	–	–
	6-MAM	0.45	2.38	2.81	1.90	1.29	0.64	0.55	0.58	0.46	0.49	0.41	0.43	0.38	0.37	0.33	0.35
	M	0.09	0.14	0.58	0.41	0.28	0.16	0.19	0.14	0.15	0.13	0.11	0.12	0.12	0.10	0.08	0.10
	AC	0.07	0.38	0.35	0.21	0.18	0.04	0.02	0.01	–	–	–	–	–	–	–	–
	C	–	0.06	0.16	0.11	0.09	0.03	0.02	0.02	0.01	0.02	0.01	0.02	0.02	0.02	0.01	0.01
B	H	0.58	1.14	1.02	0.79	0.47	0.11	–	–	–	–	–	–	–	–	–	–
	6-MAM	0.73	1.60	1.47	1.38	1.52	0.74	0.61	0.63	0.58	0.59	0.55	0.37	0.36	0.38	0.32	0.31
	M	0.22	0.42	0.84	1.12	0.78	0.32	0.29	0.27	0.24	0.2	0.16	0.12	0.12	0.13	0.11	0.1
	AC	0.15	0.23	0.21	0.16	0.09	0.04	0.03	0.01	–	–	–	–	–	–	–	–
	C	0.02	0.06	0.12	0.09	0.07	0.05	0.04	0.06	0.05	0.04	0.05	0.03	0.02	0.01	0.02	0.01
C	H	1.26	2.76	1.66	1.37	0.42	0.10	–	–	–	–	–	–	–	–	–	–
	6-MAM	1.56	2.93	2.83	2.70	1.99	0.86	0.72	0.51	0.35	0.31	0.28	0.29	0.25	0.26	0.21	0.19
	M	0.31	0.41	0.77	1.07	0.51	0.17	0.15	0.11	0.10	0.11	0.09	0.08	0.07	0.08	0.06	0.05
	AC	0.27	0.49	0.39	0.34	0.18	0.06	0.04	0.03	0.01	–	–	–	–	–	–	–
	C	0.04	0.15	0.19	0.15	0.13	0.06	0.06	0.04	0.05	0.04	0.04	0.03	0.02	0.01	0.02	0.01

(Continued)

TABLE 3.2 (Continued)

Subject	Analytes	2	6	8	15	22	29	36	43	50	57	64	71	78	85	92	99
D	H	0.84	2.80	0.99	0.49	0.33	0.17	–	–	–	–	–	–	–	–	–	–
	6-MAM	1.03	3.58	2.77	2.60	2.08	1.81	1.81	1.75	1.61	1.54	1.37	1.34	1.31	1.25	1.28	1.18
	M	0.21	1.25	0.94	0.67	0.65	0.49	0.45	0.37	0.31	0.29	0.33	0.25	0.23	0.24	0.21	0.18
	AC	0.25	0.86	0.45	0.38	0.26	0.19	0.12	0.10	0.06	0.03	–	–	–	–	–	–
	C	0.07	0.28	0.18	0.17	0.14	0.11	0.13	0.12	0.10	0.11	0.09	0.09	0.08	0.07	0.06	0.04
E	H	0.18	1.34	0.76	0.40	0.31	0.15	–	–	–	–	–	–	–	–	–	–
	6-MAM	0.37	3.46	2.93	2.44	1.10	0.78	0.56	0.50	0.44	0.37	0.33	0.28	0.25	0.21	0.18	0.15
	M	0.09	0.41	1.32	0.94	0.63	0.42	0.31	0.19	0.13	0.13	0.12	0.10	0.11	0.09	0.08	0.07
	AC	0.08	0.25	0.21	0.19	0.15	0.08	0.05	0.01	–	–	–	–	–	–	–	–
	C	0.01	0.11	0.16	0.17	0.10	0.07	0.04	0.05	0.03	0.02	0.03	0.02	0.01	0.01	–	–
F	H	0.41	1.75	0.87	0.43	0.28	0.12	–	–	–	–	–	–	–	–	–	–
	6-MAM	0.63	3.38	2.75	2.37	1.73	1.57	1.43	1.31	1.38	1.22	1.18	1.03	0.79	0.83	0.72	0.67
	M	0.14	0.71	1.07	0.81	0.68	0.50	0.53	0.36	0.29	0.21	0.23	0.18	0.19	0.15	0.11	0.12
	AC	0.12	0.34	0.29	0.15	0.11	0.08	0.02	0.03	0.01	–	–	–	–	–	–	–
	C	0.02	0.18	0.15	0.16	0.12	0.10	0.11	0.08	0.05	0.07	0.06	0.04	0.03	0.02	0.01	0.01

Source: Reproduced from Romano et al. [50]. With permission.

decontamination procedures were sufficient to remove drug contamination; the authors concluded that cutoff values and metabolite-to-drug ratio were not useful to evaluate passive contamination. This conclusion was also reached by Stout et al. [44] following an evaluation of the dynamics of external contamination of hair with cocaine. A negative result can rule out either chronic or previous use, and also contact with drugs, or circumstances in which they are used. A positive result should be interpreted with extreme caution; in the case of heroin and cocaine in particular, the result from the hair test should always be confirmed with a urine test.

Perhaps one of the most commonly encountered prescription drugs that can be found in the household of drug users or drug-dependent individuals is the morphine substitute, methadone, which is commonly used for short-term detoxification and maintenance treatment of narcotic addition. This drug is not licensed for use in children, though it can be employed for the management of neonatal opiate withdrawal syndrome. Methadone and its primary metabolite 2-ethylidene-1,5-dimethyl-3,3-diphenylpyrrolidine (EDDP) are frequently found in the hair of children living with parents on methadone maintenance therapy, with physical parent/child intimacy often proposed as a possible source for positive results.

Chatterton et al. [82] reported three separate cases of administration of prescription drugs to children, namely methadone, amitriptyline, and tramadol. In one case, methadone was detected in the hair of a deceased 14-month-old child in the concentration range 0.65–0.99 ng/mg; EDDP was also detected at a concentration above the limit of detection (0.05 ng/mg) but below the limit of quantification (0.1 ng/mg). The detected concentration of methadone was similar to that previously detected by Kintz et al. [83] in 2010. This group presented the results of six cases (two fatalities) of methadone poisoning in children and emphasized the importance of considering background information when offering a final interpretation regarding drug use as opposed to external/environmental contamination. It was not possible to conclude in these six cases, that the children were deliberately administered methadone. The results of the segmental analysis of hair could indicate that they were in an environment where methadone was being used and where the drug was not being handled and stored with appropriate care.

Although EDDP was detected in four of the aforementioned six cases, the confirmed presence of this metabolite was deemed insufficient evidence for Kintz et al. to unequivocally establish that internal drug exposure had occurred in these four cases. Boomgaarden-Brandes et al. [84] would likely have offered a different interpretation following their more recent study. This group concluded that the low content of EDDP in skin wipes, collected from the palms and armpits of 10 patients on long-term methadone programs, coupled with the absence of EDDP in hair that had been exposed intermittently to methadone via direct contact (palm to hair), indicates that detection of EDDP in human hair of children of parents on methadone maintenance therapy suggests body passage rather than external contamination by physical parent/child intimacy. This study did not consider the possibility of intrauterine exposure to methadone and EDDP or exposure via the ingestion of breast milk as alternate propositions which could offer a potential explanation for a positive hair test result in a child of a methadone-taking parent [85,86].

3.3 ALCOHOL: FAEE AND ETG

Alcohol is a legal compound in many countries and is consumed in much higher amounts in comparison to other drugs of abuse and by a much higher portion of the population. The demand for reliable alcohol markers from the forensic as well as the medical points of view to discriminate between social drinking and alcohol abuse or to verify claims of abstinence after previous harmful drinking will be discussed in detail elsewhere in this book. This section will briefly discuss contamination as a source of positive results regarding FAEE and EtG in hair.

FAEE are formed after alcohol consumption by different enzymes in blood and human tissues; they are insoluble in water and stable at neutral pH, however they are sensitive to hair treatment at alkaline pH. In 2000, the detection of FAEE in hair for the diagnosis of heavy alcohol consumption was proposed [87]. Auwarter et al. [88] found that FAEE are trapped in hair matrix mainly via sebum, after their synthesis from ethanol and fatty acids in sebaceous glands.

De Giovanni et al. [89] have demonstrated *in vitro* synthesis of FAEE, albeit under experimental conditions which did not reflect a "real-life" situation. This group noted a constant and progressive increase of FAEE following the exposure of hair to an ethanol-saturated environment. During this investigation, a lock of hair (obtained from a teetotal subject) was placed in a confided space and stored, at room temperature, for 15 days in a small glass vessel in an ethanol-containing atmosphere. Hair was tested (following a washing step to eliminate ethanol deposited on the hair surface) in duplicate at different times during the storage. Figure 3.4 shows chromatograms concerning hair analysis performed after 2 days' and 15 days' saturation. Figure 3.5 shows the production of ethyl myristate, palmitate, and stearate during the 15 days' storage in the atmosphere saturated with ethanol. It is interesting to note that the reported concentrations of these three esters were higher than those detected in "real" samples that had been obtained from alcoholics [90].

De Giovanni et al. hypothesized that esterification of fatty acids, layered on the external surface of the hair—possibly under the action of bacterial enzymes—occurs followed by incorporation of the esters into the hair matrix. Alternatively, ethanol enters the hair shaft and reacts with free fatty acids.

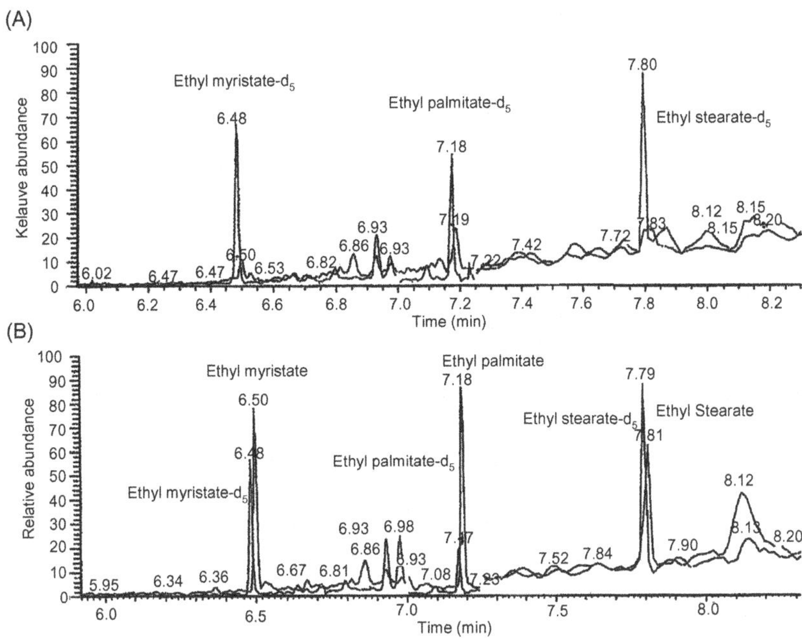

FIGURE 3.4 Chromatograms from hair analyses performed after 2 days' (A) and 15 days' saturation (B). *From De Giovanni et al. [89]. With permission.*

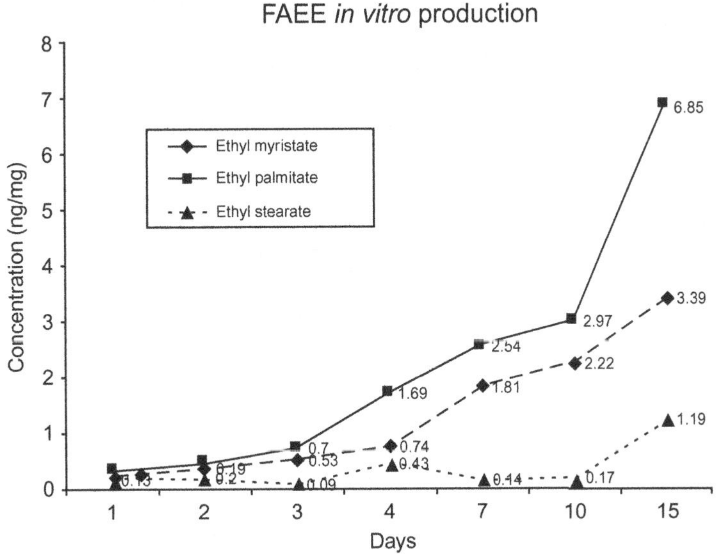

FIGURE 3.5 *In vitro* production of ethyl myristate, palmitate, and stearate during 15 days' storage in an atmosphere saturated with ethanol. *From De Giovanni et al. [89]. With permission.*

While the work of De Giovanni et al. created an unrealistic environment with regard to "exposure" to alcohol vapor, Hartwig et al. [91] have demonstrated that positive FAEE results can be achieved by simply exposing hair to a number of hair-care products. These evidentiary false positives were found after daily treatment with a hair lotion containing 62.5% ethanol, with a deodorant, and with a hair spray. Table 3.3 shows the results of hair-care products analyzed for FAEE, or applied in hair treatment experiments. This group recommended that in doubtful cases, pubic hair should be analyzed for comparison; they further hypothesized that a strong deviation from the usual concentration ratio between the four esters [88] may point to hair-care products as the source of the FAEE.

The sensitivity of FAEE to cosmetic treatment can make interpretation of analytical results somewhat problematic. The direct determination of ethanol itself in hair however is not possible due to its volatility and its potential absorption from external sources. Therefore, the minor metabolite EtG is measured as a direct marker of alcohol consumption. Although FAEE or EtG can be used independently for chronic excessive alcohol consumption assessment, for mutual confirmation and for exclusion of false positives or false-negative results, the determination of both parameters can be useful.

EtG is a stable, polar, nonvolatile and water soluble direct metabolite of ethanol that is formed by biotransformation of ethanol to EtG (ethyl-β-D-6 glucuronic acid, EtG) via conjugation with activated glucuronic acid during phase II metabolism. This minor metabolite (representing only 0.5% of complete alcohol elimination) has been determined in hair by GC-MS-EI [92,93] and by LC-MS/MS [94,95]. It is accepted that the use of bleaching products can lead to a significant decrease of the EtG concentration [96], however given that this compound is basically formed in the liver, lung, and intestine, the question of whether cosmetic treatment could yield false-positive results (as with FAEE) has generally been dismissed; this was certainly the belief until Sporkert et al. [97] reported an unusual case of external contamination of hair with EtG by an EtG-containing hair lotion. This case emphasizes the importance of gathering as much information as possible concerning hair-care products and their constituent ingredients and, if necessary, analysis of these products for the presence of EtG.

3.4 CONCLUDING REMARKS

So, is external contamination still a debate? External contamination is a scientific fact; it happens; it is not for debate. We, the forensic scientists, must recognize this fact and consider all possible hypotheses and scenarios when offering an interpretation of the analytical results of hair analysis.

As discussed, the most crucial issue regarding hair analysis and subsequent interpretation of analytical results is the avoidance of technical or evidentiary false positives. This chapter has highlighted and discussed the

TABLE 3.3 Results of the Analysis of 49 Hair-Care Products for FAEE by HS-SPME and GC-MS

Product Type or Name	Product No.	Manufacturer(s)[a]	FAEE Concentrations or Concentration Range (ng/mg)				
			Ethyl Myristate	Ethyl Palmitate	Ethyl Oleate	Ethyl Stearate	ΣFAEE
Foam setting lotions	1–3	W	0.07–0.25	0.03–0.07	0.09–0.25	0.04–0.14	0.23–0.56
Hair conditioners	1–7	L, NE, W	0.04–0.36	0.01–0.28	0.10–0.23	0.04–0.10	0.31–0.66
Hair cremes	1–3	L	0.00–0.09	0.03–0.84	0.00–0.23	0.00–0.04	0.26–0.84
Hair fixation	Londa-fix	L	0.05	0.01	0.17	0.05	0.28
Hair gels	1–3	L, W	0.01–0.04	0.02–0.08	0.19–0.21	0.06–0.17	0.32–0.47
Hair gel spray	4	C	0.46	0.49	0.40	0.25	1.6
Hair lotions	1–2	F, L	0.00–0.05	0.04–0.12	0.04–0.18	0.06–0.11	0.19–0.41
Shaving lotion	After shave	H	0.00	0.68	0.51	0.85	2.04
Hair rinsing 1	Vivality	W	0.01	0.07	0.16	0.05	0.29
Hair rinsing 2	Londa balance	L	0.09	0.19	0.17	1.26	1.71
Hair shadings	1–4	L	0.02–0.44	0.06–0.21	0.00–0.46	0.02–0.14	0.16–1.25
Shampoo 1	Happy Baby	DR	1.4	0.16	0.14	0.14	1.8
Shampoos	2–8	HE, L, N, P, T, W	0.00–0.36	0.10–0.31	0.12–0.48	0.00–0.31	0.20–1.05
Splice inhibitor	Kera-logie	LO	0.02	0.26	0.00	0.02	0.30

(Continued)

TABLE 3.3 (Continued)

Product Type or Name	Product No.	Manufacturer(s)[a]	FAEE Concentrations or Concentration Range (ng/mg)				
			Ethyl Myristate	Ethyl Palmitate	Ethyl Oleate	Ethyl Stearate	ΣFAEE
Hair spray 2	Londa style	L	0.03	0.00	2.59	0.00	2.6
Hair sprays	1, 3	C, LO	0.00–0.49	0.29–0.34	0.10–0.54	0.39–0.49	0.94–1.37
Deo spray 1	FA24h	S	0.00	0.84	1.45	0.30	2.6
Deo spray 2	Axe	LF	0.00	0.17	0.20	0.07	0.44
Hair stabilizers	1–2	L	0.00	0.21–0.29	0.16–0.48	0.16–0.48	1.0–1.1
Hair waver		L	0.03	0.01	0.33	0.03	0.40
Hair wax 1		M	0.5	7.9	8.2	1.0	17.6
Hair waxes	2–4	L, W	0.00–0.16	0.03–0.16	0.00–0.69	0.00–0.08	0.16–0.86
Hair wax 5		D	0.5	2.4	4.5	22.2	29.6
Hair wax 6	Super Light	M	0.03	0.13	6.7	3.5	10.4

[a]*Manufacturers: C, Carin; D, Dusy; DR, Drospa; F, Florena; H, Hattric; HE, Hexal; L, Londa; LF, Lever Fabergé; LO, L'Oreal; N, Nivea (Beiersdorf AG); NE, Neowell; P, Penaten; S, Schwarzkopf & Henkel; T, Toofer; M, Murray; W, Wella.*

Source: Reproduced from Hartwig et al. [91], p. 91. With permission.

various forms of external contamination which have the potential to be misinterpreted and cause evidentiary false positives. False-positive results have been shown, on occasion, to be similar (in quantitative terms) to those obtained following direct drug intake. This is problematic for the forensic toxicologist, who is challenged to provide a definitive interpretation regarding drug/alcohol intake use (or abstinence).

It is internationally accepted that routes other than direct drug intake can result in positive test results, even when the hair has been subjected to vigorous washing and decontamination steps. As always, more research is required to better understand the biological mechanisms involved with drug incorporation; the development of a decontamination technique that can differentiate, unequivocally, passive *versus* active drug use would be of huge benefit to the hair community as a whole.

REFERENCES

[1] Kintz P. Hair analysis. In: Moffat AC, Osselton MD, Widdop B, Watts J, editors. Clarke's analysis of drugs and poisons. 4th ed. London: Pharmaceutical Press; 2011. p. 323–33.

[2] Pötsch L. A discourse on human hair fibers and reflections on the conservation of drug molecules. Int J Legal Med 1996;108:285–93.

[3] Borges CR, Wilkins DG, Rollins DE. Amphetamine and N-acetylamphetamine incorporation into hair: an investigation of the potential role of drug basicity in hair color bias. J Anal Toxicol 2001;25:221–7.

[4] Gygi SP, Joseph Jr. RE, Cone EJ, Wilkins DG, Rollins DE. Incorporation of codeine and metabolites into hair: role of pigmentation. Drug Metab Dispos Biol Fate Chem 1996;24:495–501.

[5] Gygi SP, Joseph Jr. RE, Cone EJ, Wilkins DG, Rollins DE. Distribution of codeine and morphine into rat hair after long-term daily dosing with codeine. J Anal Toxicol 1995;19:387–91.

[6] Gygi SP, Joseph Jr. RE, Cone EJ, Wilkins DG, Rollins DE. A comparison of phenobarbital and codeine incorporation into pigmented and nonpigmented rat hair. J Pharm Sci 1997;86:209–14.

[7] Nakahara Y, Kikura R. Hair analysis for drugs of abuse, XIII: effect of structural factors on incorporation of drugs into hair: the incorporation rates of amphetamine analogs. Arch Toxicol 1996;70:841–9.

[8] Nakahara Y, Ochiai T, Kikura R. Hair analysis for drugs of abuse, V: the facility in incorporation of cocaine into hair over its major metabolites, benzoylecgonine and ecgonine methyl ester. Arch Toxicol 1992;66:446–9.

[9] Nakahara Y, Takahashi K, Kikura R. Hair analysis for drugs of abuse, X: effect of physiochemical properties of drugs on the incorporation rates into hair. Biol Pharm Bull 1995;18:1223–7.

[10] Pötsch L, Moeller MR. On pathways for small molecules into and out of human hair fibers. J Forensic Sci 1996;41:121–5.

[11] Pötsch L, Skopp G, Moeller MR. Influence of pigmentation on the codeine content of hair fibers in guinea pigs. J Forensic Sci 1997;42:1095–8.

[12] Pötsch L, Skopp G, Rippin G. A comparison of 3H-cocaine binding on melanin granules and human hair *in vitro*. Int J Legal Med 1997;110:55–62.

[13] Stout PR, Claffey DJ, Ruth JA. Incorporation and retention of radiolabeled S-(+)-and R-(−)-methamphetamine and S(+)- and R(−)-N-(n-butyl)-amphetamine in mouse hair after systemic administration. Drug Metab Dispos Biol Fate Chem 2000;28:286–91.

[14] Stout PR, Ruth JA. Deposition of [3H]cocaine, [3H]nicotine, and [3H]flunitrazepam in mouse hair melanosomes after systemic administrations. Drug Metab Dispos Biol Fate Chem 1999;27:731–5.

[15] Baumgartner WA, Hill VA, Blahd WH. Hair analysis for drugs of abuse. J Forensic Sci 1989;34:1433–53.

[16] Nakahara Y, Kikura R. Hair analysis for drugs of abuse, VII: the incorporation rates of cocaine, benzoylecgonine and ecgonine methyl ester into rat hair and hydrolysis of cocaine in rat hair. Arch Toxicol 1994;68:54–9.

[17] Tanaka M, Ono C, Yamada M. Absorption, distribution and excretion of 14C-levofloxacin after single oral administration in albino and pigmented rats: binding characteristics of levofloxacin-related radioactivity to melanin in vivo. J Pharm Pharmacol 2004;56:463–9.

[18] Ferko AP, Barbieri EJ, DiGregorio GJ, Ruch EK. The accumulation and disappearance of cocaine and benzoylecgonine in rat hair following prolonged administration of cocaine. Life Sci 1992;51:1823–32.

[19] Forman R, Schneiderman J, Klein J, Graham K, Greenwald M, Koren G. Accumulation of cocaine in maternal and fetal hair: the dose response curve. Life Sci 1992;50:1333–41.

[20] Robbins CR. Chemical and physical behavior of human hair. Berlin: Springer Verlag; 1994.

[21] Kronstrand R, Forstberg-Peterson S, Kagedal B, Ahlner J, Larson G. Codeine concentrations in hair after oral administration is dependent on melanin content. Clin Chem 1999;45:1485–94.

[22] Kronstrand R, Andersson MC, Ahlner J, Larson G. Incorporation of selegiline metabolites into hair after oral selegiline intake. J Anal Toxicol 2001;25:594–601.

[23] Testorf MF, Kronstrand R, Svensson SP, Lundstrom I, Ahlner J. Characterization of [3H] flunitrazepam binding to melanins. Anal Biochem 2001;298:259–64.

[24] Gautam L, Scott KS, Cole MD. Amphetamine binding to synthetic melanin and Scatchard analysis of binding data. J Anal Toxicol 2005;29:339–44.

[25] Claffey DJ, Stout PR, Ruth JA. 3H-nicotine, 3H-flunitrazepam and 3H-cocaine incorporation into melanin: a model for the examination of drug-melanin interactions. J Anal Toxicol 2001;25:607–11.

[26] Claffey DJ, Ruth JA. Amphetamine adducts of melanin intermediates demonstrated by matrix-assisted laser desorption/ionization time-of-flight mass spectrometry. Chem Res Toxicol 2001;14:1339–44.

[27] Polettini A, Cone EJ, Gorelick DA, Huestis MA. Incorporation of methamphetamine and amphetamine in human hair following controlled oral methamphetamine administration. Anal Chim Acta 2012;726:35–43.

[28] Borges CR, Martin SD, Meyer LJ, Wilkins DG, Rollins DE. Influx and efflux of amphetamine and N-acetylamphetamine in keratinocytes, pigmented melanocytes, and nonpigmented melanocytes. J Pharm Sci 2002;91:1523–35.

[29] Appenzeller BM, Schuman M, Yegles M, Wennig R. Ethyl glucuronide concentration in hair is not influenced by pigmentation. Alcohol and Alcoholism 2007b, 42:326–7.

[30] Morini L, Politi A, Polettini A. Ethyl glucuronide in hair. A sensitive and specific marker of chronic heavy drinking. Addiction 2009;104:915–20.

[31] Politi L, Leone F, Morini L, Polettini A. Bioanalytical procedures for determination of conjugates or fatty acid esters of ethanol as markers of ethanol consumption: a review. Anal Biochem 2007;368:1–16.

[32] Kulaga V, Velazquez-Armenta Y, Aleksa K, Vergee Z, Koren G. The effect of hair pigment on the incorporation of fatty acid ethyl esters (FAEE). Alcohol and Alcoholism 2009;44:287–92.

[33] Kidwell DA, Blank DL. Mechanisms of incorporation of drugs into hair and interpretation of hair analysis data. In: Cone EJ, Welch MJ. editors. Hair testing for drugs of abuse: international research on standards and technology. Rockville, MD: U.S. Department of Health and Human Services; 1995. p. 19–90.

[34] Cone EJ. Mechanisms of drug incorporation into hair. Ther Drug Monit 1996;18:438–43.

[35] Henderson GL. Mechanisms of drug incorporation into hair. Forensic Sci Int 1993;63:19–29.

[36] Ruth JA, Stout PR. Mechanisms of drug deposition into hair and issues for hair testing. Forensic Sci Rev 2004;16:116–33.

[37] Cone EJ, Hillsgrove MJ, Jenkins AJ, Keenan RM, Darwin WD. Sweat testing for heroin, cocaine, and metabolites. J Anal Toxicol 1994;18:298–305.

[38] Kacinko SL, Barnes AJ, Schwilke EW, Cone EJ, Moolchan ET, Huestis MA. Clin Chem 2005;51:2085–94.

[39] Kintz P. Excretion of MBDB and BDB in urine, saliva, sweat following single oral administration. J Anal Toxicol 1997;21:570–5.

[40] Kintz P, Tracqui A, Mangin P. Sweat testing for benzodiazepines. J Forensic Sci 1996;41:851–4.

[41] Cone EJ. Testing human hair for drugs of abuse, 1: individual dose and time profiles of morphine and codeine in plasma, saliva, urine, and beard compared to drug-induced effects on pupils and behavior. J Anal Toxicol 1990;14(1):.

[42] Henderson GL, Harkey MR, Zhou C, Jones RT, Jacob P. Incorporation of isotopically labeled cocaine and metabolites into human hair, part 3: 1, dose-response relationships. J Anal Toxicol 1996;20:1–12.

[43] Raul JS, Cirimele V, Ludes B, Kintz P. Detection of physiological concentrations of cortisol and cortisone in human hair. Clin Biochem 2004;37:1105–11.

[44] Stout PR, Ropero-Miller JD, Baylor MR, Mitchell JM. External contamination of hair with cocaine: evaluation of external cocaine contamination and development of performance-testing materials. J Anal Toxicol 2006;30:490–500.

[45] Romano G, Barbera N, Lombardo I. Hair testing for drugs of abuse: evaluation of external cocaine contamination and risk of false positives. Forensic Sci Int 2001;123:119–29.

[46] Stout PR, Ruth JA. Comparison of in vivo and in vitro deposition of rhodamine and fluorescein in hair. Drug Metab Dispos 1998;26:943–8.

[47] DeLauder SF, Kidwell DA. The incorporation of dyes into hair as a model for drug binding. Forensic Sci Int 2000;107:93–104.

[48] Cooper GAA, Kronstrand R, Kintz P. Society of Hair Testing guidelines for drug testing in hair. Forensic Sci Int 2012;218:20–4.

[49] Blank DL, Kidwell DA. External contamination of hair by drugs of abuse: an issue for forensic interpretation. Forensic Sci Int 1993;63:145–56.

[50] Romano G, Barbera N, Spadaro G, Valenti A. Determination of drugs of abuse in hair: evaluation of external heroin contamination and risk of false positives. Forensic Sci Int 2003;131:98–102.

[51] Thorspecken J, Skopp G, Potsch L. In vitro contamination of hair by marijuana smoke. Clin Chem 2004;596–602.

[52] Tsanaclis L, Wicks J. Differentiation between drug use and environmental contamination when testing for drugs in hair. Forensic Sci Int 2008;176:19–22.

[53] Auwarter V, Wohlfarth A, Traber J, Thieme D, Weinmann W. Hair analysis for Δ9-tetrahydrocannabinolic acid A—new insights into the mechanism of drug incorporation of cannabinoids into hair. Forensic Sci Int 2010;196:10–13.

[54] Gerostamoulos D, Staikos V, Vo T, Dood M, Drummer OH. Detection of oxycodone in hair, consumption or contamination? In: Presented at the 18th scientific meeting of the Society of Hair Testing. Geneva, Switzerland; 2013.

[55] Nutt J, Tsanaclis L, Bagley K, Bevan S, Wicks J. Differentiating between consumption and external contamination when testing for cocaine and cannabis in hair samples. In: Presented at the 18th scientific meeting of the Society of Hair Testing. Geneva, Switzerland; 2013.

[56] Mieczkowski T. Distinguishing passive contamination from active cocaine consumption: assessing the occupational exposure of narcotics officers to cocaine. Forensic Sci Int 1997;84(1–3):87–111.

[57] Villain M, Muller J-F, Kintz P. Heroin markers in hair of a narcotic police officer: active or passive exposure? Forensic Sci Int 2010;196(1–3):128–9.

[58] Spiandore M, Piram A, Lacoste A, Josse D, Doumenq P. External contamination of hair as a marker of exposure to chemical warfare agents. In: Presented at the 18th scientific meeting of the Society of Hair Testing. Geneva, Switzerland; 2013.

[59] Blank DL, Kidwell DA. Environmental Exposure - The Stumbling Block of Hair Testing. In: Kintz P, editor. Drug Testing in Hair. Boca Raton, FL: CRC Press; 1996. p. 17.

[60] Verstraete A, Peat M. Workplace drug testing. In: Moffat AC, Osselton MD, Widdop B, Watts J, editors. Clarke's analysis of drugs and poisons. 4th ed. London: Pharmaceutical Press; 2011. p. 73–86.

[61] Bush DM. The U.S. mandatory guidelines for federal workplace testing programs: current status and future considerations. Forensic Sci Int 2008;174:111–19.

[62] Haley NJ, Hoffmann D. Analysis for nicotine and cotinine in hair to determine cigarette smoker status. Clin Chem 1985;31:1598.

[63] Kintz P, Ludes B, Mangin P. Evaluation of nicotine and cotinine in human hair. J Forensic Sci 1992;37:72.

[64] Okoli CTC, Kelly T, Hahn EJ. Secondhand smoke and nicotine exposure. Addict Behav 2007;32:1977–88.

[65] Dimich-Ward H, Gee G, Brauer M, Leung V. Analysis of nicotine and cotinine in the hair of hospitality workers exposed to environmental tobacco smoke. J Occup Environ Med 1997;39:946–8.

[66] Al-Delaimy W, Fraser T, Woodward A. Nicotine in hair or bar and restaurant workers. N Z Med J 2001;114:80–3.

[67] Cirimele V, Sachs H, Kintz P, Mangin P. Testing human hair for cannabis. III. Rapid screening procedure for the simultaneous identification of Δ9-tetrahydrocannabinol, cannabinol, and cannabidiol. J Anal Toxicol 1996;20:13–16.

[68] Kauert G, Röhrich J. Concentrations of Δ9-tetrahydrocannabinol, cocaine and 6-monoacetylmorphine in hair of drug abusers. Int J Legal Med 1996;108:294–9.

[69] Moosmann B, Roth N, Auwarter V. Hair analysis for THCA-A, THC and CBN after passive "in vivo" exposure to marijuana smoke, http://dx.doi.org/10.1002/dta.1474 Drug testing and analysis. Wiley; 2013

[70] Baptista MJ, Monsanto PV, Pinho Marques EG, Bermejo A, Avila S, Castanheira AM, et al. Hair analysis for Δ9-THC, Δ9-THC-COOH, CBN and CBD, by GC/MS-EI: comparison with GC/MS-NCI for Δ9-THC-COOH. Forensic Sci Int 2002;128: 66–78.

[71] Kim JY, Cheong JC, Lee JI, In MK. Improved gas chromatography-negative ion chemical ionization tandem mass spectrometric method for determination of 11-nor-Δ9-tetrahydrocannabinol-9-carboxylic acid in hair using mechanical pulverization and bead assisted liquid-liquid extraction. Forensic Sci Int 2011;206:e99–e102.

[72] Mieczkowski T. Assessing the potential of a "color effect" for hair analysis of 11-nor-9-carboxy-Δ9-tetrahydrocannabinol: analysis of a large sample of hair specimens. Life Sci 2003;74:463–9.

[73] Uhl M, Sachs H. Cannabinoids in hair: strategy to prove marijuana/hashish consumption. Forensic Sci Int 2004;145:143–7.

[74] Wilkins D, Haughey H, Cone E, Huestis M, Foltz R, Rollins D. Quantitative analysis of THC, 11-OH-THC, and THCCOOH in human hair by negative ion chemical ionization mass spectrometry. J Anal Toxicol 1995;19:483–91.

[75] Baumgartner W, Hill V. Hair analysis for drugs of abuse: decontamination issues. In: Sunshine I, editor. Recent developments in therapeutic drug monitoring and clinical toxicology. New York, NY: Marcel Dekker; 1992. p. 577–97.

[76] Wang WL, Cone EJ. Testing human hair for drugs of abuse. IV. Environmental cocaine contamination and washing effects. Forensic Sci Int 1995;70:39–51.

[77] Society of Hair Testing. Statement of the Society of Hair Testing concerning the examination of drugs in human hair. Forensic Sci Int 1997;84:3–6.

[78] Society of Hair Testing. Recommendations for hair testing in forensic cases. Forensic Sci Int 2004;145:83–4.

[79] Smith FP, Kidwell DA. Cocaine in hair, saliva, skin, swabs, and urine of cocaine users' children. Forensic Sci Int 1996;83:179.

[80] Janzen K. Concerning norcocaine, ethylbenzoylecgonine, and the identification of cocaine use in human hair. J Anal Toxicol 1992;16:402.

[81] Pépin G, Gaillard Y. Concordance between self-reported drug use and findings in hair about cocaine and heroin. Forensic Sci Int 1997;84:37–41.

[82] Chatterton C, Turner K, Klinger N, Etter M, Duez M, Cirimele V. Interpretation of pharmaceutical drug concentrations in young children's head hair. J Forensic Sci 2013; http://dx.doi.org/10.1111/1556-4029.12301

[83] Kintz P, Evans J, Villain M, Cirimele V. Interpretation of hair findings in children after methadone poisoning. Forensic Sci Int 2010;196(1–3):51–4.

[84] Boomgaarden-Brandes K, Tobias N, Koc A, Sachs H, Muhlbauer B. From palm to hair: transmissibility of methadone and its metabolite EDDP. Toxichem Krimtech 2013;80 (Special Issue):354

[85] Himes SK, Goodwin RS, Rock CM, Jones HE, Johnson RE, Wilkins DG, et al. Methadone and metabolites in hair of methadone-assisted pregnant women and their infants. Ther Drug Monit 2012;34(3):337–44.

[86] Glatstein MM, Garcia-Bournissen F, Finkelstein Y, Koren G. Methadone exposure during lactation. Can Fam Physician 2008;54(12):1689–90.

[87] Pragst F, Spiegel K, Sporkert F, Bohnenkamp M. Are there possibilities for the detection of chronically elevated alcohol consumption by hair analysis? A report about the state of investigation. Forensic Sci Int 2000;107:201–23.

[88] Auwarter V, Sporkert F, Hartwig S, Pragst F, Vater H, Diefenbacher A. Fatty acid ethyl esters in hair as markers of alcohol consumption. Segmental hair analysis of alcoholics, social drinkers and teetotalers. Clin Chem 2001;47:2114–23.

[89] De Giovanni N, Donadio G, Chiarotti M. Ethanol contamination leads to fatty acid ethyl esters in hair samples. J Anal Toxicol 2008;32:156–9.

[90] De Giovanni N, Donadio G, Chiarotti M. The reliability of fatty acid ethyl esters (FAEE) as biological markers for the diagnosis of alcohol abuse. J Anal Toxicol 2007;31:93–7.

[91] Hartwig S, Auwarter V, Pragst F. Effect of hair care and hair cosmetics on the concentrations of fatty acid ethyl esters in hair as markers of chronically elevated alcohol consumption. Forensic Sci Int 2003;131:90–7.

[92] Aderjan RE, Besserer K, Sachs H, Schmitt GG, Skopp GA. Ethyl glucuronide—a nonvolatile ethanol metabolite in human hair. In: Proceedings of the joint TIAFT/SOFT international meeting. Tampa, FL; 1994.

[93] Alt A, Janda I, Seidl S, Wurst FM. Determination of ethyl glucuronide in hair samples. Alcohol Alcoholism 2000;35(3):313–14.

[94] Janda I, Alt A. Improvement of ethyl glucuronide determination in human urine and serum samples by solid phase extraction. J Chromatogr B Biomed Sci Appl 2001;758 (2):229–34.

[95] Skopp G, Schmitt G, Pötsch L, Drönner P, Aderjan R, Mattern R. Ethyl glucuronide in human hair. Alcohol 2000;35:283–5.

[96] Morini L, Zucchella A, Polettini A, Politi L, Groppi A. Effect of bleaching on ethyl glucuronide in hair: an in vitro experiment. Forensic Sci Int 2010;198:23–7.

[97] Sporkert F, Kharbouche H, Augsburger MP, Klemm C, Baumgartner MR. Positive EtG findings in hair as a result of a cosmetic treatment. Forensic Sci Int 2012;218:97–100.

Chapter 4

Alcohol Biomarkers in Hair

Fritz Pragst
Charité—University Medicine Berlin, Berlin, Germany

4.1 INTRODUCTION

Drinking of alcoholic beverages belongs to human culture within living memory and all attempts in history to ban alcohol have failed. However, beside the social and recreational use, uncontrolled drinking has the shady sides of physical and psychic impairment, addiction, and chronic damage of all essential organs, and plays a negative role in offences and crimes. Therefore, from forensic as well as clinical point of view, it is often important to know the current effect of alcohol on a person as well as his general drinking behavior. Whereas the acute degree of alcoholic impairment can be assessed from the current concentration of ethanol in blood, the chronic alcohol consumption has to be evaluated by means of biomarkers with a much longer time window than ethanol itself.

An overview of the different kinds of alcohol biomarkers, their properties, advantages, and limitations was described in several reviews [1–6] and shall not be repeated here. Traditional markers are mainly determined in blood (serum or plasma). Hair as a matrix for assessment of chronic alcohol intake has gained practical importance only in the last decade. From the different possibilities of hair for this purpose which were summarized in previous reviews [7,8] only the minor metabolites of ethanol, ethyl glucuronide (EtG), and fatty acid ethyl esters (FAEEs) have attracted practical attention (Figure 4.1). Both markers have the advantage that they contain the ethyl group of ethanol which excludes other reasons than alcohol for a positive result. Cocaethylene (CE) as a marker of combined use of cocaine and alcohol is generally included in hair testing for illegal drugs. However, the search for 11-nor-Δ^9-tetrahydrocannabinol-9-carboxylic acid ethyl ester (THC-COOEt) as an analogous combined marker of cannabis and alcohol use in blood and in hair was without success [9]. There were no other new approaches to this problem within the last 7 years since the author's contribution in the preceding book on hair [8].

Hair Analysis in Clinical and Forensic Toxicology.

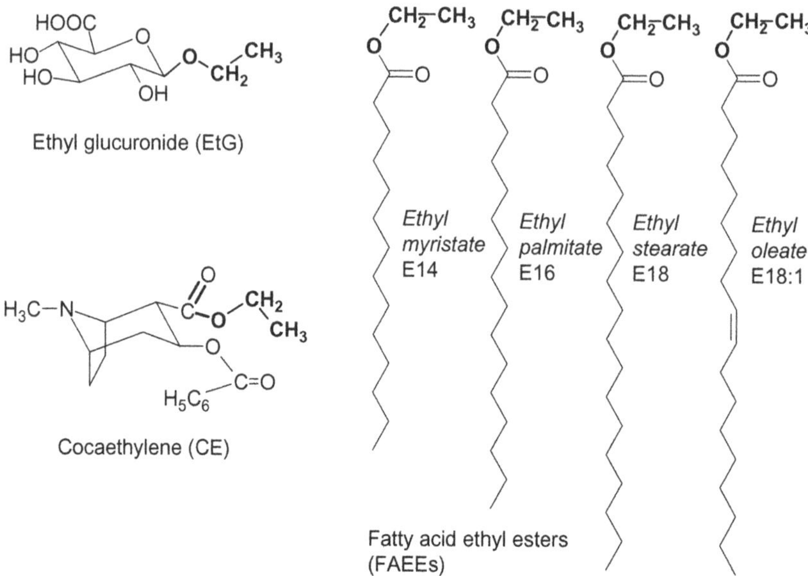

FIGURE 4.1 Structures of ethyl glucuronide (EtG), fatty acid ethyl esters (FAEEs), and cocaethylene (CE).

However, the progress in analytical techniques has enabled a tremendous improvement of analytical sensitivity as a basis of routine procedures for EtG and FAEE and has paved the way for the widespread application in practice. Both EtG and FAEE in hair, alone or in combination, are regularly used in casework with forensic background and particularly EtG in hair has been introduced in legal regulations in some European countries [10,11]. Methods and results were published in more than 150 papers, and a comprehensive review about all essential aspects of EtG in hair appeared recently [12]. For uniform performance and interpretation, guidelines and cutoffs were agreed in consensuses of the Society of Hair Testing (SoHT) [13–15] and the European Workplace Drug Testing Society [16], and proficiency tests were implemented [17]. The comparison with other alcohol consumption biomarkers showed superior properties concerning sensitivity and specificity [18,19].

In this chapter, biochemical and physiological principles, analytical methods, interpretation concerning alcohol consumption, and practical applications of EtG and FAEE shall be described and the knowledge about CE shall be updated.

4.2 ALCOHOL AMOUNT, DRINKING PATTERN, AND MINOR METABOLITES OF ETHANOL IN HAIR

The intended use of alcohol biomarkers is the discrimination between abstainers, social drinkers (=moderate drinkers), and excessive drinkers or alcoholics. There are no uniform terms and definitions of these categories with respect to consumed alcohol quantities. *Strict abstainers* differ from *nondrinkers* with a mean of $\leq$1 g ethanol per day. The further classification is difficult because of differences between men and women and the influence of body weight. According to WHO recommendations *low-risk drinkers* were defined as subjects who consume $\leq$30/20 g/day for men/women. *At-risk alcohol consumption* >30/20 g/day (male/female) is associated with a significantly increased risk of disease and premature mortality. *High-risk drinking* is considered for daily consumption of $\geq$40 g pure ethanol for women and $\geq$60 g for men. In a less-restrictive classification, risky drinking is assumed for 60–120 g/day and excessive drinking (alcohol abuse, alcoholism) for >120 g/day [20]. However, the impact of alcohol consumption is more complex than simply ascertaining the amount of alcohol consumed and depends also on the drinking patterns and is different, for example, for habitual and episodic excessive drinking (binge drinking). Nevertheless, despite these complications and in agreement with the SoHT consensus [14], in this chapter a long-term ethanol intake of, as a mean, $\geq$60 g/day is regarded as chronic excessive alcohol consumption, heavy drinking, or alcohol abuse.

The concentrations of EtG (C_{EtG}) and of FAEE (C_{FAEE}) in hair are used for characterization of the drinking behavior. The interpretation by cutoff values is based on the assumption of a clear relationship between alcohol dose and C_{EtG} or C_{FAEE} in hair. However, as shown in Figure 4.2, there are several steps between drinking of the alcoholic beverage and the incorporated minor metabolites in hair which were discussed from pharmacokinetic point of view more closely in Ref. [21]. If in a first approximation all interindividual variations by individual properties and external factors are neglected, according to the pioneering work of Nakahara et al. [22] the concentration in hair correlates best with the area of the concentration in blood versus time curve AUC (Eq. (4.1)) with the drug specific "incorporation rate" (ICR) as the proportionality factor. This is comprehensible since the growing hair quasi-integrates the concentration captured in the hair root before keratinization.

$$C_{\text{hair},x} = \text{ICR}_x \times \text{AUC}_{\text{blood},x} \tag{4.1}$$

$C_{\text{hair},x}$ = Concentration of substance x in hair;
ICR_x = Incorporation rate of the substance x;
$\text{AUC}_{\text{blood},x}$ = Area under the concentration in blood versus time curve of substance x.

This relationship is valid primarily only for incorporation from blood. However, it was assumed that despite the complicated formation and incorporation mechanisms of both markers (cf. Sections 4.3.1 and 4.4.1) this holds true also for EtG and FAEE, and that AUC_{EtG} as well as AUC_{FAEE} in blood should be proportional to the area under the ethanol concentration in blood versus time curve AUC_{EtOH}. That means that, in a first

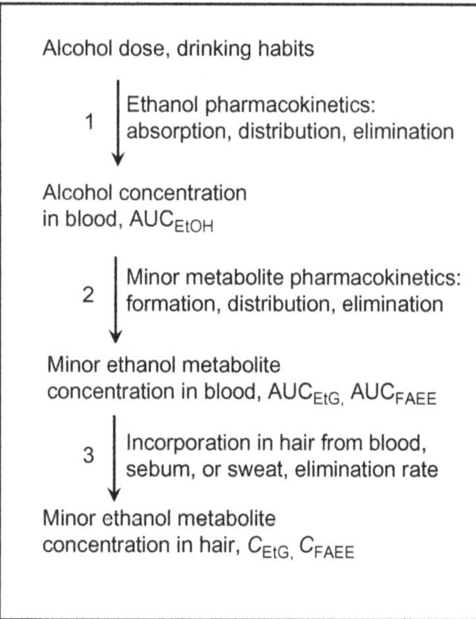

FIGURE 4.2 Steps from alcohol dose to concentrations of EtG and FAEE in hair. AUC = area under the concentration versus time curve in blood.

approximation, the concentrations of EtG and FAEE in hair are proportional to AUC_{EtOH} (Eqs. (4.2) and (4.3)) but not to the medium alcohol dose as generally assumed.

$$C_{EtG,hair} = F_{EtG} \times AUC_{EtOH} \tag{4.2}$$

$$C_{FAEE,hair} = F_{FAEE} \times AUC_{EtOH} \tag{4.3}$$

$C_{EtG,hair}$ and $C_{FAEE,hair}$ = Concentrations of EtG and FAEE in hair;
F_{EtG} and F_{FAEE} = Proportionality factors describing the efficiencies of formation and incorporation of EtG and FAEE;
AUC_{EtOH} = Area under the ethanol concentration in blood versus time curve.

It was further shown in Ref. [21] that, due the zero-order metabolism of ethanol in man, AUC_{EtOH} depends in addition to the alcohol dose also strongly on the drinking pattern. As an example it is shown in Figure 4.3 that drinking of 120 g ethanol within 3 h leads to an about eightfold higher AUC_{EtOH} than drinking the same amount in six portions of 20 g ethanol every 2.5 h. Systematic calculation of AUC_{EtOH} per month for different alcohol doses and drinking habits from occasional intake of two single drinks per month to severe alcoholics with 380 g ethanol per day (Figure 4.4) showed

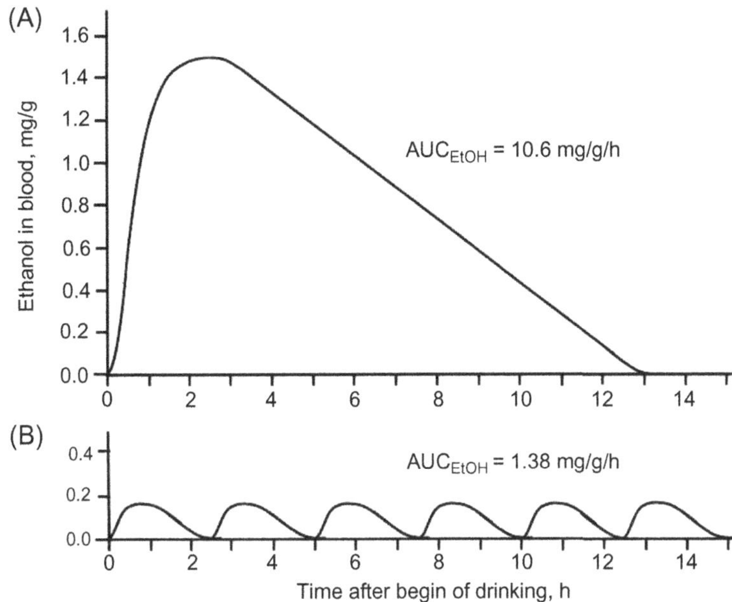

FIGURE 4.3 Effect of drinking pattern on AUC_{EtOH}. A total of 120 g EtOH was consumed (A) in one event with a drinking and absorption time of 3 h and (B) in six single drinks of 20 g EtOH with a time difference of 2.5 h and a drinking and absorption time of 1 h. Parameters for calculation: Male, body mass 70 kg, absorption deficit 0.20, volume of distribution 0.70 L/kg, and ethanol elimination rate 0.15 mg/g/h. *Modified from Ref. [21].*

that AUC_{EtOH} remains low in the range of social drinkers up to about 60 g/day and then strongly increases to high levels above 120 g/day. This should correspondingly be reflected in C_{EtG} and C_{FAEE} in hair.

It was concluded from these considerations that

- the selectivity of EtG and FAEE in hair for discrimination between social drinking and alcohol abuse is highly favored by this over-proportional increase of AUC_{EtOH} in the range between 60 and 120 g ethanol/day, and
- the drinking pattern, for example, low amounts every day or binge drinking of the same total amount at weekends (Figure 4.4, points e, f, and g), has a very strong effect on the concentrations found in hair with increased sensitivity for binge drinking.

Although this effect of drinking habits is strongly superimposed by individual variations in alcohol metabolism and incorporation of EtG or FAEE in hair as well by external effects, which will be described below separately for EtG and FAEE, it should be taken into account in the interpretation of these markers.

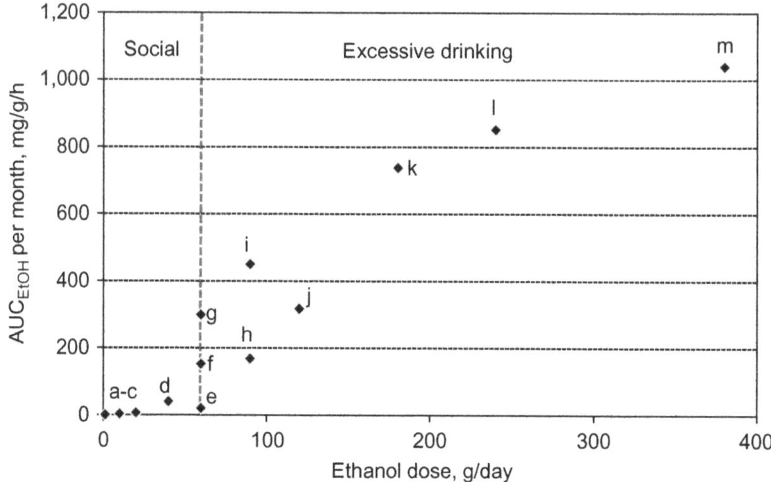

FIGURE 4.4 Areas under the ethanol concentration in blood versus time curves AUC_{EtOH} per month calculated for different daily ethanol dose and drinking pattern by Eq. (4.1). (a) 2 single drinks per months; (b) 1 × 0.5 L beer every second day; (c) 1 × 0.5 L beer per day; (d) 1.0 L beer within 1 h, per day; (e) 3 × 0.5 L beer per day, at least 3 h interval; (f) 1.5 L beer within 2 h, per day; (g) beer and brandy, 210 g EtOH within 5 h, 2 × per week; (h) 1 L wine (11 vol%), every evening within 3 h; (i) beer and brandy, 210 g EtOH within 5 h, 3 × per week; (j) brandy and beer, 120 g EtOH, every evening within 3 h; (k) brandy and beer, 180 g EtOH daily within 5 h; (l) brandy, 1 × 60 g + 6 × 30 g, every 2 h; (m) brandy, 1 × 100 g + 7 × 40 g, every 2 h. Parameters: Male, body mass 70 kg, absorption deficit 0.20, volume of distribution 0.70 L/kg, and ethanol elimination rate 0.15 mg/g/h with the exception of case "m" (0.20 mg/g/h). *Modified from Ref. [21].*

4.3 ETHYL GLUCURONIDE

4.3.1 Formation and Incorporation of EtG in Hair

The conjugation of ethanol by activated glucuronic acid UDP-GA and UDP glucuronolsyl transferase (UGT) occurs in the endoplasmic reticulum of liver cells and, to a minor extent, in cells of the intestine mucosa and of the kidney [23–25]. Interindividual variations can be explained by the polymorphism of this enzyme superfamily. According to a recent study including UGT1A1, 1A3, 1A4, 1A6, 1A9, 2B7, 2B10, and 2B15, EtG formation was observed for all enzymes under investigation with the highest activity of UGT1A9 and UGT2B7 [25]. In a previous study UGT1A1 and 2B7 were found to be the most prevalent isoforms [23]. After co-incubation with the flavonoids, quercetin and kaempferol, formation of EtG was significantly reduced to <60% for all enzymes except for UGT2B15 [25]. Obviously, nutritional components that are largely cleared by glucuronidation affect conversion of ethanol to EtG. This observation may serve as a further explanation of variable EtG formation in man. It is not excluded that decreased

EtG values or even false-negative results can be caused by genetically low UGT activity like Gilberts syndrome [26] or by inhibiting effects of food constituents or certain regularly ingested medicines.

EtG is almost completely excreted in urine with a renal clearance of 5.3 to 20.8 L/h. It followed from the total excreted amount that, on molar basis, between 0.013% and 0.022% of the drunken ethanol is transformed into EtG [27–29]. The elimination of EtG in heavy drinkers does not significantly differ from social drinkers [30]. Pharmacokinetic studies showed that EtG reaches its maximum concentration in blood between 2 and 5 h after ethanol and decreases linearly until the ethanol elimination is complete and after that exponentially according to a first-order elimination rate with a half-life of 1.7–3.3 h [29,31,32]. The maximum concentrations in blood are between 0.3 and 5 mg/L, depending on the alcohol dose. The serum/whole blood ratio was determined between 1.33 and 1.90 (median 1.69) [33]. The volume of distribution was calculated at 0.12–0.38 L/kg [23].

A relatively low variation of the areas under the curve was found for 10 volunteers after a fixed ethanol dose of 0.5 g/kg (AUC_{EtOH} = 1.34–2.23 g/L h and AUC_{EtG} = 1.73–3.19 mg/L h) [29], and a good correlation between AUC_{EtOH} and AUC_{EtG} was described for 13 volunteers in another study [32]. This supports the relevance of AUC_{EtOH} for the concentration of EtG in hair as considered above in Section 4.2. A strong relationship between AUC_{EtOH}, AUC_{EtG} and the concentration in hair was also confirmed in a study with ethanol fed rats [34].

Because of the high hydrophilicity and the anionic state of EtG at physiological pH which does not bind to melanin, a very low ICR was generally predicted for deposition of EtG in hair from blood, and its incorporation from sweat was assumed to be the main route of deposition in hair. However, the excretion of EtG in sweat was found to be very inefficient with strongly varying concentrations about two orders of magnitude below the blood levels [35]. Furthermore, this route of incorporation should depend to a high degree on surrounding temperature, physical exercise, and individual inclination to sweating and should lead to large variations in hair results.

Different from these assumptions, the determination of EtG in hair of rats after intragastric administration of ethanol led to the conclusion that the bloodstream is likely to play a major role in EtG incorporation in hair since rats do not sweat [34]. This result was confirmed by investigation of daily shaved beard hair of three human volunteers after single high alcohol doses [36]. After drinking of 153–200 g ethanol, small concentrations of EtG were already detected 9 h after end of drinking. The concentrations increased to maxima of 182, 242, and 74 pg/mg on days 2–4, and then gradually decreased to the limit of quantification (LOQ; 2 pg/mg) on days 8–10. The values are arranged in Figure 4.5 for one volunteer in the approximate position in the hair root at the time of alcohol consumption. It was concluded that the predominant portion of EtG is incorporated from blood in the upper

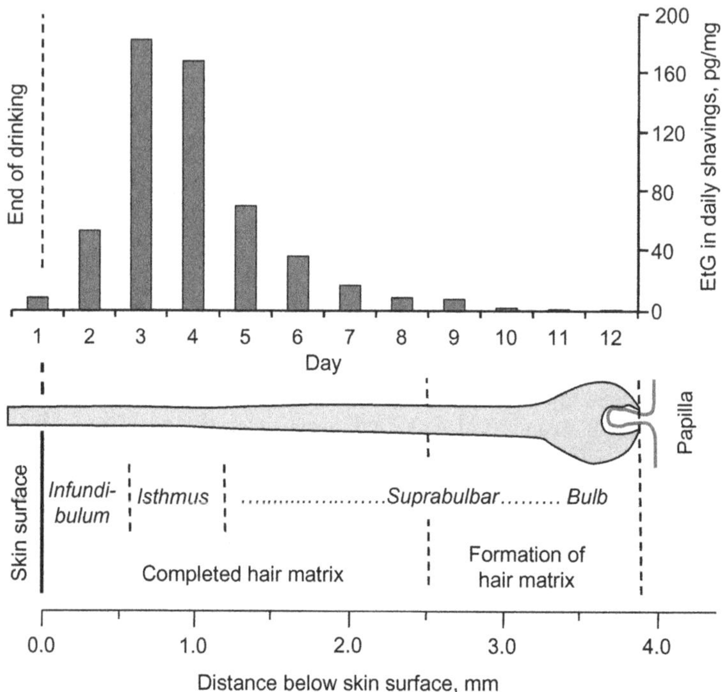

FIGURE 4.5 EtG concentrations in daily shavings after a single ethanol dose of 153 g arranged in the approximate position of the corresponding hair root sections at the time of alcohol consumption. *Data from Ref. [36].*

part of the hair root between suprabulbar region and isthmus leading to a positive zone in the hair shaft of about 3 mm (8–9 days) after a single drinking event. Deposition from sweat which is only possible into the residual hair stubble after shaving and in the infundibulum down to the sebaceous gland mouth was found to be of minor importance but could play a greater role in long hair.

It follows from these experiments that EtG in hair fulfills the prerequisites for time-resolved interpretation of segmental concentrations and that a single excessive drinking can be well detected in sufficiently short hair segments.

4.3.2 Analytical Determination of EtG in Hair

Methods for determination of EtG in hair were described in several papers and are contained in the experimental sections of almost all publications about this topic. They differ often only marginally by chromatographic parameters or used instruments. In this section the methodical progress of the essential steps shall be summarized. The experimental characteristics of some validated and sufficiently sensitive procedures are given in Table 4.1.

4.3.2.1 Sampling, Segmentation, and Storage

Scalp hair is preferentially collected in the vertex posterior region. It is recommended to cut two locks: one for analysis and the second for a possible reanalysis in case of any doubt. However, the EtG concentrations in such hair samples may be different even if the same hair length is analyzed [54]. For instance, EtG in 10 simultaneously collected locks (2 frontal, 4 temporal, 3 occipital, and 1 coronal; length 3.5 cm) of a social drinker (41 g ethanol/day) ranged from 7.7 to 24 pg/mg (mean 17.3 pg/mg). As the coefficient of variation for the three included volunteers (14.2–27.6%) is much higher than of the control (6.6%) analytical reasons were excluded and varying ICR depending on the sampling site was assumed.

According to the 2009 SoHT consensus the proximal segment 0–3 cm should be analyzed [13]. A strong decrease of C_{EtG} from proximal to distal by normal hair hygiene was found by analysis of a larger number of routine samples in three 1 cm segments [55]. But, as agreed in the reviewed consensus 2011, a longer segment up to 0–6 cm can generally be interpreted by the same cutoff in order to cover a longer time period [14]. This is justified since the decrease by washout effects was shown to concern mainly the first proximal centimeters and the superficial part, probably recently from sweat deposited portion of EtG whereas the concentration remains more stable in the distant segment up to 12 cm [56]. The hair samples should be stored at room temperature under dry and dark conditions. Control samples were found to be stable under these conditions for more than 4 months [42].

4.3.2.2 External Decontamination

Since external contamination by EtG is improbable, washing of hair is performed in order to remove impurities which could disturb the analysis. There is no generally accepted decontamination procedure and water, acetone, methanol, or dichloromethane are used alone or in different sequences and with different intensity for this purpose (Table 4.1). Nonswelling solvents such as acetone and dichloromethane are favorable to prevent extraction of EtG in this step.

4.3.2.3 Grinding or Cutting to Small Pieces

It was shown that grinding the sample leads to 0.95- to 1.8-fold higher extraction yields after overnight incubation and to 1.4- to 2.3-times higher extraction yields after 2 days incubation with water than cutting to short hair pieces [48,57]. It followed from another study about the extraction kinetics that incubation of powdered hair reached a plateau already after 4 h whereas for cut hair pieces the extraction was not completed after 18 h [49]. Cutting

TABLE 4.1 Procedures for Quantitative Determination of EtG in Hair

Reference	Washing	Milling	Sample Amount, Extraction	Clean-Up[a], Derivatization[b]	Analysis, LC Column[c]	LOD/LOQ, pg/mg
[37]	CH_2Cl_2, CH_3OH	No	100 mg, 0.7 mL H_2O, overnight incubation + 2 h ultrasonication	Direct injection	LC-MS/MS Chrompack Inertsil ODS-3	2/3
[38,39]	CH_3OH	No	10 mg, 1 mL H_2O overnight ultrasonication	SPE, OASIS MAX, BSTFA	GC-EI-MS/MS	5/10
[40, 41]	H_2O, acetone	Yes	20–30 mg, 2 mL H_2O, 2 h ultrasonication	SPE, OASIS MAX, PFPA, HFBA	GC-NCI-MS	0.7/2.3 0.5/1.5
[42]	5 min H_2O + 5 min acetone	Yes	30 mg, 1 mL H_2O, 2 h ultrasonication	SPE, OASIS MAX, PFPA	GC-NCI-MS/MS	3/8.4
[43]	H_2O, CH_2Cl_2	No	30 mg, 2 mL H_2O, 2 h ultrasonication at 50°C	SPE, Clean Screen EtG	LC-MS/MS Uptisphere-3SI[c]	4/10
[44]	H_2O, acetone	Yes	10–50 mg, 2 mL H_2O, 2 h ultrasonication	SPE, Clean Screen EtG, HFBA	HS-SPME-GC-NCI-MS/MS	0.6/2.8
[45]	CH_2Cl_2	No	100 mg, 0.7 mL H_2O, incubation overnight + 2 h ultrasonication	Direct injection, 10 μL	LC-MS/MS RP phenylpropyl[c]	1/4
[46]	H_2O, CH_2Cl_2, CH_3OH	No	20 mg, 1 mL H_2O, 1 h ultrasonication + overnight incubation	Deproteination with acetonitrile/H_2O 7:1 v/v[d], Pyridine + BSTFA	GC-EI-MS/MS, large volume injection	5/10
[47]	No data	No	25 mg, 2 mL H_2O/acetonitrile (1:1 v/v), 16 h incubation + 2 h ultrasonication	Direct injection	HILIC-MS/MS Luna HILIC[c]	—/20

(*Continued*)

TABLE 4.1 (Continued)

Reference	Washing	Milling	Sample Amount, Extraction	Clean-Up[a], Derivatization[b]	Analysis, LC Column[c]	LOD/LOQ, pg/mg
[48]	10 min ultrasonication in CH_2Cl_2, 30 s CH_3OH	Yes	75 mg, 0.5 mL H_2O incubation overnight at room temperature	Direct injection, 10 μL	LC-MS/MS Hypercarb[c]	1.7/2.3
[49]	10 min CH_2Cl_2, 2 min CH_3OH, both ultrasonication	No	30 mg, 1.5 mL H_2O incubation for 18 h at 37°C	SPE, Clean Screen EtG	UPLC-MS/MS Acquity UPLC1 HSS T3[c]	—/2.0
[50]	Acetone, H_2O	No	50 mg, 2 mL H_2O + 3.5 mL acetonitrile + 0.3 mL 1 N HCl 4 h ultrasonication at 50°C	SPE, Isolute-aminopropyl	LC-MS/MS Zorbax Eclipse XDB-C8	1/2.6
[51]	3 min CH_2Cl_2 + 3 min CH_3OH	No	50 mg, 0.5 mL H_2O/CH_3OH 35:1 (v/v) for 15 h incubation + 90 min ultrasonication	Direct injection, 3 μL	UPLC-MS/MS Acquity UPLC1 BEH C18	0.5/1.0
[52]	CH_2Cl_2, CH_3OH	No	100 mg, 0.5 mL H_2O, overnight ultrasonication at 50°C	Direct injection	LC-MS/MS XTerra® Phenyl	—/5
[53]	2 min H_2O, 1 min CH_2Cl_2	No	30 mg, 2 mL H_2O, 2 h ultrasonication at 0°C	SPE, Clean Screen EtG	LC-MS/MS Uptisphere-3SI	1.0/3.0

[a]*SPE, solid-phase extraction; cartridges:* Clean Screen1 EtG *(3 mL, 200 mg) (Amchro, Hattersheim, Germany);* Isolute NH2, *aminopropyl (1 g/6 mL) (Biotage, Uppsala, Sweden);* OASIS MAX *(3 mL, 60 mg) (Waters Corporation, Milford, MA).*
[b]*Derivatization: BSTFA, N,O-bis(trimethylsilyl)trifluoroacetamide; HFBA, heptaflourobutyric anhydride; PFPA, pentafluoropropionic anhydride.*
[c]*LC-columns:* Acquity UPLC1 HSS T3*—100 × 2.1 mm, 1.8 μm, high strength silica (HSS) T3 column (Waters Corporation, Milford, MA);* Acquity UPLC1 BEH C18*—100 × 2.1 mm, 1.7 μm (Waters Corporation, Milford, MA);* Chrompack Inertsil ODS-3, *100 × 3 mm, 3 μm (Varian, Walnut Creek, CA);* Hypercarb*—100 × 2.5 mm, 5 μm (Thermo Scientific, Waltham, MA);* Luna HILIC*—150 × 3 mm, 5 μm (Phenomenex, Torrance, CA);* Phenylpropyl*—Synergi Polar RP 250 × 2 mm, 4 μm (Phenomenex, Aschaffenburg, Germany);* Uptisphere-3SI*—100 × 2 mm, 3 μm (Interchim, France);* XTerra Phenyl*—150 × 3.9 mm, 3.5 μm (Waters Corporation, Milford, MA);* Zorbax Eclipse XDB-C8*—4.6 × 50 mm, 1.8 μm (Agilent, Waldbronn, Germany).*
[d]*Use of Sirocco Protein Precipitation Plates (Waters Corporation, Milford, MA).*

the hair below water proved to be advantageous in order to avoid problems by electrostatic charge [50].

4.3.2.4 Hair Extraction

Commercially available deuterated EtG-d5 (0.5–20 ng per sample) was generally added as internal standard before extraction. From methanol, methanol/water (1:1), water and water/trifluoroacetic acid (9:1), water was found to be the best extracting solvent [58] and is applied in most procedures (Table 4.1). Nevertheless, addition of small amounts of methanol to water (1:35 v/v) [51] and mixtures of acetonitrile and water (1:1 or 7:4, v/v) [47,50] are also applied. The extraction is supported by 1–4 h treatment in ultrasonic bath which is frequently preceded or followed by overnight incubation. Optimization for hair powder showed that the extraction with water and ultrasonication is complete after 120 min [42].

Micropulverized extraction, a combination of simultaneous milling in a ball mill and extraction, enabled quantitative extraction within 30 min [59]. By this technique 50 mg hair, internal standard, 0.5 mL water, and two steel balls (∅ 5 mm) are oscillated in a 2 mL plastic vial with a frequency of 30 s^{1}. Up to 20 samples can simultaneously be extracted. A problem appeared to be the leakage of the vials under these conditions.

It was shown in a study with rat hair that two freezing/thawing cycles plus 24 h incubation of hair pieces resulted in higher extraction yields than different combinations of ultrasonication and incubation [60].

4.3.2.5 Clean-up

Clean-up by solid-phase extraction (SPE) is frequently performed in order to increase the sensitivity by removal of impurities and for enrichment of EtG (Table 4.1). This can be based either on anion exchange cartridges (OASIS MAX®, Waters Corporation, Milford, MA, or Isolute-aminopropyl, Biotage, Uppsala, Sweden) utilizing the acidic structure of EtG, or on a carbon-based sorbent in especially for EtG developed cartridges (Clean Screen® EtG, Amchro, Hattersheim, Germany). Another possibility of clean-up is protein precipitation after evaporation of the aqueous hair extract with acetonitrile/H_2O (7:1, v/v) and subsequent filtration with Sirocco™ protein precipitation plates (Waters Corporation) [46]. The protein precipitating effect of acetonitrile plays also a role in using H_2O/acetonitrile mixtures for hair extraction [47,50].

4.3.2.6 GC-MS and GC-MS/MS Procedures

Gas chromatography-mass spectrometry (GC-MS) of EtG requires generally derivatization and was applied in electron impact (EI) as well as negative chemical ionization (NCI) mode. Derivatization with perfluorinated carboxylic acid anhydrides transforms EtG to the double-acylated lactone structure. NCI delivers generally better results. MS/MS instead of the originally used single MS [61] led to improvement of the sensitivity and accuracy particularly in EI mode [42]. Large volume injection (20 μL) increased essentially the sensitivity [46]. An increase in sensitivity was also achieved by using headspace solid-phase microextraction (HS-SPME) through additional clean-up of the derivatized EtG by headspace extraction and increase of the analyzed amount by accumulation on the fiber [44]. A GC-MS/MS method for simultaneous determination of EtG and γ-hydroxybutyric acid was developed [39].

4.3.2.7 LC-MS/MS Procedures

Many laboratories prefer currently LC-MS/MS procedures because they are faster and no derivatization is needed. Since EtG is an anion, generally electrospray ionization (ESI) in negative ion mode with the transitions 221->75 (quantifier), 221->113, and 221->85 (qualifiers) is used. Optimization of LC conditions showed that EtG was eluted on reversed phase columns in the first part of the chromatogram with interfering signals and strong matrix effects [47]. Moreover, a high water content of the mobile phase was necessary in order to retain EtG, and postcolumn addition of acetonitrile was often required to enhance the low ESI yield under these conditions [37,62]. These difficulties were solved by using hydrophilic interaction liquid chromatography (HILIC) columns [46,47,63,175] or a silica column (Uptisphere-3SI, Interchim, France) [43,53] with high content of acetonitrile in the mobile phase. An increase of the retention time was also achieved by use of a porous graphite carbon column (Hypercarb™, Thermo Scientific) [48,64]. Although in many procedures the hair extract is directly injected, clean-up by SPE is recommended particularly at low concentrations in order to avoid false-negative results caused by ion suppression.

4.3.2.8 Validation and Quality Assurance

The methods described for forensic purposes were generally validated according to international and national guidelines [52,65]. Acceptable validation data for linearity, intra- and interassay precision and accuracy, and ion suppression (LC-MS/MS) were described for the EtG methods included in Table 4.1. Limits of detection (LOD) around 1 pg/mg and lower limits of quantification (LLOQ) ≤3 pg/mg have become standard in up-to-date procedures.

An important issue is quality assurance using control samples. Quality controls for two concentration levels (approximately 50 and 500 pg/mg) were prepared by spiking hair samples with known concentrations of EtG, fortifying hair by incubation of blank hair with EtG solutions for several days or by use of authentic hair samples positive for EtG [52]. Repeated measurement by a validated method showed that EtG concentrations in authentic hair exhibited poor intraassay precision, with coefficients of variation of 25.1% and 20.9%, compared with 17.7% and 18.5% for fortified hair and 17.4% and 11.3% for spiked hair, for the lower and higher concentrations, respectively. The interassay precision for authentic hair was also poorer, 35.7% and 22.5%, compared with fortified (28.2% and 19.8%) and spiked (18.4% and 13.2%) hair for the lower and higher concentrations. In order to get a more realistic view of the method, the authors conclude that there should be regular performance monitoring of the analysis of authentic hair. If a large supply of authentic hair is lacking, fortified hair should be employed preferentially over spiked hair.

Hair reference materials containing EtG in the range indicative for moderate drinking behavior are produced by Medichem (Steinenborn, Germany) from authentic hair samples obtained from different persons and were tested for homogeneity [66]. Two hair reference materials with mean EtG contents of 8.48 and 22.0 pg/mg showed in 15 aliquots only standard deviations of 1.60 and 1.90 pg/mg, respectively.

Regular proficiency tests for EtG in hair are performed by the SoHT and the GTFCh (Proficiency Testing Organization ARVECON GmbH, Walldorf, Germany). As an example, in the proficiency test 2010 of the SoHT with two powdered samples (median concentrations A and B, 15.3 and 38 pg/mg) 21 of 25 participating laboratories delivered results [17]. The test was not passed three times for sample A and two times for sample B. In these cases, too high concentrations were reported from LC-MS/MS.

4.3.3 Alcohol Consumption and EtG Concentrations in Hair

4.3.3.1 Animal Experiments

Literature data about alcohol consumption and EtG concentrations in hair determined in animal experiments and human scalp hair are summarized in Table 4.2. There are two animal studies about a correlation between alcohol dose and EtG in hair [34,60]. Each five Longe Evans rats obtained 1, 2, or 3 g EtOH/kg/day on 4 days per week for 3 weeks in the first study [34]. Hair was shaved before and 4 weeks after the first administration. The EtG concentrations increased with rising dose and were in a good correlation with the areas under the EtG in blood versus time curve ($AUC_{EtG,blood}$) which was determined for the same animals.

In the second study, 50 Warsaw high-preferring (WHP) rats could drink 10% EtOH in water *ad libitum* from drinking bottles from which the drinking amount (0.07–13.8 g ethanol/day) was calculated [60]. The EtG concentrations in hair collected after 4 weeks (150–20,729 pg/mg) were much higher than in Ref. [34] with a correlation between drinking amount and EtG in hair only for female but not for male rats.

4.3.3.2 Meta-analysis of EtG Concentration in Human Hair

The EtG concentrations in hair of humans with different drinking behavior and published until 2012 were thoroughly reviewed and statistically

TABLE 4.2 Literature Data about Alcohol Consumption and EtG Concentrations in Animal Experiments and in Human Hair

Reference	Subjects, Alcohol Dose	Hair Length, cm	EtG, pg/mg			Remarks
			Range	Mean[c]	Median	
[34]	3 Controls	Full length	0	—	0	Intragastric administration on 4 days per week for 3 weeks, EtG concentration in hair correlates with AUC of EtG in blood
	5 Rats 1 g/kg/day		16–46	—	21	
	5 Rats 2 g/kg/day		79–158	—	104	
	5 Rats 3 g/kg/day		101–194	—	189	
[60]	50 Rats, 10% EtOH in water ad, 4 weeks, 0.07–13.8 g/kg/day	Full length	170–20,720	3,890 (female)	—	Calculated from drinking volume. Female consumed more alcohol than male
				2,180 (male)	—	
[67]	Social <0–60 g/day	0–3 to 0–6	4.7–10.2[a]	7.5	—	Meta-analysis of data from 15 publications
	Heavy ≥60 g/day		99.9–185.5[a]	142.7	—	
	Postmortem heavy		177.2–995[a]	586.1	—	
[51]	44 Teetotalers or abstainers, children	1.5–6.0	0.3–2.1	0.8 ± 0.4	—	Quantification by standard addition method
[49]	23 Abstainers	2	<2 ($n = 2$)	—	—	Prospective study over 3 months
	7 Male 32 g/day		<2–11 ($n = 6$)			
	14 Female 16 g/day		<2–3 ($n = 6$)			

(*Continued*)

TABLE 4.2 (Continued)

Reference	Subjects, Alcohol Dose	Hair Length, cm	EtG, pg/mg			Remarks
			Range	Mean[c]	Median	
[19]	43 Teetotalers	3	0–10	0.6	0	Prospective study based on daily self-report log
	44 Low risk (≤20/30 g/day)		0–32	4.9	3.5	
	38 At risk (> 20/30 g/day)		1–1,190	88.3	31.9	
[68]	22 Withdrawal patients	3–5	<2–434.7	62.8	37.4	Self-reported drinking data
	21 Social 2–60 g/day		<2–35.4	5.1	0	
	7 Teetotalers		<2	—	—	
[69]	15 Strong alcoholics in withdrawal treatment	3.5–8.5	8–261	113.3	88	Self-reported drinking data, segmental analysis, EtG of highest segments
	180–720 g EtOH/day					
[64]	6 of 17 Withdrawal patients	2–6	90–640	280	250	6 patients positive (>LOD of 25 pg/mg); 10 patients negative (<LOD)
[70]	16 Withdrawal patients	3	10–1,528	388	214	Self-reported drinking data
	60–564 g/day					
[43]	12 Heavy drinkers	No data	54–497	220	216	No quantitative drinking data
[71]	<60 g EtOH/day ($n = 23$)	3	—	9.6	0	Self-reported drinking data
	60–119 g EtOH/day ($n = 25$)		—	125	64	
	120–179 g EtOH/day ($n = 18$)		—	199	179	
	≥180 g EtOH/day ($n = 32$)		—	254	123	

(Continued)

TABLE 4.2 (Continued)

Reference	Subjects, Alcohol Dose	Hair Length, cm	EtG, pg/mg			Remarks
			Range	Mean[c]	Median	
[72]	154 Applicants for driver's license regranting	No data	<7– >300	<7 ($n = 70$)	—	Alleged abstinence of 1 year
				7–30 ($n = 26$)		
				31–100 ($n = 41$)		
				>100 ($n = 12$)		
[46]	5 Children 0 g/day	5	<LOD	—	<LOD	Self-reported drinking data, China
	4 Social 4–30 g/day		<LOD—11	—	—	
	11 Abusers 50–200 g/day		10–78	30.4	22	
[73]	50 Nonabusers	3	0–26.7	—	9	Overall medical decision
	26 Abusers		0–887	—	50.3	
[74]	19 Teetotalers	1 or 3	0–37	1/19 >30	—	Self-reported drinking data
	51 Low risk, 1–20 units per week		0–510	5/51 >30	—	
	11 Increasing risk, 21–50 units per week		0–698	5/11 >30	—	
	19 High risk, >50 units per week		0–504	11/19 >30	—	

(Continued)

TABLE 4.2 (Continued)

Reference	Subjects, Alcohol Dose	Hair Length, cm	EtG, pg/mg			Remarks
			Range	Mean[c]	Median	
[50]	317 Abstinent	3	0–493	19.8	2	Parents in child protection cases, self-reported drinking data
	65 Low moderate		0–90	9.1	3	
	672 Moderate		0–1,340	48.2	14.5	
	322 Excessive		0–1,420	122.7	66	
[75]	89 Nondrinkers and social drinkers	3	—	3.9 (±259%)	—	Self-reported drinking data
	80 Alcoholics		—	167 (±152%)	—	Traditional alcohol markers
[40]	18 Social 20–60 g/day	No data	0.7–41	11.9	8	Self-reported drinking data
	3 Risky >60 g/day		38–176	131	—	
[76]	66 Social 0.5–60 g/day	2–5	0–1,600	75.7	21.5	Self-reported drinking data
	2 Excessive 61 and 120 g/day		50, 430			
[77]	82 Abstainers	3	—	—	0	Self-reported drinking data, liver disease
	57 Moderate (<28 g/day)		0–10[b]	—	0	
	52 Drinkers		31–187[b]	—	100	
[78]	36 Severe drinkers (60–650 g/day)	3	32–662	—	—	Self-reported drinking data, linear correlation with EtG

[a]*95% confidence interval.*
[b]*Interquartile range.*
[c]*For Ref. 72 and 74 the frequency of results is given instead of the mean value.*

evaluated in a meta-analysis [67]. From overall 366 records mentioning EtG in hair, only 58 provided EtG concentrations and only 15 studies matched the analytical and statistical selection criteria of the authors. Teetotalers were not statistically evaluated because of mainly data below LOQ. Social drinkers (>0–60 g EtOH/day) resulted in a mean of 7.5 pg/mg and a 95% confidence interval (95% CI) of 4.7–10.5 pg/mg. Heavy drinkers (> 60 g EtOH/day) showed a mean of 142 pg/mg with a 95% CI of 99.9–185.5 pg/mg whereas the mean for deceased individuals with history of chronic excessive drinking was 586.1 pg/mg with the 95% CI of 177.2–995.0 pg/mg. Despite the distinct differences between social and heavy drinkers in the overall statistics, the 95% CI of the single studies overlapped often.

4.3.3.3 Basic EtG Levels in Hair of Abstainers

Hypothetical basal levels of EtG around 0.8 ± 0.4 pg/mg in hair from 26 children (teetotalers) and 16 adult abstainers were estimated using the method of standard addition and extrapolation and are believed to originate from ethanol-containing food, prescribed medicals, and cosmetic and hair care products [51]. In accordance, negative results with LOD of 1 pg/mg were also found for hair samples from 10 children [53].

4.3.3.4 EtG in Hair after Single Excessive Drinking

It is generally assumed that a single dinking episode does not lead to a positive hair test. Contrary to that, maximum EtG concentrations of 74–242 pg/mg were measured after a single drinking session of three volunteers with 153–200 g ethanol in the daily shaved beard hair after 2–4 days (Figure 4.5) [36]. These results show that a single excessive drinking can be well detected in sufficiently short hair segments. However, despite these high peak concentrations, the total concentration of the shavings of 32 days (corresponding to about 1 cm hair) was only 8.6–24.6 pg/mg, and for 3 months it was calculated as 3.1–8.7 pg/mg. It can be expected that the real concentrations in the 3-cm proximal segment would be even lower because of washout effects.

4.3.3.5 Prospective Studies

In a prospective study with red wine, 7 male volunteers drunk 32 g/day ethanol and 14 female volunteers drunk 16 g/day over a period of 3 months, whereas 23 individuals stayed abstinent over this time [49]. EtG was detected only in 6 of the 32 g consumers (<2–11 pg/mg) and in 6 of the 16 g consumers (<2–3 pg/mg). Two of the abstainers provided traces of EtG (<2 pg/mg).

Another prospective study was based on the daily alcohol self-monitoring log (report of alcohol consumption on day-by-day basis) of 125 individuals over 3 months and classification into teetotalers, low-risk drinkers (women/men ≤20/30 g EtOH/day), and at-risk drinkers >20/30 g EtOH/day) and within the latter group heavy drinkers (> 60 g EtOH/day, women and men) [19]. It was shown that EtG in the 0–3 cm hair segment enables to discriminate teetotalers and low-risk drinkers on one side and at-risk drinkers on the other with a sensitivity of 0.82 and a specificity of 0.93. Heavy drinkers could be distinguished from all the others even better with a sensitivity of 0.95 and a specificity of 0.97. However, discrimination between teetotalers and low-risk drinkers appeared not to be possible.

4.3.3.6 EtG in Hair and Retrospectively Self-Reported Drinking Data

Most data included in Table 4.2 are based on retrospectively self-reported drinking data. Although such data are generally not very reliable and drinking amounts are frequently underreported, they give some further impression about the concentration ranges observed in different groups of alcohol consumers. EtG ranges usually from 0 to 1,500 pg/mg and higher values were only measured in exceptional cases. Means and medians or frequency of positive results are clearly different for self-reported abstainers, social drinkers, and chronic excessive consumers but there are many deviations in negative and particularly positive direction.

Although EtG appears to have good prerequisites for following up the drinking history by segmental hair analysis [36], this was investigated until present only in one pilot study [69]. The cessation of drinking of strong alcoholics was indicated by low concentrations in the proximal segments if the hair sample was collected at least 1 month after starting of abstinence.

4.3.3.7 Effect of Decreased Kidney Function

Since renal excretion is the major elimination route of EtG, increased concentrations in blood and consequently also in hair were found for persons with decreased kidney function [79]. For 5 out of 11 patients with daily ethanol intake of 0.5–12 g and serum creatinine concentrations of 141–500 μmol/L (reference range <97 (female) and <115 (male)) 12–134 pg/mg EtG were determined in hair.

4.3.4 Cutoff Values, Sensitivity, and Specificity

An efficient statistical method for evaluation of the discrimination power of a parameter and for selection of a suitable cutoff with optimized sensitivity and specificity is the Receiver Operation Characteristics (ROC) analysis [80]. This technique was applied in several studies to EtG in hair and for its comparison with other markers of excessive alcohol consumption [18,19,50,71,75,77]. The results are summarized together with those for FAEE in Table 4.3 and three examples are shown in Figure 4.6. The areas

TABLE 4.3 Literature Data about Discrimination Power, Cutoff Values, and Corresponding Sensitivity and Specificity of EtG (0–3 cm) and FAEEs (0–3 or 0–6 cm) in Hair for Detection of Chronic Excessive Alcohol Consumption

Reference	Marker	Consumer Groups	AUC[a] (ROC)	Cutoff	Sensitivity	Specificity
[71]	EtG	Heavy drinker (≥60 g/day)/all others	0.94	27 pg/mg	0.92	0.96
[18]	EtG	116 Abstainers and social drinkers/59 heavy drinkers	0.982	30 pg/mg	0.915	0.97
[19]	EtG	At-risk drinker (≥30 g/day)/all others	0.95	9 pg/mg	0.82	0.93
		Heavy drinker (≥60 g/day)/all others	0.99	25 pg/mg	0.95	0.97
[50]	EtG	111 Abstainers and low moderate drinkers/150 heavy drinkers	0.95	30 pg/mg	0.75	0.97
				10 pg/mg	0.88	0.88
[77]	EtG	Abstainers/any drinking	0.79	—	—	—
		<28 g/day/≥28 g/day	0.95	8 pg/mg	0.90	0.88
[75]	EtG	89 Social and nondrinkers/80 confirmed alcoholics	0.94	28 pg/mg	0.83	0.97
[81]	FAEE (0–6)	38 Teetotalers or social drinkers/65 withdrawal patients + 21 postmortem alcoholics	0.93	0.50 ng/mg	0.90	0.90
				1.00 ng/mg	0.74	1.00
[75]	FAEE (0–3)	89 Social and nondrinkers/80 confirmed alcoholics	0.83	0.68 ng/mg	0.59	0.91

(Continued)

TABLE 4.3 (Continued)

Reference	Marker	Consumer Groups	AUC[a] (ROC)	Cutoff	Sensitivity	Specificity
[50]	FAEE (0–6)	92 Abstainers and low moderate drinkers/137 heavy drinkers	0.95	1.0 ng/mg	0.77	0.96
				0.63 ng/mg	0.88	0.88
[8]	FAEE (0–6)	Postmortem samples, 171 alcohol abusers, 61 low drinkers	0.99	0.80 ng/mg	0.95	0.95
[82]	FAEE (0– ≤ 6)	Postmortem samples, 168 social drinkers/502 alcohol abusers	0.75	1.08 ng/mg	0.56	0.80

[a] *A test is interpreted to be noninformative with* $AUC = 0.5$*, less accurate with* $0.5 < AUC \leq 0.7$*, moderate accurate with* $0.7 < AUC \leq 0.9$*, highly accurate with* $0.9 < AUC < 1.0$ *and perfect with* $AUC = 1.0$*.*

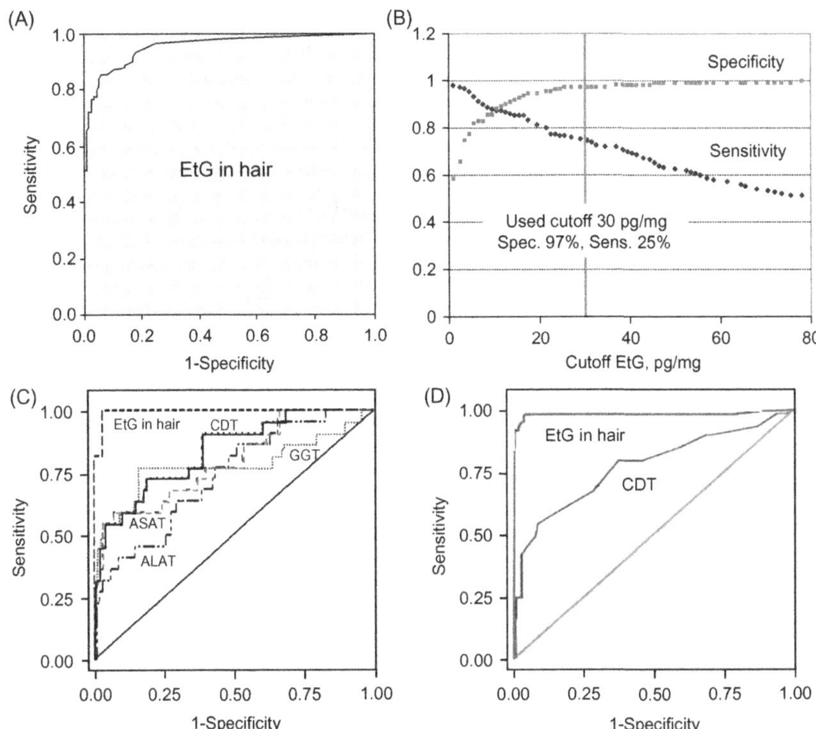

FIGURE 4.6 Evaluation of the discrimination power of EtG in hair for detection of chronic excessive alcohol consumption by ROC analysis and comparison with other alcohol biomarkers. (A) 111 abstainers or low moderate drinkers versus 150 excessive drinkers [50], AUC 0.95, highly accurate. (B) Plot of sensitivity and specificity versus cutoff for the same groups as in A. (C) 104 abstainers and social drinkers versus 21 excessive drinkers [19], AUC 0.99, highly accurate. AUCs of other alcohol markers for comparison: ALAT 0.73, ASAT 0.79, GGT 0.79, CDT 0.84. (D) 116 abstainers and social drinkers versus 59 heavy drinkers [18], AUC 0.98, highly accurate. For comparison CDT: AUC 0.77.

under the ROC curve AUC were between 0.94 and 0.99 in these studies and indicate EtG in hair as a highly accurate test for discrimination between abstainers and social drinkers on one side and excessive alcohol consumers on the other.

However, the outcome of the ROC analysis depends to a high degree on the studied population. For instance, the discrimination of social drinkers with 40 g ethanol/day from excessive drinkers with 80 g/day would be considerably less efficient because of the many other individual factors which have an effect on the EtG concentration and are discussed in other sections below.

The sensitivity (1-portion of false-negative results) and specificity (1-portion of false-positive results) can be calculated for the chosen cutoff from the

ROC data. The plot of sensitivity and specificity versus the cutoff as in Figure 4.6B demonstrates the influence of the cutoff selection on the rates of false-positive and false-negative results. It shows in agreement with the data of Table 4.3 that the cutoff 30 pg/mg is chosen in favor of higher specificity (low rate of false-positive results) in order to avoid disadvantages for the accused drinker. But it has to be kept in mind that false-negative results may have equally disastrous consequences, for example, in child custody cases.

Based on these studies the cutoff for excessive drinking of 30 pg/mg was recommended in the current SoHT consensus [14]. No real evaluation was performed concerning abstinence assessment. The currently used cutoff of 7 pg/mg [15] is probably too high and has to be optimized if more data are available about the EtG level in hair of abstainers. Both cutoffs were confirmed in the revised consensus 2014 [83].

4.3.5 EtG in Nonhead Hair

4.3.5.1 General Differences as Compared to Scalp Hair

Hair from other anatomic sites can be investigated as an alternative if scalp hair is not available or in addition to scalp hair for confirmation and as a source of further information. The main differences are a smaller growth rate and a much higher portion of telogen hair which makes it difficult to assign a hair segment to a certain time period. As a rule of thumb to estimate the maximum time window, the Eq. (4.4) was suggested, as given in Ref. [76]:

$$\text{Time window} = \text{hair length}/[\text{growth rate} \times (1 - \%\text{telogen}/100)] \qquad (4.4)$$

Since the length of body hair is not uniform, the 75 percentile of the length distribution of the sample under investigation was proposed to use in this equation. Growth rate and percentage of telogen hair are taken from a table of literature data.

Advantages of nonhead hair are that it usually is not submitted to cosmetic treatment and that it is less exposed to environmental influences. EtG in nonhead hair was compared with head hair in several studies [40,50,63,76,84,85]. By now there are no sufficient data known for beard hair.

4.3.5.2 EtG in Pubic Hair

Pubic hair can display extremely higher EtG concentrations than head hair. As an example, for eight volunteers with EtG in scalp hair below 10 pg/mg, in pubic hair between 12 and 1,370 pg/mg were measured [63]. In other studies the concentration ratio between pubic hair and head hair ranged from 1.42 to 69.1 ($N = 20$, mean 14.7, median 6.9 [40]) and 1.3 to 255 ($N = 19$, mean 86, median 48.5 [84]) with negative results in pubic hair for teetotalers and increase of this ratio with drinking amount. In agreement with these findings a much higher portion of positive EtG results in pubic hair (62%)

than in scalp hair (16%) was found in the total statistics of 2,347 cases [84]. Contamination by urine is assumed to be the main reason.

For this reason, EtG in pubic hair cannot be used for discrimination between social and excessive drinking but can be helpful in abstinence assessment. According to the SoHT consensus about the use of alcohol markers in hair for abstinence assessment a negative EtG result in pubic hair strongly confirms abstinence [15]. However, false-positive results may occur from EtG in urine caused by ethanol-containing food products or hand sanitizers.

4.3.5.3 *EtG in Axillary Hair*

Axillary hair contained frequently less EtG than head hair. The concentration ratio axillary/head hair ranged from 0.03 to 2.5 ($N=8$, mean 0.78, median 0.36) [40] or from 0.00 to 0.63 ($N=10$, mean 0.19, median 0.13 [76]). Accordingly, the percentage of positive results in axillary hair (3.5% and 2.6%) was significantly lower than in scalp hair (16% and 11.4%) in the statistics of 2,347 and 22,825 cases, respectively [84,85]. It is proposed that the low EtG results in axillary hair arise from decomposition by deodorants or from leaching by sweat [76]. Because of this lower sensitivity, EtG in axillary hair is considered unsuitable for assessment of alcohol consumption.

4.3.5.4 *EtG in Chest, Arm, and Leg Hair*

Chest hair (usually only from male subjects) appeared to be most similar to head hair and is recommended as the best alternative. The EtG concentration ratio of chest hair/scalp hair in two studies ranged from 0.41 to 1.83 ($N=10$, mean 1.16, median 1.12) [40] or from 0.14 to 4.0 ($N=45$, mean 1.07, median 0.88) [76]. In the total statistics of 2,347 and 22,825 cases the portion of positive results was 21.5% and 9.9% respectively for chest hair as compared to 16% and 11.4% for scalp hair [84,85].

Arm and leg hair exhibits frequently higher EtG concentrations than head hair. The ratio ranged from 0.58 to 4.11 ($N=11$, mean 2.40, median 2.11) [40], or from 0.28 to 7.3 ($N=49$, mean 1.45, median 1.00) for arm hair and 0.26 to 10.9 ($N=59$, mean 1.77, median 1.05) for leg hair [76].

The interpretation of EtG in chest, arm, and leg hair with respect to the cutoff values led in the majority of cases to the same result as scalp hair despite large intraindividual differences of the concentration often found between the different kinds of hair.

4.3.6 Effect of Hair Color, Hair Care, and Cosmetic Treatment

It was shown in animal experiments [34] as well as by comparison of white and pigmented hair from grizzle subjects [86] that EtG in hair is not biased against pigmentation. This is in agreement with general knowledge about incorporation in hair since EtG as an anionic substance at physiological pH does not bind to melanin.

Because of its structure as a polyhydroxy compound, EtG is sensitive to oxidizing agents like H_2O_2 used in bleaching and permanent dyeing processes. It was shown in an *in vitro* study that application of a commercial bleaching kit to hair samples from 4 social drinkers and 21 excessive drinkers (EtG 7.7–149 pg/mg) totally removed EtG from hair [87]. Besides chemical degradation, increased leaching from the cosmetically damaged hair was also assumed. In a similar *in vitro* study a decrease of EtG by, as a mean, 73.5% was found after bleaching whereas incubation of EtG alone with H_2O_2 (15%) led only to a decomposition of 45% [88]. Accordingly, the overall statistics of a large number of samples showed significantly lower EtG concentrations in bleached and dyed hair as compared to untreated hair (Table 4.4) [50,89]. Contrary to that, no effect on dyeing was reported in another statistic study [90]. But, this statistics has to be seen with caution since false-negative results are not included (portion of positive results in cosmetically treated samples was only 4.1% [91] as compared to 21% and 12.7% [92,93] in the same whole clientele).

Perm treatment of 23 samples decreased the concentration in hair, as a mean, by 95.7%, and the agent ammonium thioglycolate totally decomposed EtG [88]. On the other hand, treatment of 10 samples with coloring products did not show any important change in EtG results in the same study. Thermal straightening of hair by heating with a curling iron to 200°C for 60 s in an *in vitro* experiment decreased the EtG concentration in 20 out of 41 samples by as a mean 20% (range 0.7–79%) whereas in 21 of 41 samples an increase by as a mean 15% (range 2–51%) was found [94]. Since hair straightening is sometimes daily applied degradation or increase of extractability (washout) of EtG in practical cases is possible.

The effect of different cosmetic treatments on EtG evaluated from a larger number of routine samples is shown in Table 4.4 [50]. Different from FAEE (Section 4.4.5), ethanol in hair lotions or hair sprays has no effect on the EtG concentration [41]. However, in some commercial herbal hair tonics 0.07–2.7 mg/L of EtG were detected [95,96] and can lead to false-positive results after continuous application as high as 910 pg/mg in one case [95]. In such lotions, EtG is formed during extraction of plants such as fresh thyme, dried birch, oak, or plantain with ethanol [97]. Analysis of the lotion for EtG and additional testing of nonhead hair can help to clear such an error in rare cases.

Four special cleansing shampoos offered in internet to "detox" and to reduce EtG in hair were applied to 21 hair samples with EtG between 7 and 288 pg/mg and did not lead to any significant decrease of the concentration after a single application [98].

Until present, information about hair care and cosmetic treatment is based mainly on reports of the tested persons and to a limited extent on visual

TABLE 4.4 Effect of Cosmetic Hair Treatment on EtG Concentrations [50]

Treatment	Other Treatment Excluded	*N*	EtG, pg/mg		Effect	ANOVA Test		
			Mean	Median		α	F	Significance
No treatment known	—	1,079	67.9	19	—	—	—	—
Bleaching	Yes	164	13.9	5	Strong decrease	0.000	20.361	Yes
Dyeing	Yes	96	25.3	3	Strong decrease	0.007	7.274	Yes
Bleaching and/or dyeing	No	341	18.9	5	Strong decrease	0.000	33.525	Yes
Hair spray	Yes	79	73.9	11	No effect	0.737	0.113	No
Hair gel	Yes	99	71.2	22	No effect	0.835	0.043	No
Hair wax, oil, or grease	Yes	34	58.8	23	No effect	0.728	0.121	No

examination of the hair sample, for example, differences in color with newly grown hair near the root. In light of the strong effects of cosmetic treatment particularly on the alcohol markers in hair, one should be aware of a possible manipulation of hair by such treatments. Therefore, an objective test to prove bleaching, dyeing, or permanent wave should belong to the routine procedures of hair laboratories. It was shown recently that attenuated total reflectance Fourier transform infrared spectroscopy (ATR-FTIR) of pulverized hair is a nondestructive possibility to detect oxidative treatment [99]. This is based on the S–O stretching band at 1,040 cm^{-1} of the sulfonic group of cysteic acid residues which are formed by action of H_2O_2 from cysteine or cystine moieties in hair proteins. Since the IR spectrum of the hair specimen does not significantly differ between donors, the spectrum can be normalized in order to quantify the effect.

4.3.7 Effect of Other Parameters and Habits on EtG in Hair

The influence of other specific characteristics such as age, gender, body mass index (BMI), smoking, or preferred beverage (beer or wine) of the individuals on the performance of EtG in hair was studied by ROC analysis in Ref. [71]. None of the factors examined was found to influence significantly the marker performance if the relatively low number of cases is taken into account.

However, in the evaluation of 272 FAEE positive samples by measurement of EtG, a gender-related decrease in EtG sensitivity of female subjects was observed independent of artificial hair coloring [89]. Therefore, the authors propose to use the lower cutoff 20 pg/mg for women. In the interpretation of group-level factors from a large population dataset (18,920 males and 1,373 females), the application of 30 pg/mg led to positive results in 12.3% of the male and only 6.6% of the female subjects [85]. If the cutoff of 20 pg/mg was applied for the women, the positive portion increased to 9.6%. Different from that, the correlation of EtG in hair and self-reported drinking amounts from 25 males and 11 females showed that the lower results for women were caused by lower alcohol consumption and that the same cutoff can be used for both genders [78].

Concerning age of the subjects, in the large population study, an increase of the percentage of positive results from 18–30 years ($n = 1{,}764$; 4%) to 31–40 years ($n = 5{,}878$; 8%) to over 40 years ($n = 9{,}731$; 18%) was found [85]. This is obviously caused by higher alcohol consumption with increasing age combined with different drinking behavior (e.g., binge-drinking behavior versus daily substantial alcohol consumption at mealtime), tolerance phenomenon, and change in ethanol metabolism or less willing of elderly people to refrain from alcohol consumption in driving ability examination cases. No significant influence of the BMI was seen in this study.

Interestingly, the authors found a seasonal effect of the sampling date in the 4-year study with highest mean and median values for samples collected in winter and lowest collected in summer [85]. Most probable explanations are drinking of more and higher concentrated alcoholic beverages during the cold season and increased elimination of EtG from hair by sweating, frequent showers, and bathing in summer.

It was shown by EtG in hair that drug users obviously consume less alcohol than nondrug users [38]. From 57 hair samples which were positive for one or more illegal drugs (opiates, cocaine, amphetamines, methamphetamines, benzodiazepines, and cannabinoids) EtG was detected in 12 samples (21%, mean 11 pg/mg) whereas from 42 drug-negative samples 17 were positive for EtG (40.5%, mean 107 pg/mg). Sources of bias such as age, gender, and living area were excluded.

4.4 FATTY ACID ETHYL ESTERS

4.4.1 Formation and Incorporation of FAEEs in Hair

FAEEs are a group of more than 20 minor metabolites of ethanol and are formed in presence of alcohol from free fatty acids, triglycerides, lipoproteins, or phospholipids by specific cytosolic FAEE synthases as well as by nonspecific enzymes such as carboxylesterase, lipoprotein lipase, carboxylester lipase, or cholesterol esterase in blood and almost all tissues. Biochemistry, physiology, metabolism, and its possible role in the pathogenesis of alcohol-induced tissue damage were thoroughly reviewed previously [8,100–102] and there is no really new knowledge about the fundamentals of these compounds in the last 10 years.

Ethyl palmitate and ethyl oleate are the most abundant FAEEs in blood [103]. Depending on the alcohol dose, 0.36–0.9 μg/mL was determined in whole blood [101] which decreased to 7–25 ng/mL within about 30 h. This is believed to be the metabolic base concentration in absence of exogenic alcohol [104,105]. The elimination from blood occurs by redistribution into tissues and by FAEE hydrolases in at least two-phase kinetics with a primary and a terminal half-life of 3 and 11 h, respectively [106].

More than 15 FAEEs including unsaturated and branched FAEEs as well as saturated FAEEs with an odd number of carbon atoms were detected by a GC-MS method in hair samples from alcoholics [8]. From these, ethyl myristate (E14:0), ethyl palmitate (E16:0), ethyl oleate (E18:1), and ethyl stearate (E18:0) were chosen for routine analysis of hair. It was concluded from the high concentrations in the external lipid layer of hair and from the segmental distribution along the hair shaft that FAEEs found in hair are mainly synthesized in the sebum glands and are incorporated into hair via the steadily renewed sebum layer [8,107]. This mechanism was supported by a systematic analysis of FAEEs in newly excreted sebum which was collected from the forehead of teetotalers, social drinkers, and alcohol withdrawal

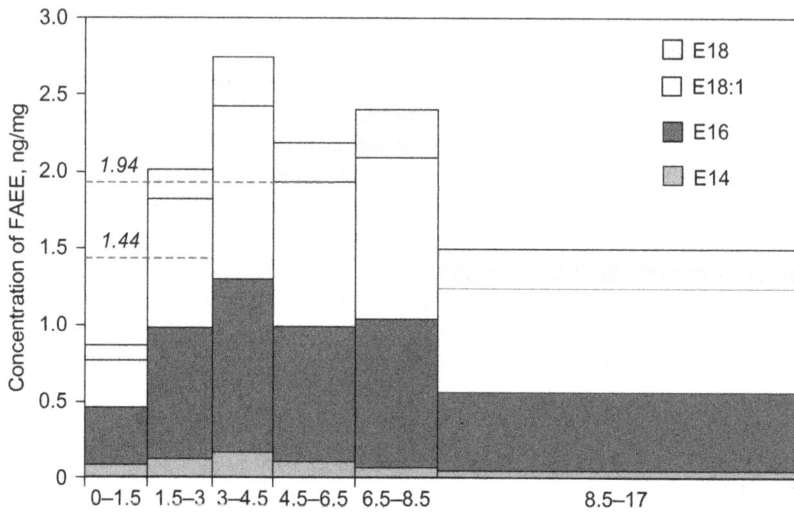

FIGURE 4.7 FAEE concentrations in hair segments of a 35-year-old woman in withdrawal treatment with long-term daily ethanol intake of 200 g and abstinence begins 6 days before sampling. Mean concentration in 0–3 cm segment 1.44 ng/mg and in 0–6 cm segment 1.94 ng/mg. The concentration ratio 0–3 cm/0–6 cm is 0.74. *Data from Ref. [107].*

patients with sebutapes (CuDerm, Dallas, TX) [108]. High concentrations of >60–850 ng/sebutape were only measured within 2–3 days after abstinence begin. After that it slowly decreased to a basic level of 20–30 ng/sebutape which was also found in abstinent volunteers.

Caused by this mechanism the incorporation occurs over the whole hair length and the FAEEs accumulate with increasing concentrations from proximal to distal. A typical segmental distribution is shown in Figure 4.7 and can originate from permanent excessive drinking as well as from a short but heavy relapse after long lasting abstinence [107]. Consequently, FAEEs in hair are not suitable for exploring the drinking history and the concentration measured at constant drinking behavior in the proximal hair segment increases with its length, for example, from 0–3 to 0–6 cm. Nevertheless, cessation of drinking can be recognized by a low concentration in the proximal segment.

4.4.2 Analytical Determination of FAEEs in Hair by HS-SPME and GC-MS

Although LC-MS/MS methods with ESI were published for analysis of FAEEs from meconium [109–111], the original procedure with HS-SPME and GC-MS described in 2001 [107,112] seems to be still the most efficient way for hair. The steps of this procedure were optimized and revalidated in later studies [113] and by other authors [114–116].

4.4.2.1 Sample Pretreatment and External Decontamination

The hair samples are washed with water and air dried. According to the consensuses, generally the proximal hair segment 0–3 or 0–6 cm is analyzed if available. Plastic material should be avoided and all steps should be performed in glass vials or aluminum dishes. External lipids originating from recent sebum excretion or from hair cosmetics are removed by washing two times with *n*-heptane. This is a critical step and must be performed in a constant and reproducible way in order to remove FAEEs in the external lipids but to avoid their extraction from the hair matrix. Alternatively, two washings with methanol (2 mL for about 2 min each) were described [117,118] but cannot be recommended because the swelling effect on hair and the lower solubility of external lipids in this polar solvent. Always hair pieces of 1–2 mm length were analyzed. Grinding is accounted to be unsuitable because of losses of FAEEs by evaporation. It was used only in one study but its effect on extraction yield was not examined [118].

4.4.2.2 Hair Extraction

The hair pieces are extracted for 15 h at 25°C by a two-phase mixture of 0.5 mL dimethylsulfoxide (DMSO) and 2 mL *n*-heptane containing each 40 ng of the deuterated FAEE (R-$COOC_2D_5$) as internal standards which are commercially available and are imperative for the HS-SPME method. Ethyl heptadecanoate which is recommended in Ref. [118] is unsuitable as internal standard because of its presence in sebum and hair after alcohol consumption. The aprotic solvent mixture appeared to be most suitable because of the highest extraction yields and the avoidance of ester hydrolysis or transesterification in presence of methanol or other alcohols. After extraction the mixture is cooled to below 0°C. The heptane layer can be easily decanted from the frozen DMSO and is evaporated to dryness in 10 mL headspace vials.

4.4.2.3 Headspace Solid-Phase Microextraction

HS-SPME appears to be an essential clean-up step of the procedure. To the hair extract in the headspace vials, 1 mL phosphate buffer (0.1 M, pH 7.4) and 0.5 g NaCl are added in order to retain free fatty acids from evaporation and to use the salting-out effect. However, it was shown that salt-free analysis does not reduce sensitivity or accuracy [114]. All steps of HS-SPME are automatically performed with a 65-μm polydimethylsiloxane/divinylbenzene (PDMS/DVB) fiber as optimized and described in previous papers [112,113,115].

4.4.2.4 Gas Chromatography-Mass Spectrometry

FAEEs are well analyzed by a single quadrupole GC-MS in EI mode with molecular ions as quantifiers and highly abundant fragment ions from the McLafferty rearrangement (m/z = 88) and of the β-cleavage (m/z = 101) as qualifiers. Positive chemical ionization and MS/MS (88, 101, and M + 1 for FAEEs, and 93, 106, and M + 1 for FAEEs-D5) were also reported but without data about the performance [119] or with unreliable results because of other experimental reasons [118].

4.4.2.5 Calibration and Validation

The method was validated according to forensic guidelines [120] and showed good results for sensitivity, linearity, precision and accuracy, recovery, and stability [114]. The calibration was linear between 0.02 and 1.2 ng/mg for each of the four esters. LOD and LOQ were determined between 0.005 and 0.009 ng/mg and between 0.016 and 0.025 ng/mg, respectively [113,114] or better [117]. Caused by saturation effects of the SPME fiber the calibration curves are slightly bent at higher concentrations. Therefore, in another study a low range (0.05–0.5 ng/mg) and a high range calibration (0.5–5 ng/mg) were conducted in order to get sufficient linearity in each range [113].

4.4.2.6 Attempts for Improvement and Alternative Methods

There were several attempts to simplify the HS-SPME procedure. It was tried to save the liquid extraction and to submit the washed hair pieces directly with phosphate buffer to HS-SPME at 90°C (M. Hastedt, F. Pragst, unpublished results). Despite excellent extraction yields from the hair matrix, no reproducible results could be obtained because of the different HS-SPME behavior of the deuterated standards added by spiking.

Experiments to find a common procedure for FAEE and EtG led to the proposal of washing the hair sample with dichloromethane and methanol followed by addition of 1 mL of methanol and sonication for 30 min to extract the FAEEs and EtG from hair [119]. However, this procedure was not validated and not really tested in forensic practice and, in consideration of previous results [58,112], it should suffer from incomplete extraction for both EtG and FAEE because of the short extraction time and from increased transesterification of FAEE at longer extraction time.

As an alternative to HS-SPME, the *n*-hexane layer of the hair extract was cleaned-up by SPE on aminopropyl cartridges with elution by *n*-hexane and dichloromethane and subsequent GC-MS of the four esters with alpha-cholestane as internal standard and LOQs of 0.01 ng/mg in a validated procedure [121].

4.4.3 FAEE Concentrations in Hair and Alcohol Consumption

4.4.3.1 Concentration Ratio of the Four FAEEs

The sum of the four ester concentrations ranged from <0.1 to more than 80 ng/mg in the different reports from routine and research measurements. The ratio between the four esters was quite variable with E16 and E18:1 around 40% and E14 and E18 around 10%. As a typical example the distribution for 644 samples with a mean ratio E14:E16:E18:1:E18 of 8:45:38:9 is shown in Figure 4.8 [115]. A ratio of 8:37:47:10 was described in Ref. [117]. Large deviations from this ratio or almost exclusive presence of only one ester could indicate an external source, for instance cosmetic products. But, this has not gained practical importance until now and no examples were described. Therefore, it is considered to restrict the FAEEs on E16 and E18:1 or even only on E16. This would require new definition of the cutoffs but should also decrease analytical uncertainty of the summary parameter.

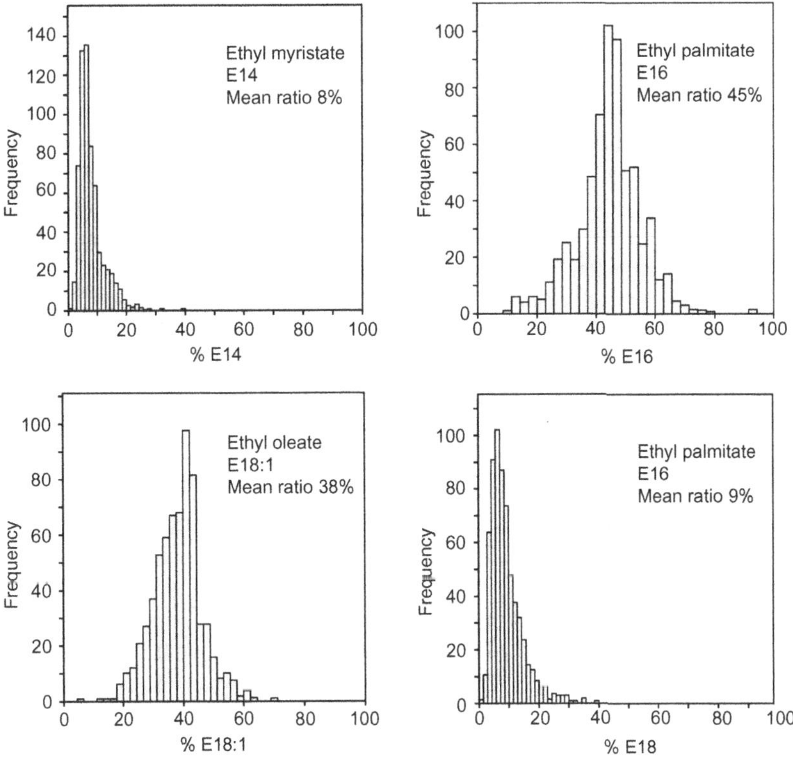

FIGURE 4.8 Distribution of the concentration ratios of ethyl myristate, ethyl palmitate, ethyl oleate, and ethyl stearate to the sum of the four concentrations in 644 hair samples. *Data from Ref. [115].*

4.4.3.2 FAEE Concentrations in Hair and Self-Reported Drinking Data

FAEE concentrations found in hair of different alcohol consumer groups are summarized in Table 4.5. Further studies are discussed in Section 4.4.4 and combined with EtG or in context of application in the Sections 4.5 and 4.7. Only self-reported drinking data were available and no prospective studies are known.

Concentrations in hair of abstainers (children, reliable adult teetotalers, ethanol-containing hair cosmetics excluded) in the 0–6 cm hair segment ranged from 0.06 to 0.37 ng/mg [8,116]. Values below 0.2 ng/mg in the 0–3 cm segment were shown to be typical for abstainers in abstinence assessment [114,122]. This appears to be the range of basic levels of FAEE in hair of children and strict teetotalers.

It was concluded from the hair samples of volunteers that social drinking does not lead to FAEE concentrations in the 0–6 cm hair segment above 1.0 ng/mg [8,116]. This was confirmed also in practical applications [50,82,115,123]. In the same way, 0.5 ng/mg in the 0–3 cm segment has proved itself as the upper limit for social drinking in practice [89,113].

However, the data in Table 4.5 show that drinking amounts are frequently underreported, particularly if higher quantities are in disadvantage for the investigated person. From 553 subjects tested in context of child custody the FAEE concentrations of the self-reported subgroups of teetotalers ($N = 156$, no alcohol consumption for at least 6 months), moderate drinkers ($N = 252$, occasional drinking of alcohol up to 60 g ethanol/day), and excessive alcohol consumers ($N = 145$, equal or more than 60 g ethanol/day) were statistically well separated [115]. But, there was a high portion of positive results in the groups of self-reported teetotalers and moderate drinkers.

Similar results were obtained for 1,381 individuals who were divided into the subgroups abstinent, low moderate, moderate, and excessive drinking according to their self-reports [50]. The group "low moderate" included individuals who reported only low doses at special occasions (mean <1 g/day). This group provided lower FAEE results than the group of self-reported abstainers. The low reliability of these drinking data was in the same way obvious from the EtG concentrations measured in the same samples (Table 4.2). From these data, "reliable" cases were selected for evaluation of cutoffs using a series of exclusion and inclusion criteria such as EtG in the same hair sample, carbohydrate-deficient transferrin (CDT) in blood, and no cosmetic treatment (Section 4.4.4).

Based on self-reported drinking data, a high rate of 46 apparently false-positive FAEE results from the 0–3 cm segment and using the 0.5 ng/mg cutoff was found for 160 volunteers (77 male, 83 female) [117]. A closer evaluation showed that this concerned mainly female participants and that 11 could be explained by using ethanol-containing hair lotions whereas for 22 the regular intake of contraceptives (ethinylestradiol with a progestin, either drospirenone, gestodene, or levonorgestrel) was considered as a possible reason for increased

TABLE 4.5 Literature Data about Alcohol Consumption and FAEE Concentrations in Human Hair

Reference	Subjects, Drinking Behavior	Hair Length, cm	FAEE, ng/mg			Remarks
			Range	Mean	Median	
[8]	17 Teetotalers	≤6	0.06–0.37	0.17	0.14	Volunteers, self-reported drinking data
	20 Moderate (2–20 g/day)		0.08–0.87	0.40	0.29	
	49 Withdrawal patients		0.20–20.5	2.70	1.40	
[8]	61 Alcohol abuse excluded	≤6	0.03–0.89	0.32	0.29	Postmortem cases; Data from case history and autopsy
	171 Known alcohol abuse		0.4–42	5.0	3.0	
	96 Without drinking data		0.8–18.9	1.38	0.66	
[116]	10 Teetotalers	5	<0.24			Only E14 and E16
	10 Social, 1–4 U/day		0.15–0.85			
	12 Alcoholics, 2–35 U/day		0.06–6.92			
[115]	156 Abstinent	6 or <6 in full length	0.11–10.8	0.97	0.43	Parents in child protection cases, self-reported drinking data of 6 months before sampling
	252 Moderate (<60 g/day)		0.11–11.7	1.57	0.81	
	145 Excessive (>60 g/day)		0.21–31.3	2.03	3.16	

(Continued)

TABLE 4.5 (Continued)

Reference	Subjects, Drinking Behavior	Hair Length, cm	FAEE, ng/mg			Remarks
			Range	Mean	Median	
[50]	242 Abstinent	6 or <6 in full length	0.04–23	1.12	0.30	Parents in child protection cases, self-reported drinking data
	28 Low moderate		0.06–2.5	0.36	0.25	
	765 Moderate		0.03–33	1.48	0.62	
	346 Excessive		0.06–55	4.27	2.07	
[82]	168 Social drinkers	3 or 6	0.008–14.3	0.875	0.302	Postmortem samples, data from police reports, and autopsy findings
	502 Alcohol abusers		0.01–83.7	3.439	1.346	
[75]	89 Nondrinkers and social drinkers	3	—	0.42 (±114%)	—	Self-reported drinking data
	80 Confirmed alcoholics		—	1.41 (±186%)	—	Traditional alcohol markers
[117]	160 Volunteers	3	0.02–10.9	1.16	0.60	Self-reported drinking data, possible effect of contraceptives
[89]	73 Male, FAEE ≥0.5	3	0.58–29.4	2.75	1.49	Parents in child protection, cases with FAEE positive
	199 Female, FAEE ≥0.5		0.52–49.5	2.85	1.42	

FAEE formation. Since this effect was not observed in the studies of female volunteers before, it needs further confirmation. One explanation which should apply to EtG in the same way is that women taking oral contraceptives demonstrate a significantly decreased ethanol elimination rate [124].

4.4.3.3 FAEEs in Postmortem Hair Samples

FAEE concentrations found in hair of death cases are also included in Table 4.5. The usability for detection of antemortem alcohol abuse was demonstrated in 328 death cases [8]. The results were in the range of teetotalers and social drinkers for 61 death cases in which alcohol abuse was excluded (0.03–0.89 ng/mg, mean 0.32 ng/mg). However, the FAEE concentrations of 171 cases with known excessive drinking (0.4–42 ng/mg, mean 5.0 ng/mg) were on average clearly higher than for withdrawal patients. This was explained by the even higher alcohol consumption or by neglected hair care of these individuals who came to a higher portion from antisocial environment.

The concentrations of FAEEs in hair samples from 1,057 autopsy cases which were later examined in the same institute ranged from 0.008 to 83.7 ng/mg [82]. The antemortem drinking behavior was evaluated using police investigation reports, medical history, and macroscopic and microscopic alcohol-typical results from autopsy. The correlation with the FAEE concentrations was less significant than in the first study. A total of 168 social drinkers and 502 alcohol abusers were selected and evaluated using ROC analysis. As an applicable cutoff for postmortem samples 1.08 ng/mg was proposed but the discriminating performance of the test appeared to be relatively low with a ROC area of 0.75, a sensitivity of 56%, and specificity of 80% at this cutoff. The authors assume the unknown hair cosmetics to be the main reason of the low performance but uncertainties in the drinking behavior and variation in the analyzed hair length might have contributed too.

4.4.4 Cutoff Values, Sensitivity, and Specificity

ROC analysis was used several times to evaluate the discrimination power and to determine a suitable cutoff for FAEEs in hair. The results are given in Table 4.3 together with EtG. As explained above, it is necessary to standardize the length of the assessed segment (0–3 or 0–6 cm) because of the increasing concentration profile from proximal to distal (Figure 4.7). Furthermore, cases of cosmetic treatment had to be excluded. The discrimination between abstainers and social drinkers is not possible since social drinkers have frequently FAEE concentration in the range of abstainers. Therefore, the discrimination concerns only abstainers and social drinkers on one side and alcoholics or chronic excessive drinkers on the other.

The AUC of the ROC analysis of 0.83–0.99 indicates moderately accurate or mainly highly accurate performance of the test. Figure 4.9 shows an

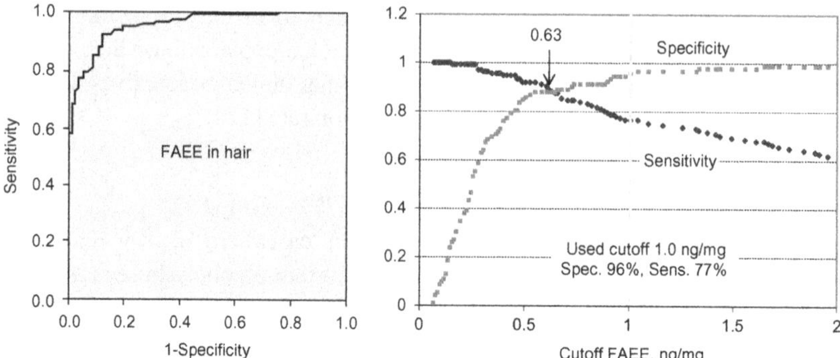

FIGURE 4.9 ROC analysis for FAEE in the proximal 6 cm hair segment of selected cases with reliable drinking data in the categories "abstinent and low moderate" (true negative, $N = 92$) and "excessive" (true positive, $N = 137$). *From Ref. [50].*

example for the segment 0–6 cm with AUC 0.95 together with a plot of sensitivity and specificity versus cutoff [50]. The current cutoff 1.0 ng/mg enables 96% specificity but only 77% sensitivity and is in favor of false-negative results. Equal sensitivity and specificity of 88% would be attained with a cutoff of 0.63 ng/mg. However, from the other ROC studies and according to the present experience, the less-restrictive cutoff 1.0 ng/mg appears to be most suitable for the 0–6 cm hair segment.

The only ROC examination for FAEE in the 0–3 cm segment proposed a cutoff of 0.68 ng/mg with 59% sensitivity and 91% specificity [75]. The congruence of the current cutoffs 0.5 and 1.0 ng/mg for 0–3 and 0–6 cm respectively was only recently examined [125]. For this purpose both 0–3 and 0–6 cm of 159 hair samples were analyzed in parallel. The mean FAEE concentration ratio 0–3/0–6 cm was 0.83 (median 0.82, range 0.31–1.52, interquartile range, 0.70–0.96). Reevaluation of the segmental concentrations from 42 samples in Ref. [107] yielded in almost the same results (concentration ratio 0–3/0–6 cm: mean 0.80, median, 0.82, range 0.38–1.23, interquartile range 0.72–0.91). As an example, the ratio is 0.74 for the hair segments of the case shown in Figure 4.7. The variation of the ratio should be caused to an essential degree by the different time period covered by the two segment lengths with a possible change of alcohol consumption. However, this should be equalized in mean and median.

These results were not yet reflected in the 2014 revision of the SoHT consensus [83] and the following unchanged cutoffs were agreed for scalp hair:

Alcohol abuse (≥60 g ethanol/day)	0–3 cm: 0.5 ng/mg
	0–6 cm: 1.0 ng/mg
No contradiction to alleged abstinence:	0–3 cm: 0.2 ng/mg
	0–6 cm: 0.4 ng/mg

No cutoffs were recommended for nonhead hair in the consensus [83]. However, according to literature data, the values for the 0–6 cm scalp hair segment can also be used as a good approximation for nonhead hair (cf. Section 4.4.6).

4.4.5 Effect of Hair Color, Hair Care, and Cosmetic Treatment

Natural hair color: It was found in animal experiments [126] that FAEE incorporation into hair is not affected by hair color and pigmentation. No results from white and pigmented hair in grizzled samples were found in literature but evaluation of a larger number of individuals showed no bias of natural hair color.

FAEE content in hair care products: In a systematic investigation by HS-SPME/GC-MS, traces of the four FAEEs were detected in all of 49 frequently applied hair care products with the highest concentration only 0.003% in a hair wax [127]. However, treatment of a hair sample from an alcoholic with hair wax containing 10.4 ng/mg FAEE decreased the concentration in hair from 3.6 to 1.6 ng/mg, obviously by extraction from hair into the lipophilic wax.

Shampooing: It was shown also by *in vitro* experiments with hair samples from alcoholics that shampooing partly removes the external lipid layer and in this way decreases the FAEE concentration in the heptane washings, but there was no substantial decrease of the FAEE extracted from the hair matrix after 20 and 38 times shampooing [127]. Although lower FAEE results can be expected after regular removing the sebum layer by frequent shampooing, no influence of the shampooing frequency in the range between daily and once per week could be separated from the effect of other parameters for 75 individuals.

Permanent wave, bleaching, dyeing, and shading: Only small decreases of the FAEE concentrations by 4–14% were found in *in vitro* experiments by permanent wave, bleaching, and shading [127]. However, dyeing with alkaline medium decreased FAEE by 64%. Hair samples from alcoholics, which were bleached or dyed some months before sampling, displayed no obvious difference between the treated distal segment and the nontreated proximal segment which could be explained by the cosmetics. This is obviously caused by the continued deposition of FAEE from sebum after treatment.

Ethanol in hair lotions, hair sprays, deodorants, and hair stabilizers: It was experimentally shown that regular application of ethanol-containing hair lotions leads to FAEE concentrations in the range typical for alcohol abuse [127]. It was assumed that the topically applied ethanol diffuses to the sebum glands, and is metabolized there to FAEEs which are distributed and incorporated in hair as described above. However, FAEEs can also directly be formed from lipids or free fatty acids in the hair matrix or external lipid layer if the sample is stored in ethanol-containing atmosphere [127]. The spontaneous formation of up to 11.5 ng/mg E14 + E16 + E18 in hair samples stored for 15 days in the atmosphere above 96% ethanol was confirmed in a later study [128]. Although these experiments were carried out under extreme conditions which could not represent real life, it shows the possibility of increased FAEE results for persons working in ethanol-containing atmosphere, for example, in pubs. Hair sprays and "mists" frequently contain between 10% and 95% by volume ethanol. Their use led to FAEE concentrations between 0.5 and 5.0 ng/mg in hair of nine subjects with moderate or no alcohol consumption [129]. The risk for false-positive results for this reason appears particularly high when monitoring a female population. Individuals under monitoring should be advised to discontinue use of any ethanol-containing hair care products.

The effect of different kinds of hair cosmetics under real conditions was evaluated for a larger number of samples and is shown in Table 4.6 [50].

4.4.6 FAEEs in Nonhead Hair

In a first investigation including 28 scalp, 28 pubic, 10 axillary, 16 beard, 5 leg, and 5 chest hair samples large differences in the FAEE concentrations in hair from different sites in the same individual were found [130]. However, cases of chronic excessive alcohol consumption were characterized by FAEEs >1.0 ng/mg in almost all samples. Also the statistical comparison of 541 scalp hair and 84 body hair concentrations for FAEEs and for the individual esters by ANOVA showed that there is no significance to the differences between both hair types [115]. Likewise, the overall comparison of 214 body hair samples and 689 scalp hair samples from male subjects revealed no significant difference [50]. Armpit hair ($N = 41$) and leg hair ($N = 11$) had lower and chest hair ($N = 29$) had higher FAEE concentrations than scalp hair without significant difference.

4.4.7 Other Parameters with Effect on FAEE Concentration in Hair

Statistical examination of a possible relationship between FAEE concentration and body weight or adipose state showed that heavy and light people equally contribute to high and low levels of FAEEs, so the test is not biased against physically fit or obese people [115]. Further parameters of influence will be dealt together with EtG in Section 4.5.4.

4.5 COMBINED USE OF ETG AND FAEES

Both EtG and FAEEs can separately be used as markers for chronic alcohol consumption. However, due to the lacking quantitative relationship with the

TABLE 4.6 Effect of Cosmetic Hair Treatment on FAEE [50]

Treatment	Other Treatment Excluded	N	FAEE, ng/mg		Effect	ANOVA Test		
			Mean	Median		α	F	Significance
No treatment known	—	1,079	1.99	0.69	—	—	—	—
Bleaching	Yes	164	1.67	0.73	No effect	0.299	1.078	No
Dyeing	Yes	96	1.47	0.67	No effect	0.190	1.712	No
Bleaching and/or dyeing	No	341	1.68	0.77	No effect	0.165	1.959	No
Hair spray	Yes	79	3.89	1.45	Strong increase	0.000	15.311	Yes
Hair gel	Yes	99	2.05	0.56	No effect	0.886	0.021	No
Hair wax, oil, or grease	Yes	34	1.51	0.50	Slight decrease	0.547	0.460	No

alcohol dose for both markers and the influence of hair cosmetics, there remains a risk of false-positive or false-negative results. Therefore, it is reasonable to use the same sample for determination of both markers. This seems particularly advantageous since, as shown in the previous sections, EtG and FAEEs complete each other in an optimal way because of different biochemical formation, different incorporation mechanism in hair, different sensitivity to hair cosmetics, and different analytical determination. The combined use was investigated in context of abstinence assessment [8,114,122] as well as of detection of chronic excessive alcohol consumption [8,21,50,61,75,81,89,113]. The results are summarized in Table 4.7.

4.5.1 EtG and FAEEs in Abstinence Assessment

The cutoffs for any alcohol intake are currently 7 pg/mg EtG, and 0.2 ng/mg FAEE in the proximal 0–3 cm hair segment and 0.4 ng/mg FAEE in the proximal 0–6 cm segment [83]. A result below these cutoffs is not in contradiction to alleged abstinence and a result equal to or above these cutoffs strongly suggests repeated alcohol consumption. Because of the possibility of false-positive FAEE results by ethanol-containing hair lotions and hair sprays, EtG should be first choice in abstinence assessment. Nevertheless, besides nonjustified disadvantages of false-positive results for the tested person, false-negative EtG results may have serious consequences as well, for example, for other involved persons or for public safety, and have to be avoided as far as possible. Therefore, the additional determination of FAEE is advised particularly in case of bleached or dyed hair or because of the washout effect of EtG.

The additional utility of FAEE for this purpose was examined in the context of driving ability examination by determination of EtG and FAEE in the 0–3 cm hair segments from 160 persons with known history of alcohol abuse who were now claiming abstinence [122]. Agreeing EtG and FAEE results (pos-pos and neg-neg) were found in 77.5% of the cases (Table 4.7). The concentrations of the residual 36 cases (22.5%) are compared in Figure 4.10A. From the disagreeing results, 8 FAEE values and 2 EtG values are even in the ranges of excessive drinking (≥0.5 ng/mg and ≥30 pg/mg). EtG between 4 and 7 pg/mg was detected in five of the ambiguous cases confirming the positive FAEE result.

For further verification of the contribution of FAEE to abstinence assessment 73 hair samples with EtG between 4 and 10 pg/mg were analyzed for FAEE [114]. The concentrations are shown in Figure 4.10B. Agreeing results related to the cutoffs were obtained in 72.6% of the cases. EtG between 4 and 7 pg/mg and FAEE ≥0.2 ng/mg were found in 13 cases (17.8%) whereas the positive EtG result ≥7 pg/mg was not confirmed in 7 cases (9.6%) by FAEE.

TABLE 4.7 Combined Use of EtG and FAEE in Hair

Application	Individuals	Hair Length, cm	Cutoff		Results EtG–FAEE, %				Reference
			EtG, pg/mg	FAEE, ng/mg	Pos-Pos	Pos-Neg	Neg-Pos	Neg-Neg	
Abst. Ass.	40	6	8	0.40	22.5	15	15	47.5	[8]
Abst. Ass.	160	3	7	0.20	9.4	5.6	16.9	68.1	[122]
Abst. Ass.[a]	73	3	7	0.20	13.7	9.6	17.8	58.9	[114]
Alc. Abuse	180	≤6	25	0.50	15.6	8.9	14.4	61.1	[81]
Alc. Abuse	174	≤6	25	0.50	16.1	7.5	17.2	59.2	[21]
Alc. Abuse	78	3	30	0.50	10.3	7.7	14.1	67.9	[113]
Alc. Abuse	1,657	3/≤6	30	0.50/1.00[b]	25.5	10.1	18.5	45.9	[50]
Alc. Abuse	*946 male*	*3/≤6*	*30*	*0.50/1.00*[b]	*35.5*	*13.7*	*12.2*	*38.7*	[50]
Alc. Abuse	*689 female*	*3/≤6*	*30*	*0.50/1.00*[b]	*18.7*	*6.3*	*24.8*	*50.7*	[50]
Alc. Abuse	272	3	20	0.50	52.2	0[c]	47.8	0[c]	[89]
Alc. Abuse	97	3–6	30	0.50/1.00	22.7	4.1	27.8	45.4	[131]

[a]*Only samples with EtG between 4 and 10 pg/mg were included in this study.*
[b]*The following FAEE cutoffs were used: 0.50 ng/mg for the proximal 0–3 cm segment and 1.00 ng/mg for the 0–3 to 0–6 cm segment.*
[c]*Only samples with FAEE ≥ 0.5 ng/mg (73 male, 199 female) were included in this study.*

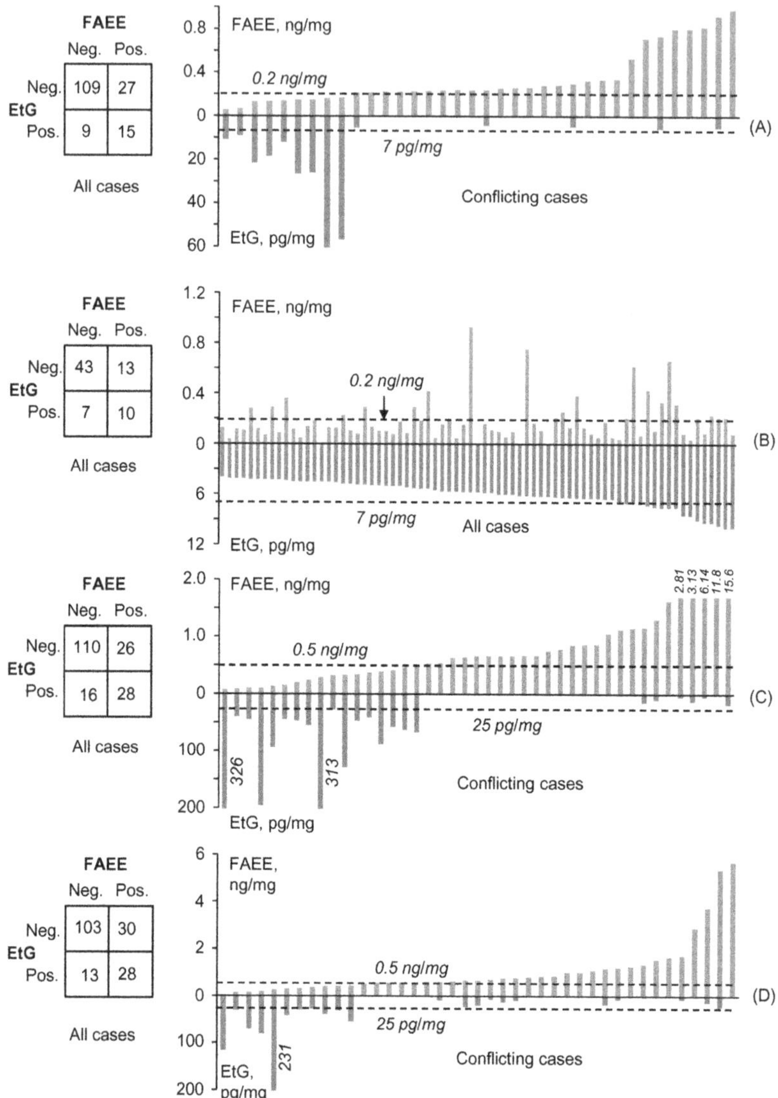

FIGURE 4.10 Combined use of EtG and FAEE. (A) 160 cases of driving ability examination with alleged abstinence. (B) 73 cases from abstinence assessment with EtG between 4 and 10 pg/mg. (C) 180 forensic samples tested for excessive alcohol abuse. (D) 174 forensic samples tested for excessive alcohol consumption. *Data from: Part (A) Ref. [122]; Part (B) Ref. [114]; Part (C) Ref. [81]; Part (D) Ref. [21].*

These examples show that FAEE can confirm ambiguous EtG findings in the context of abstinence tests. According to the consensus, a negative FAEE result cannot overrule an EtG result ≥7 pg/mg [83] but it makes heavy drinking more improbable. EtG concentrations below 7 pg/mg with FAEE

≥ 0.2 ng/mg are not consistent with alleged abstinence. In this way, the rate of false-negative results can be decreased. In case of not detected EtG and positive FAEE result, particularly in cases of FAEE ≥ 0.5 ng/mg, a possible effect of hair cosmetics should be checked and additional evidence from other markers should be consulted.

4.5.2 EtG and FAEE for Detection of Chronic Alcohol Abuse

The discrimination between social drinking and chronic alcohol abuse using a fixed cutoff of only one hair marker is bound to a high risk of errors because of the missing quantitative relationship between alcohol dose and marker concentration in hair. This risk can be decreased by combined use of EtG and FAEE. The results of the comparison between both markers which were described in five papers [21,50,81,89,113] are also shown in Table 4.7. The rate of agreeing results was similar to abstinence assessment between 71% and 78% and the portion of differing results was higher for EtG–FAEE neg-pos (14.1–18.5%) than vice versa (7.5–10.1%). Figure 4.10C and D shows the concentrations of the conflicting results for two of these studies. The examination of subpopulations in Ref. [50] showed that the higher portion of neg-pos results is mainly caused by the effect of hair cosmetics on both markers and therefore particularly pronounced in the female subgroup (24.8%) as compared to the male subgroup (12.2%).

Further reasons for the lower portion of EtG positive results can be the washout effect and the less-restrictive cutoff of EtG. The evaluation of 272 FAEE-positive samples by EtG measurement led to an almost twofold portion of concordant results if 20 pg/mg instead of 30 pg/mg was used as the cutoff [89]. Even at this cutoff, 26% of the discordant EtG-negative samples had evidence of recent alcohol abuse from consultation histories or positive CE findings, and 29% were dyed. The standard procedure of this laboratory dictates that hair samples testing positive for FAEE are further analyzed for the presence of EtG above 20 pg/mg in order to control for influence of external ethanol contamination on the FAEE result [89].

4.5.3 Combined Interpretation of EtG and FAEE in Hair

Whereas agreeing EtG and FAEE results increase the reliability of the interpretation by mutual confirmation, it is often not clear how to deal with differing results of both markers. It makes the interpretation apparently more complicated and confusing for the toxicologist. This may be a reason to stick to only one marker, for example, EtG. However, particularly in these cases of conflicting results the combined use of both markers contributes to a more profound judgment by questioning hair cosmetics, including other alcohol markers or repeating the measurement. The complexity of such cases with deviating results was demonstrated in two case reports [132,133].

A scheme for systematic interpretation of EtG and FAEE results was proposed in a previous paper [21] of the author of this chapter and is given in Table 4.8 using the current cutoffs. Because of the many and varying effects on the relationship between drinking behavior and EtG or FAEE

TABLE 4.8 Proposal of Combined Interpretation of EtG and FAEE in the 0–3 cm Hair Segment for Diagnosis of Alcohol Abuse [21]

Concentration in Hair		Interpretation
FAEE, ng/mg[a]	EtG, pg/mg	
<0.20	<7	Abstinence or moderate drinking
<0.20	7–29	Moderate drinking, abstinence excluded
<0.20	*30–60*	*Repeat analysis of both parameters. If confirmed:* Social drinking, weak indication of abuse
<0.20	*>60*	*Repeat analysis of both parameters. If confirmed: look for reasons in aggressive hair cosmetics*
		No diagnosis. Collect new sample and/or analyze body hair other than pubic hair
0.20–0.49	<7	Moderate drinking, abstinence improbable
0.20–0.49	7–29	Moderate drinking, abstinence excluded
0.20–0.49	30–60	Social drinking, weak indications of abuse
0.20–0.49	>60	Indications of abuse
0.50–1.00	*<7*	*Repeat analysis of both parameters. If confirmed, look for reasons in hair cosmetics.* Social drinking, weak indication of abuse
0.50–1.00	7–29	Social drinking, weak indications of abuse
0.50–1.00	30–60	Indications of abuse
0.50–1.00	>60	Strong indication of alcohol abuse
>1.00	*<7*	*Repeat analysis of both parameters. If confirmed look for EtOH in hair cosmetics, or extreme shampooing. No diagnosis. Collect new sample and/or analyze body hair*
≥1.00	7–29	Indication of alcohol abuse
≥1.00	30–60	Strong indication of alcohol abuse
≥1.00	>60	Strong indications of alcohol abuse

Cutoffs: FAEE 0.5 ng/mg and EtG 30 pg/mg.
[a]*If the 0–6 cm hair segment was used for FAEE determination, 0.40, 1.00, and 2.00 ng/mg should be used as the critical values.*

concentrations in hair, interindividually differing results, for example, up to the factor of two, are quite natural and must be taken into account. Therefore, the levels of abstinence (<7 pg/mg and <0.20 ng/mg) and the double of the cutoff values (60 pg/mg and 1.00 ng/mg) were used as additional criteria. Smaller differences were regarded to be in the range of biological variability and interpreted in favor of a lower degree of alcohol consumption. A strong contradiction exists only if one marker is in the range of abstinence and the other exceeds the double of the cutoff. In these rather rare cases the measurement of both markers should be repeated. In case of the same analytical outcome it should be looked for an explanation in hair cosmetics. If no diagnosis is possible, a new scalp hair sample and/or a body hair sample should be tested and additional evidence from other alcohol markers should be included. In general, hair results should not be applied schematically and isolated from the circumstances of the respective case and should be critically seen in context of the medical diagnosis.

4.5.4 Time Period Between Initiation of Abstinence and Collection of Hair Sample

It is often important for the interpretation to know how long after the beginning of abstinence a previous alcohol abuse can be detected by positive EtG or FAEE results in the routinely analyzed 3 or 6 cm long proximal hair segment. Dilution by newly grown hair and elimination of the markers from hair by various described mechanisms decrease the EtG and FAEE concentration with increasing time after cessation of drinking. The effect of the self-reported time of abstinence before sampling on EtG in the 0–3 cm proximal segment and on FAEE in the 0–6 cm proximal segment was evaluated for a larger number of previous excessive drinkers and is shown in Figure 4.11 [50]. Cases with cosmetic hair treatment were excluded. The mean and median EtG concentrations of samples collected between 1 and 3 months after initiation of abstinence are clearly below the cutoff of 30 pg/mg (Figure 4.11A). The same is true for FAEE (Figure 4.11B). The mean and median concentrations in the 6 cm hair segment decrease with increasing duration of abstinence and are below the cutoff (1.0 ng/mg) already 2 months after initiation of abstinence. That means that EtG below 30 pg/mg in the 3 cm segment or FAEE below 1.0 ng/mg in 6 cm segment do not fully exclude that there may have been chronic excessive drinking within the time represented by the hair segment which stopped, for instance, 1 or 2 months before sampling. On the other hand, in distinct cases FAEE concentrations above the cutoff were still observed even in samples from previous excessive drinkers collected more than 6 months after abstinence began (Figure 4.11C). Possible explanations are enzyme induction or delayed incorporation from fat deposits.

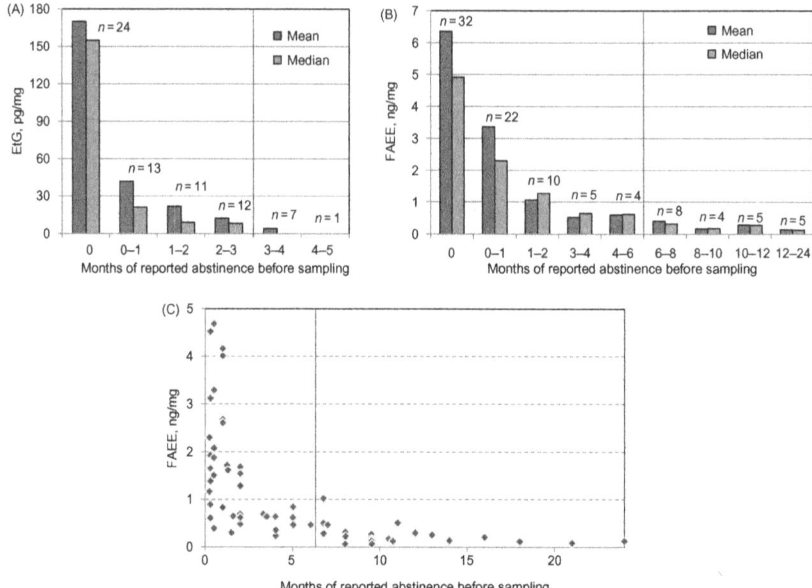

FIGURE 4.11 Effect of the time period between self-reported initiation of abstinence after excessive drinking and collection of hair sample on alcohol marker concentrations in hair. (A) Mean and median of EtG concentrations in the 3 cm proximal hair segment. (B) Mean and median of FAEE concentrations in the proximal 6 cm hair segment. (C) FAEE concentrations of distinct cases in proximal 6 cm hair segment. Cases with cosmetic hair treatment and without reported data about cosmetic treatment were excluded. *From Ref. [50].*

4.6 COMPARISON OF ETG AND FAEE IN HAIR WITH OTHER ALCOHOL MARKERS

EtG and to a lower extent also FAEE in hair were compared several times with other markers of chronic excessive alcohol consumption such as CDT, gamma-glutamyltransferase (GGT), aspartate aminotransferase (AST), alanine aminotransferase (ALT), mean corpuscular volume (MCV), or phosphatidyl ethanol (PEt) [18,19,70,72,73,75,113,115,134,135]. Compared criteria were the area under the ROC curve, sensitivity and specificity, or more simply the percentage of correctly identified alcohol abusers. The results are shown in Table 4.9.

The data show clearly a better performance of the alcohol markers in hair, which is based on the higher specificity as direct alcohol markers, the higher sensitivity, and the longer time window. Nevertheless, as shown in previous sections, both EtG and FAEEs in hair have their limitations should be used with caution and cannot be regarded as golden standards for detection of chronic alcohol abuse. Therefore, it is a common tenor of these comparative studies that CDT and PEt as well as the nonexpensive traditional alcohol biomarkers continue to have practical importance. They can be

TABLE 4.9 Performance of EtG and FAEE in Hair in Comparison to Other Markers of Chronic Excessive Alcohol Consumption[a]

Reference	Subjects	Criterion	EtG in Hair	FAEEs in Hair	CDT	GGT	ALT	AST	MCV	PEt
[134][b]	70 Postmortem cases	% Positive	56	—	—	—	—	—	—	41
[70]	16 Withdrawal patients	% Positive	94	—	64	93	67	67	—	—
[72]	154 Driving ability examination	% Positive	34[c]	—	9.7	—	—	—	—	—
[115]	151 Forensic cases	% Positive	—	59	28	30	20	20	—	—
[73]	76 Forensic cases	% Sensitivity/% Specificity	69/100	—	27/100	46/47	42/61	42/21	27/100	—
[18]	175 Patients	Area ROC	0.98	—	0.77	0.83	0.61	0.73	0.73	—
		% Sensitivity/% Specificity	92/97	—	51/90	66/81	36/80	48/82	17/97	—
[19]	125 Volunteers	Area ROC	0.99	—	0.84	0.79	0.73	0.79	—	—
[113]	78 Workplace testing	% Positive	14	28	—	10	15	1.7	—	—
[75]	80 Withdrawal patients + 89 Volunteers	Area ROC	0.94	0.83	0.90	0.94	0.81	0.80	0.80	—
		% Sensitivity/% Specificity	83/97	59/91	80/91	83/92	46/91	61/91	19/100	—
[135]	562 Forensic cases	% Positive	13.5	—	7.4	—	—	—	—	—

[a] *The applied reference ranges or cutoffs of the markers are given in the original papers and are not uniform in the different studies.*
[b] *PEt was detected in 51% and EtG ≥ 7 pg/mg in 70% of the cases.*
[c] *55% are ≥ 7 pg/mg.*

applied in combination with one or both hair markers and can be used as additional evidence in doubtful cases. Critical cases should be reviewed in context of all available information including these markers, case history, and self-report.

4.7 PRACTICAL APPLICATIONS OF ETG AND FAEE IN HAIR

4.7.1 Driving Ability Examination and Driver's License Regranting

Abstinence assessment in context of driving license regranting is one of the most frequent applications of EtG determination in hair [72,92,93,136] and has been introduced in national guidelines [10,11]. In Germany, the candidates can demonstrate 1 year abstinence either by six random (24 hours' notice) urine tests in 12 months or by four separate 3 cm long hair strands collected every 3 months (EtG cutoffs 0.1 mg/L in urine and 7 pg/mg in hair). The superiority of hair was shown in a study where only 92 out of 4,248 urine samples (2%) were tested positive, whereas 81 out of 386 hair samples (21%) were positive in the same time period [92]. This was confirmed in a later paper from the same laboratory for 13,441 urine and 3,952 hair samples with 2.1% and 12.7% positives respectively [93]. In hair, the EtG concentration was 7–30 pg/mg for 7.8% and >30 pg/mg for 4.9% of the samples. To a lower extent, the advantage of hair was still true for cosmetically treated hair with a positive rate of 4.09% versus 2.08% in urine [91].

Similarly, alleged abstinence was disproved in 55% of 154 fitness-to-drive control cases with only 9.7% positive by CDT [72]. In Switzerland, the cutoff value of 30 pg/mg in hair is used in order to control imposed "zero tolerance when driving" (ZTD) which means that a driver is not allowed to drink any alcohol before driving (0.0‰ in blood, usually <0.5‰) and to consume alcohol above the level of social drinking at any other time [136]. The hair test is performed every 6 months. From 234 cases, EtG was not detected in 126 cases (53.8%, LOD 5 pg/mg), was in the range of social drinking in 93 cases (39.7%, <30 pg/mg), and showed excessive alcohol consumption in 15 cases (6.4%, ≥30 pg/mg).

In an investigation of 146 drivers charged with alcohol impairment (DUI) and with installed ignition interlock devices in their car in Alberta, Canada, EtG in hair emerged as a top overall predictor for discriminating new recidivism events, for entry alcohol dependence, and for the highest interlock blood alcohol concentrations (BACs) recorded [137].

4.7.2 Workplace and Preemployment Alcohol Testing

Both EtG and FAEE with the cutoffs 30 pg/mg and 0.5 ng/mg for chronic excessive alcohol consumption and 7 pg/mg EtG for refuting a claim of abstinence are also recommended in the Guidelines for European workplace drug and alcohol testing in hair [16]. In practice, drug and alcohol testing is regarded useful especially, for example, for pilots, lorry drivers, train drivers, or operators of heavy machinery such as cranes. The legal status is complicated and requires approval of the tested person even in these cases [138]. In most European countries it is performed in case of suspicion or within the preemployment medical examination or yearly health check. A workplace alcohol testing program by combined use of EtG and FAEE in hair which was performed under these conditions for 78 employees and by using the gradual system for combined interpretation of both markers (Table 4.8) no indications of alcohol abuse were obtained in 50 cases (64%), slight indications were seen in 13 cases (17%), and clear indications in 11 cases (14%) [113]. GGT, AST, ALT, and MCV confirmed the hair results only partly and displayed altogether a lower portion of positive results.

4.7.3 Prenatal Alcohol Exposure

Alcohol use during pregnancy is bound to prenatal ethanol exposure and to a high risk of fetal alcohol spectrum disorder (FASD), the most prevalent cause of neurocognitive handicap among children. The damaging effect of alcohol occurs during all three trimesters. Therefore, retrospective detection of alcohol consumption during pregnancy is an important part of the diagnosis of FASD. Hair analysis was applied for this purpose in several studies [123,139–141]. Preferentially hair from the mother is analyzed but alcohol abuse of the male partner is also investigated as a contributing factor since this has been shown to be a strong determinant of maternal alcohol use during pregnancy. In a study about FAEEs in neonate hair no difference between cases with nondrinking and socially drinking mothers were found [142]. FAEEs were measured in a cohort of 324 parents from British Columbia (Canada) [123]. The rate of positive samples (≥0.5 ng/mg) was 33.3% (32.4% among 225 women and 35.4% among 96 men) and 19% were above 1.0 ng/mg. From 26 women in pregnancy two had FAEE values between 0.5 and 1.0 ng/mg and three above 1.0 ng/mg. This high rate of positive results appeared to be comprehensible since mainly suspicious cases were selected by the social workers for the test. The relationship between social worker reports and the FAEE test was examined for 119 mothers and 53 fathers from the same area by odds ratio (OR) analysis [141]. Forty of 119 women tested above 0.5 ng/mg and 26 (19%) pregnant women were found positive for heavy alcohol exposure. Reported factors directly related to alcohol use were significantly associated with testing positive for FAEEs. For instance, factors associated with testing positive for hair FAEE in mothers alone were: knowledge of a specific instance of problem drinking within the past 6 months (OR = 8.51) and third party reports alleging alcohol abuse (OR = 3.30). Mothers who admitted to heavy drinking were also seven times more likely to test positive for hair FAEE (OR = 6.74) than those who did not. Furthermore, it was shown that mothers testing positive for FAEE had a threefold increased risk of testing positive for drugs [140].

In two studies with 99 and 151 mother–infant pairs, maternal and fetal hair or nails were tested for EtG and compared with FAEEs and EtG in meconium and EtG in nails [139,143]. Neither maternal hair nor neonatal hair or nails appeared to be good predictors of gestational ethanol consumption because of insufficient sensitivity to detect social alcohol consumption and too low neonate sample amount. A systematic investigation of maternal hair with three 3 cm segments for EtG which could characterize drinking habits in the three trimesters of pregnancy is still missing.

4.7.4 Family Courts and Child Custody Cases

Testing hair for alcohol abuse on order of family courts and social authorities in context of child custody is one of the most frequent applications. Children of alcoholic parents are frequently threatened by neglect, maltreatment, poverty, and social discrimination [144]. Purpose of the hair alcohol test is either to characterize the drinking habit of one or both parents in case of suspected or blamed excessive drinking or to control imposed abstinence in case of previous abuse. Although several of the reports with large cohorts referred in this chapter include mainly samples from this background [50,115], no overview about the role of hair markers was described. The interpretation problems became obvious in a forensic case of abstinence determination which was dealt with in a hearing before the Family Division of the High Court of London [132]. The mother with a longstanding history of alcohol misuse had to show absolute abstinence for 1 year in order to regain custody of her child. Altogether five hair samples were subsequently analyzed by two laboratories with EtG 22 pg/mg in the proximal 0–1 cm segment and <2.4–3.3 pg/mg in the 0–3 cm segment and negative or low FAEE results (0.03 and 0.27–0.53 ng/mg in 0–6 cm segment) and were controversially interpreted. Despite these findings it was concluded in the judgment that the evidence did not indicate that the mother had consumed alcohol in the period tested by the hair samples [145]. Hair testing for family courts is frequently performed together for drug and alcohol abuse. The legal background in Germany, practical experiences, and advantages and limitations of hair testing in family cases including alcohol markers were also illustrated from judiciary viewpoint [146].

4.7.5 Liver Transplantation and Patients with Liver Disease

A high percentage of liver transplants are performed for patients with end-stage alcoholic liver disease and alcohol use disorders are one of the leading causes of liver transplantation. Therefore, an abstention period of 6 months prior to listing is required in most transplant programs. Reasons for monitoring abstinence in these patients by alcohol biomarkers are the stabilization of their medical condition before surgery and the improvement of their prospects for long-term sobriety afterward [147]. EtG in hair has been shown to be superior to other markers for this purpose [77,148–150]. For instance, in a group of 63 transplantation candidates 39 patients tested positive for drinking by at least one of the applied alcohol markers. From these patients, EtG in hair (cutoff 30 ng/mg) was positive in 87%, EtG in urine in 46%, CDT in 18%, ethanol in blood in 2%, and methanol in blood in 8% of the cases [148]. EtG in hair was the only positive marker in 44% of the cases. In a similar way, for 42 patients who had been denied a transplant due to recent substance abuse, 64% of the positive EtG results in hair (cutoff 2 pg/mg) were not self-reported and 88% were not confirmed by breath testing [149].

The relationship between hair EtG and self-reported drinking during the last 3 months in patients with liver disease was studied for 82 abstainers, 57 patients drinking <28 g ethanol/day (two standard drinks in United States), and 52 patients drinking ≥ 28 g ethanol/day [77]. ROC analysis showed a modest discrimination power (AUC 0.79) for any drinking, but a high AUC (0.93) for drinking at least 28 g/day over the prior 3 months. The results indicate that a cutoff 8 pg/mg performed particularly well in detecting subjects averaging at least 28 g ethanol/day (sensitivity 90%, specificity 88%), and illustrate the trade-off between sensitivity and specificity (81% versus 93%) when using the alternative 30 pg/mg. The results confirmed the utility of hair EtG in patients with liver disease. Severity of liver disease may modestly influence the accuracy of hair EtG, but hair EtG performed well regardless of cirrhosis.

No impact of the liver disease was found in other studies [148]. In summary, EtG in hair is a sensitive and useful tool for assessment of alcohol abstinence in liver transplant candidates. However the authors agreed that no patient should be denied transplantation because of a positive marker result alone.

4.7.6 Postmortem Cases and Historical Study

Hair is a preferable autopsy material for forensic analysis, since it is well protected against postmortem changes in a decaying cadaver. EtG captured in hair under dry and dark conditions appeared to be stable over centuries and was used to explore alcohol use in history. Hair analysis of 38 mummies from the Capuchin Catacombs of Palermo (nineteenth century, 6 children, 23 male, and 9 female adults) showed 17 positive samples of adults with EtG in the concentration range 2.3–517 pg/mg while all samples from children were negative [151]. The results were discussed in context of high wine production and drinking in Sicily in sixteenth to nineteenth centuries.

EtG in hair of 70 consecutive medicolegal autopsies was demonstrated in 49 cases at concentrations ranging from 7.5 to 10,400 pg/mg [134]. For seven death cases with known alcohol abuse or unknown drinking history, the EtG concentration in hair ranged from 61 to 819 pg/mg [42]. EtG was also detected in three hair segments of a mummified corpse 27 years after death but, due to severe ion suppression, quantification could not be performed [152]. About 46 and 54 pg/mg were measured in the two segments of the 4 cm long hair of a 5-year-old male child who died after ingestion of 50–60 g ethanol (120–150 mL vodka) and who obviously had drunken larger amounts of alcohol several times before [153]. In three 4 cm long segments of the postmortem hair sample of a 26-year-old man who died from a combined clomipramine/alcohol poisoning and had abused alcohol for a few years the following EtG concentrations were determined from proximal to distal: 440 pg/mg, 70 pg/mg, and not detected [154].

FAEEs were routinely determined in postmortem hair samples for several years [82]. The concentrations of 502 identified alcohol abusers ranged from 0.01 to 83.7 ng/mg (median 1.35 ng/mg).

4.8 COCAETHYLENE

The prevalence of concurrent use of and dual dependence on cocaine and ethanol among cocaine abusers is very high [155]. It has been estimated that 60–85% of cocaine users also co-abuse alcohol. Combining cocaine and alcohol enhances and prolongs the euphorigenic effects of cocaine while also diminishing the acute paranoia and agitation that accompanies cocaine withdrawal [156,157]. Results of a study about the motivations for simultaneous using of cocaine and alcohol suggested that this has higher social importance to those who primarily snort cocaine than to those who smoke crack [158].

It was shown that this change is caused mainly by synergetic action and by the effect of alcohol on the pharmacokinetics of cocaine and not by the mixed metabolite of both substances, CE [157,159–162]. CE displayed the same pharmacodynamic profile as cocaine but appeared to be less potent and had a longer elimination half-life (e.g., 1.68 h versus 1.07 h for COC [160]).

Cocaine (COC) is esterified to CE by the same nonspecific hepatic cocaine carboxylesterase that hydrolyzes cocaine into benzoylecgonine (BE) [163]. In a pharmacokinetic study with the deuterated drug 0.3–1.2 mg/kg cocaine were intravenously administrated to 10 volunteers 1 h after ingestion of 1 g/kg ethanol [164]. A total of $17 \pm 6\%$ of the drug was converted to CE at the expense of BE which attained significantly lower concentrations in blood and urine. The half-life of COC was 1.91 ± 0.51 h in the cocaine plus ethanol condition compared with 1.29 ± 0.15 h in the cocaine alone condition. Furthermore, ethanol significantly increased the urinary portion of COC and ecgonine methyl ester (EME), and decreased urinary portion of BE. Later, it was shown by the same group under similar conditions that due to the first-pass metabolism the fraction of CE formed from the bioavailable COC was significantly higher for oral administration (34 ± 20%) than for intravenous injection (24 ± 11%) or smoking (18 ± 11%) [165].

From the viewpoint of hair analysis, it is an important result of these kinetic studies [158,159] that the ratio CE/COC is essentially determined by the BAC during cocaine use and only to a much lower extent by the cocaine dose and consumption frequency. The ICR in hair of CE should be similarly high as for COC because of the small structural difference. Therefore, the ratio CE/COC in hair should approximately represent the mean ratio of both compounds in blood during consumption which, as shown above, is essentially determined by the BAC and is more characteristic for the drinking behavior than the CE concentration alone.

Data about CE concentrations and concentration ratios CE/COC in hair were previously reviewed [8] and are given including newer results [166–168] in Table 4.10. CE ranged from 0.01 to 328 ng/mg. CE/COC is mainly between 0.004 and 0.6 and only in some exceptional cases higher than 1.0. In a recent comparative investigation coca chewers displayed both higher CE concentrations and higher CE/COC ratios than cocaine abuser [167]. This may be caused by the higher portion of gastrointestinal absorption in case of coca chewing although other differences between both consumer groups are possible. In none of these studies drinking data were reported. However, in three studies CE was compared with EtG [168] and FAEE [121,168–170] in hair. The results are shown in Figure 4.12.

Cocaine-positive hair samples (0.475–361 ng/mg) from 68 individuals (64 male, 4 female) were tested for CE and EtG [168]. Reevaluation of the results with the cutoffs 7 pg/mg for EtG and 0.1 ng/mg for CE leads to 37 samples positive and 12 samples negative for both markers. Five samples were CE positive and EtG negative, and 14 samples were EtG positive and CE negative (Figure 4.12A). There was no quantitative correlation between EtG and CE as well as between EtG and CE/COC (Figure 4.12B) or CE/(COC + BE + EME). However, the portion of positive EtG increased with increasing ratio CE/(COC + BE + EME) from 60% at <0.005 to 82% at 0.01–0.03 to 95% at >0.03. A similar degree of agreement was obtained in comparing FAEE (cutoff 0.5 ng/mg) with CE (cutoff 0.05 ng/mg [121] or LOD [169], Figure 4.12C and E). The correlation slightly improved when using CE/(COC + BE + EME) instead of CE (Figure 4.12D). In one study a positive predictive value of 0.66 was estimated which indicates that a sample is 66% likely to be positive for excessive alcohol consumption if CE is detected [169].

There are several explanations for disagreeing results in COC-positive hair samples. A high EtG or FAEE concentration without detection of CE could mean that alcohol and COC are not or only occasionally used together, that COC results only from external contamination or that COC and CE are less efficiently incorporated in nonpigmented hair. On the other hand, a

TABLE 4.10 Literature Data about Concentrations of Cocaine (COC) and Cocaethylene (CE) in Hair after Concurrent Use of Cocaine and Alcohol[a]

Number of Individuals	CE, ng/mg	COC, ng/mg	CE/COC, %	Remarks	Reference
	Range (Mean, Median)	Range (Mean, Median)	Range (Mean, Median)		
126	0.01–30.3	0.01–268.6	0.01–247	Summarized data from drug users in 7 studies[b]	[8]
	(1.94, —)	(34.1, —)	(14.9, —)		
33	0.011–3.66	0.66–81.8	1–23	Self-reported cocaine users	[166]
	(0.48, 0.21)	(16.3, 10.8)	(5.8, 2.0)		
24	0.27–60.1	0.008–13.8	0.02–72	Argentinean coca chewers	[167]
	(24.4, 18.65)	(3.29, 0.72)	(12.7, 7.7)		
20	2.0–38.7	0.002–2.10	0.044–20.1	German cocaine users	[167]
	(11.4, 6.6)	(0.47, 0.077)	(3.3, 0.83)		
51	1.4–328	0.1–5.38	0.0004–0.266	Cocaine abuser, comparison with EtG	[168]
	(57.4, 19.3)	(0.89, 0.48)	(0.053, 0.024)		

[a]*Only samples detected with CE were included from all studies.*
[b]*Original references given in Ref. [8].*

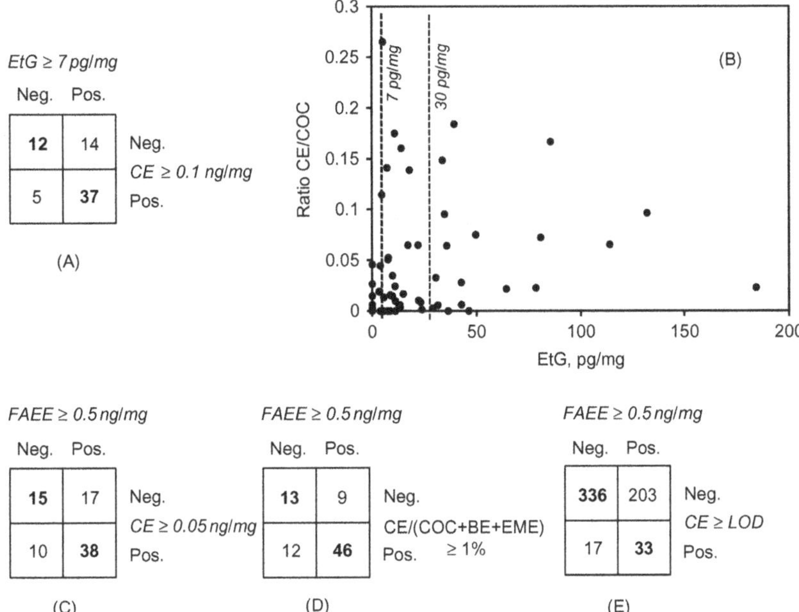

FIGURE 4.12 Concentrations of CE, EtG, and FAEEs in hair of cocaine users. (A) 68 volunteers with 0.47–361 ng/mg cocaine in hair. (B) Plot of CE/COC versus EtG for the same data in [168]. (C) and (D) 80 forensic samples tested positive for cocaine in context of regranting driver's license. (E) 588 individuals tested for cocaine and metabolites with 105 identified cocaine users. *Data from: Parts (A) and (B) Ref. [168]; Parts (C) and (D) Ref. [121]; Part (E) Ref. [169].*

positive CE result despite negative EtG or FAEE can occur in cases of moderate alcohol use concurrent to frequent cocaine use or if the illicit cocaine contained CE which is formed in certain manufacturing procedures with alcohol as the solvent [171]. Furthermore, hair care and hair cosmetics may affect the markers to a different degree.

With respect to interpretation, it can be concluded from the present state of results that:

- CE in hair can only be interpreted if it was not contained in the illicit drug preparations abused by the tested person. This can be presumed in most cases and is presupposed in the next points.
- Detection of CE proves concurrent use of cocaine and alcohol and contradicts alleged abstinence from alcohol. It corroborates corresponding results from other alcohol markers.
- Both the concentration of CE and the ratio CE/COC should be considered for interpretation of the abuse behavior. CE/COC ratios >0.10 indicate that the drug was used regularly in combination with high alcohol levels. In this case the probability of chronic excessive alcohol abuse increases with increasing CE (or COC) concentration.

- On the other hand, CE/COC in the lower range should show that alcohol was present only occasionally or in low concentrations during cocaine use.
- Differences to EtG or FAEE in hair can be explained by nonconcurrent use of cocaine and alcohol, positive cocaine result caused by external contamination, or different effects of hair care and cosmetics.

Combined use of alcohol and cocaine as a part of human sacrifice ceremonies in the Inca capacocha rite was also evidenced by CE in hair [172]. Determination of COC, BE, and CE in hair segments from the frozen bodies of three entombed Inca children (aged 4, 5, and 13 years) who were sacrificed before the Spanish conquest near the Andean summit of Volcán Llullaillaco, Argentina, showed escalating coca and alcohol ingestion in the lead-up to death. In case of the 13-year-old girl, CE increased from an average baseline of 0.6 ng/mg to 1.8 ng/mg 12 months before death, a small peak of 2.6 ng/mg 6 months before death, and to 8.4 ng/mg in the weeks immediately before death. The high CE/COC ratio of 0.53 in the last 1.5 months before death suggests extensive alcohol intake coupled with frequent coca chewing.

4.9 OUTLOOK

Attempts to ban or even to reduce alcohol in human life have failed in history and are unrealistic to be successful in near future. Diagnosis of alcohol abuse is one important step to minimize damage by this substance. Hair analysis for biomarkers of alcohol has become one of the most efficient methods for this purpose in the last 10 years. The advantages but also the limitations of EtG and FAEE in hair have been shown in many reports from research work and routine applications. However, further investigations are necessary for optimal exploiting the potential of these markers in analysis as well as in interpretation. The biannual reviews of the SoHT consensus on alcohol markers in hair will be helpful to implement new knowledge into practice.

Harmonization of the analytical methods, further improvement of quality control and regular proficiency tests with real samples will be important steps for a uniform performance. Progress of MS techniques can be expected to increase sensitivity and accuracy for both markers down to the range of abstinence. For the FAEEs it has to be examined whether the confinement on the most abundant esters, ethyl palmitate and ethyl oleate, enables the same sensitivity and specificity as the sum of the presently analyzed four esters. Besides methodical simplification, this would also decrease analytical errors caused by error addition from the single esters.

A key problem for interpretation of EtG and FAEE is the effect of hair cosmetics. This is based currently only on self-reports of the tested person. Analytical tests for detection of bleaching, dyeing, or markers of hair sprays have to be developed and regularly used in order to bring this to an objective and more reliable level.

Finally, it should always be kept in mind that hair results may have drastic consequences on the persons concerned. False-positive or false-negative results can be equally detrimental. In view of the many and highly variable influencing factors, alcohol markers in hair should not be applied in isolation but should be critically interpreted in context of all available information from case history or data from blood and urine.

In addition to further development and improvement of biomarkers in hair and body fluids, new electronic ways of controlling alcohol abuse may become serious competition to hair analysis. Breath alcohol ignition locks are already regularly used to prevent drunken driving [173]. Alcohol monitoring bracelets based on steady transdermal measurement of ethanol in insensible sweat [174] could find broader application for abstinence assessment if the rather uncomfortable devices can be miniaturized. The toxicologist should always keep an open mind to such progress.

REFERENCES

[1] Helander A. Biological markers in alcoholism. J Neural Transm 2003;66(Suppl.):15–32.

[2] Hannuksela ML, Liisanantti MK, Nissinen AE, Savolainen MJ. Biochemical markers of alcoholism. Clin Chem Lab Med 2007;45:953–61.

[3] Niemelä O. Biomarkers in alcoholism. Clin Chim Acta 2007;377:39–49.

[4] Jones AW. Biomarkers of acute and chronic alcohol ingestion. In: Garriott JC, editor. Medicolegal aspects of alcohol. 5th ed. Tucson, AZ: Lawyers and Judges Publishing Company; 2008. p. 157–203.

[5] Pragst F. Toxicological markers of chronic alcohol abuse. In: Madea B, editor. Handbook of forensic medicine. Oxford: Wiley-Blackwell; 2014. p. 1099–107.

[6] Maenhout TM, De Buyzere ML, Delanghe JR. Non-oxidative ethanol metabolites as a measure of alcohol intake. Clin Chim Acta 2013;415:322–9.

[7] Pragst F, Spiegel K, Sporkert F, Bohnenkamp M. Are there possibilities for the detection of chronically elevated alcohol consumption by hair analysis? A report about the state of investigation. Forensic Sci Int 2000;107:201–23.

[8] Pragst F, Yegles M. Alcohol markers in hair. In: Kintz P, editor. Analytical and practical aspects of drug testing in hair. Boca Raton, FL: CRC Taylor & Francis; 2006. p. 287–323.

[9] Nadulski T, Bleeck S, Schräder J, Bork WR, Pragst F. 11-nor-delta9-tetrahydrocannabinol-9-carboxylic acid ethyl ester (THC-COOEt): unsuccessful search for a marker of combined cannabis and alcohol consumption. Forensic Sci Int 2010;196:78–84.

[10] Schubert W, Dittmann V, Brenner-Hartmann J. Urteilsbildung in der Fahreignungsbegutachtung—Beurteilungskriterien. Kirschbaum—Fachverlag für Verkehr und Technik, Bonn; 2013.

[11] Schweizerische Gesellschaft für Rechtsmedizin, Arbeitsgruppe Haaranalytik. Bestimmung von Ethylglucuronid (EtG) in Haarproben. <http://www.sgrm.ch/uploads/media/EtG_FINAL_2014.pdf>; 2014 [accessed 17.09.14].

[12] Crunelle CL, Yegles M, Nuijs AL, Covaci A, De Doncker M, Maudens KE, et al. Hair ethyl glucuronide levels as a marker for alcohol use and abuse: a review of the current state of the art. Drug Alcohol Depend 2014;134:1–11.

[13] Kintz P. Consensus of the Society of Hair Testing on hair testing for chronic excessive alcohol consumption 2009. Forensic Sci Int 2009;196:2.

[14] Kintz P. Consensus of the Society of Hair Testing on hair testing for chronic excessive alcohol consumption 2011. Forensic Sci Int 2011;218:2.

[15] Society of Hair Testing. Consensus of the Society of Hair Testing: use of alcohol markers in hair for abstinence assessment. <http://www.soht.org>; 2012 [accessed 12.05.14].

[16] Agius R, Kintz P, European Workplace Drug Testing Society. Guidelines for European workplace drug and alcohol testing in hair. Drug Test Anal 2010;2:367–76.

[17] Sporkert F. Inter-laboratory proficiency test: ethyl glucuronide in hair. Examination period: 16 December 2010–16 January 2011. <http://www.soht.org//pdf/EtG_PT_2010.pdf>; 2010 [accessed 13.05.14].

[18] Pirro V, Valente V, Oliveri P, De Bernardis A, Salomone A, Vincenti M. Chemometric evaluation of nine alcohol biomarkers in a large population of clinically-classified subjects: pre-eminence of ethyl glucuronide concentration in hair for confirmatory classification. Anal Bioanal Chem 2011;401:2153–64.

[19] Kharbouche H, Faouzi M, Sanchez N, Daeppen JB, Augsburger M, Mangin P, et al. Diagnostic performance of ethyl glucuronide in hair for the investigation of alcohol drinking behavior: a comparison with traditional biomarkers. Int J Legal Med 2012;126:243–50.

[20] World Health Organization (WHO). Global status report on alcohol and health. <www.who.int/.../global_alcohol_report/msbgsruprof>; 2011 [accessed 12.05.14]

[21] Pragst F, Rothe M, Moench B, Hastedt M, Herre S, Simmert D. Combined use of fatty acid ethyl esters and ethyl glucuronide in hair for diagnosis of alcohol abuse: interpretation and advantages. Forensic Sci Int 2010;196:101–10.

[22] Nakahara Y, Takahashi K, Kikura R. Hair analysis for drugs of abuse. X. Effect of physicochemical properties of drugs on the incorporation rates into hair. Biol Pharm Bull 1995;18:1223–7.

[23] Foti RS, Fisher MB. Assessment of UDP-glucuronosyltransferase catalyzed formation of ethyl glucuronide in human liver microsomes and recombinant UGTs. Forensic Sci Int 2005;153:109–16.

[24] Al Saabi A, Allorge D, Sauvage FL, Tournel G, Gaulier JM, Marquet P, et al. Involvement of UDP-glucuronosyltransferases UGT1A9 and UGT2B7 in ethanol glucuronidation, and interactions with common drugs of abuse. Drug Metab Dispos 2013;41:568–74.

[25] Schwab N, Skopp G. Identification and preliminary characterization of UDP-glucuronosyltransferases catalyzing formation of ethyl glucuronide. Anal Bioanal Chem 2014;406:2325–32.

[26] Miners JO, McKinnon RA, Mackenzie PI. Genetic polymorphisms of UDP-glucuronosyltransferases and their functional significance. Toxicology 2002;181–182:453–6.

[27] Dahl H, Stephanson N, Beck O, Helander A. Comparison of urinary excretion characteristics of ethanol and ethyl glucuronide. J Anal Toxicol 2002;26:201–4.

[28] Goll M, Schmitt G, Ganssmann B, Aderjan RE. Excretion profiles of ethyl glucuronide in human urine after internal dilution. J Anal Toxicol 2002;26:262–6.

[29] Høiseth G, Bernard JP, Karinen R, Johnsen L, Helander A, Christophersen AS, et al. A pharmacokinetic study of ethyl glucuronide in blood and urine: applications to forensic toxicology. Forensic Sci Int 2007;172:119–24.

[30] Høiseth G, Morini L, Polettini A, Christophersen A, Mørland J. Blood kinetics of ethyl glucuronide and ethyl sulphate in heavy drinkers during alcohol detoxification. Forensic Sci Int 2009;188:52–6.

[31] Droenner P, Schmitt G, Aderjan R, Zimmer H. A kinetic model describing the pharmacokinetics of ethyl glucuronide in humans. Forensic Sci Int 2002;126:24–9.

[32] Halter CC, Dresen S, Auwaerter V, Wurst FM, Weinmann W. Kinetics in serum and urinary excretion of ethyl sulfate and ethyl glucuronide after medium dose ethanol intake. Int J Legal Med 2008;122:123–8.

[33] Høiseth G, Morini L, Polettini A, Christophersen AS, Johnsen L, Karinen R, et al. Serum/whole blood concentration ratio for ethylglucuronide and ethyl sulfate. J Anal Toxicol 2009;33:208–11.

[34] Kharbouche H, Steiner N, Morelato M, Staub C, Boutrel B, Mangin P, et al. Influence of ethanol dose and pigmentation on the incorporation of ethyl glucuronide into rat hair. Alcohol 2010;44:507–14.

[35] Schummer C, Appenzeller BM, Wennig R. Quantitative determination of ethyl glucuronide in sweat. Ther Drug Monit 2008;30:536–9.

[36] Schräder J, Rothe M, Pragst F. Ethyl glucuronide concentrations in beard hair after a single alcohol dose: evidence for incorporation in hair root. Int J Legal Med 2012;126:791–9.

[37] Morini L, Politi L, Groppi A, Stramesi C, Polettini A. Determination of ethyl glucuronide in hair samples by liquid chromatography/electrospray tandem mass spectrometry. J Mass Spectrom 2006;41:34–42.

[38] Paul R, Kingston R, Tsanaclis L, Berry A, Guwy A. Do drug users use less alcohol than non-drug users? A comparison of ethyl glucuronide concentrations in hair between the two groups in medico-legal cases. Forensic Sci Int 2008;176:82–6.

[39] Paul R, Tsanaclis L, Kingston R, Berry A, Guwy A. Simultaneous determination of GHB and EtG in hair using GCMS/MS. Drug Test Anal 2011;3:201–5.

[40] Kerekes I, Yegles M, Grimm U, Wennig R. Ethyl glucuronide determination: head hair versus non-head hair. Alcohol Alcohol 2009;44:62–6.

[41] Martins Ferreira L, Binz T, Yegles M. The influence of ethanol containing cosmetics on ethyl glucuronide concentration in hair. Forensic Sci Int 2012;218:123–5.

[42] Kharbouche H, Sporkert F, Troxler S, Augsburger M, Mangin P, Staub C. Development and validation of a gas chromatography-negative chemical ionization tandem mass spectrometry method for the determination of ethyl glucuronide in hair and its application to forensic toxicology. J Chromatogr B Analyt Technol Biomed Life Sci 2009;877:2337–43.

[43] Lamoureux F, Gaulier JM, Sauvage FL, Mercerolle M, Vallejo C, Lachâtre G. Determination of ethyl-glucuronide in hair for heavy drinking detection using liquid chromatography-tandem mass spectrometry following solid-phase extraction. Anal Bioanal Chem 2009;394:1895–901.

[44] Agius R, Nadulski T, Kahl HG, Schräder J, Dufaux B, Yegles M, et al. Validation of headspace solid-phase microextraction-GC-MS/MS for the determination of ethyl glucuronide in hair according to forensic guidelines. Forensic Sci Int 2010;196:3 9.

[45] Albermann ME, Musshoff F, Madea B. A fully validated high-performance liquid chromatography-tandem mass spectrometry method for the determination of ethyl glucuronide in hair for the proof of strict alcohol abstinence. Anal Bioanal Chem 2010;396:2441–7.

[46] Shi Y, Shen B, Xiang P, Yan H, Shen M. Determination of ethyl glucuronide in hair samples of Chinese people by protein precipitation (PPT) and large volume injection-gas chromatography-tandem mass spectrometry (LVI-GC/MS/MS). J Chromatogr B Analyt Technol Biomed Life Sci 2010;878:3161–6.

[47] Tarcomnicu I, van Nuijs AL, Aerts K, De Doncker M, Covaci A, Neels H. Ethyl glucuronide determination in meconium and hair by hydrophilic interaction liquid chromatography-tandem mass spectrometry. Forensic Sci Int 2010;196:121–7.

[48] Albermann ME, Musshoff F, Aengenheister L, Madea B. Investigations on the influence of different grinding procedures on measured ethyl glucuronide concentrations in hair determined with an optimized and validated LC-MS/MS method. Anal Bioanal Chem 2012;403:769–76.

[49] Kronstrand R, Brinkhagen L, Nyström FH. Ethyl glucuronide in human hair after daily consumption of 16 or 32 g of ethanol for 3 months. Forensic Sci Int 2012;215:51–5.

[50] Suesse S, Pragst F, Mieczkowski T, Selavka CM, Elian A, Sachs H, et al. Practical experiences in application of hair fatty acid ethyl esters and ethyl glucuronide for detection of chronic alcohol abuse in forensic cases. Forensic Sci Int 2012;218:82–91.

[51] Pirro V, Di Corcia D, Seganti F, Salomone A, Vincenti M. Determination of ethyl glucuronide levels in hair for the assessment of alcohol abstinence. Forensic Sci Int 2013;232:229–36.

[52] Turfus SC, Beyer J, Gerostamoulos D, Drummer OH. A comparison of the performance of quality controls prepared from spiked, fortified and authentic hair for ethyl glucuronide analysis. Forensic Sci Int 2013;232:60–6.

[53] Imbert L, Gaulier JM, Dulaurent S, Morichon J, Bevalot F, Izac P, et al. Improved liquid chromatography-tandem mass spectrometric method for the determination of ethyl glucuronide concentrations in hair: Applications to forensic cases. Int J Legal Med 2014;128:53–8.

[54] Dussy F, Carson N, Hangartner S, Briellmann T. Is one hair lock really representative? Drug Test Anal 2014;6(Suppl. 1):5–8.

[55] Tsanaclis L, Kingston R, Wicks J. Testing for alcohol use in hair: is ethyl glucuronide (EtG) stable in hair? Ann Toxicol Anal 2009;21:67–71.

[56] Agius R, Ferreira LM, Yegles M. Can ethyl glucuronide in hair be determined only in 3 cm hair strands? Forensic Sci Int 2012;218:3–9.

[57] Mönch B, Becker R, Nehls I. Quantification of ethyl glucuronide in hair: effect of milling on extraction efficiency. Alcohol Alcohol 2013;48:558–63.

[58] Jurado C, Soriano T, Giménez MP, Menéndez M. Diagnosis of chronic alcohol consumption. Hair analysis for ethyl-glucuronide. Forensic Sci Int 2004;145:161–6.

[59] Mönch B, Becker R, Nehls I. Determination of ethyl glucuronide in hair: a rapid sample pretreatment involving simultaneous milling and extraction. Int J Legal Med 2014;128:69–72.

[60] Małkowska A, Szutowski M, Dyr W. Deposition of ethyl glucuronide in WHP rat hair after chronic ethanol intake. Pharmacol Rep 2012;64:586–93.

[61] Yegles M, Labarthe A, Auwärter V, Hartwig S, Vater H, Wennig R, et al. Comparison of ethyl glucuronide and fatty acid ethyl ester concentrations in hair of alcoholics, social drinkers and teetotallers. Forensic Sci Int 2004;145:167–73.

[62] Cabarcos P, Hassan HM, Tabernero MJ, Scott KS. Analysis of ethyl glucuronide in hair samples by liquid chromatography-electrospray ionization-tandem mass spectrometry (LC-ESI-MS/MS). J Appl Toxicol 2013;33:638–43.

[63] Kintz P, Villain M, Vallet E, Etter M, Salquebre G, Cirimele V. Ethyl glucuronide: unusual distribution between head hair and pubic hair. Forensic Sci Int 2008;176:87–90.

[64] Concheiro M, Cruz A, Mon M, de Castro A, Quintela O, Lorenzo A, et al. Ethylglucuronide determination in urine and hair from alcohol withdrawal patients. J Anal Toxicol 2009;33:155–61.

[65] Peters FT, Hartung M, Herbold M, Schmitt G, Daldrup T, Mußhoff F. Anhang B zur Richtlinie der GTFCh zur Qualitätssicherung bei forensisch-toxikologischen Untersuchungen: Anforderungen an die Validierung von Analysenmethoden. Toxichem Krimtech 2009;76:185–208.

[66] Mönch B, Becker R, Jung C, Nehls I. The homogeneity testing of EtG in hair reference materials: a high-throughput procedure using GC-NCI-MS. Forensic Sci Int 2013;226:202–7.

[67] Boscolo-Berto R, Viel G, Montisci M, Terranova C, Favretto D, Ferrara SD. Ethyl glucuronide concentration in hair for detecting heavy drinking and/or abstinence: a meta-analysis. Int J Legal Med 2013;127:611–19.

[68] Politi L, Morini L, Leone F, Polettini A. Ethyl glucuronide in hair: is it a reliable marker of chronic high levels of alcohol consumption? Addiction 2006;101:1408–12.

[69] Appenzeller BM, Agirman R, Neuberg P, Yegles M, Wennig R. Segmental determination of ethyl glucuronide in hair: a pilot study. Forensic Sci Int 2007;173:87–92.

[70] Høiseth G, Morini L, Polettini A, Christophersen A, Mørland J. Ethyl glucuronide in hair compared with traditional alcohol biomarkers—a pilot study of heavy drinkers referred to an alcohol detoxification unit. Alcohol Clin Exp Res 2009;33:812–16.

[71] Morini L, Politi L, Polettini A. Ethyl glucuronide in hair. A sensitive and specific marker of chronic heavy drinking. Addiction 2009;104:915–20.

[72] Liniger B, Nguyen A, Friedrich-Koch A, Yegles M. Abstinence monitoring of suspected drinking drivers: ethyl glucuronide in hair versus CDT. Traffic Inj Prev 2010;11:123–6.

[73] Morini L, Varango C, Filippi C, Rusca C, Danesino P, Cheli F, et al. Chronic excessive alcohol consumption diagnosis: comparison between traditional biomarkers and ethyl glucuronide in hair, a study on a real population. Ther Drug Monit 2011;33:654–7.

[74] Lees R, Kingston R, Williams TM, Henderson G, Lingford-Hughes A, Hickman M. Comparison of ethyl glucuronide in hair with self-reported alcohol consumption. Alcohol Alcohol 2012;47:267–72.

[75] Hastedt M, Büchner M, Rothe M, Gapert R, Herre S, Krumbiegel F, et al. Detecting alcohol abuse: traditional blood alcohol markers compared to ethyl glucuronide (EtG) and fatty acid ethyl esters (FAEEs) measurement in hair. Forensic Sci Med Pathol 2013;9:471–7.

[76] Pianta A, Liniger B, Baumgartner MR. Ethyl glucuronide in scalp and non-head hair: an intra-individual comparison. Alcohol Alcohol 2013;48:295–302.

[77] Stewart SH, Koch DG, Willner IR, Randall PK, Reuben A. Hair ethyl glucuronide is highly sensitive and specific for detecting moderate-to-heavy drinking in patients with liver disease. Alcohol Alcohol 2013;48:83–7.

[78] Crunelle CL, Cappelle D, Covaci A, van Nuijs AL, Maudens KE, Sabbe B, et al. Hair ethyl glucuronide as a biomarker of alcohol consumption in alcohol-dependent patients: role of gender differences. Drug Alcohol Depend 2014;141:163–6.

[79] Høiseth G, Morini L, Ganss R, Nordal K, Mørland J. Higher levels of hair ethyl glucuronide in patients with decreased kidney function. Alcohol Clin Exp Res 2013;37(Suppl. 1): E14–16.

[80] Greiner M, Pfeiffer D, Schmidt RD. Principles, practical application of the receiver-operating characteristic analysis for diagnostic tests. Prev Vet Med 2000;45:23–41.

[81] Pragst F, Yegles M. Determination of fatty acid ethyl esters (FAEE) and ethyl glucuronide (EtG) in hair: a promising way for retrospective detection of alcohol abuse during pregnancy? Ther Drug Monit 2008;30:255–63.

[82] Hastedt M, Bossers L, Krumbiegel F, Herre S, Hartwig S. Fatty acid ethyl esters in hair as alcohol markers: estimating a reliable cut-off point by evaluation of 1,057 autopsy cases. Forensic Sci Med Pathol 2013;9:184–93.

[83] Society of Hair Testing. 2014 consensus for the use of alcohol markers in hair for assessment of both abstinence and chronic excessive alcohol consumption, <http://www.soht.org/images/pdf/2014>; 2014 [accessed 18.09.14].

[84] Pirro V, Di Corcia D, Pellegrino S, Vincenti M, Sciutteri B, Salomone A. A study of distribution of ethyl glucuronide in different keratin matrices. Forensic Sci Int 2011;210:271–7.

[85] Salomone A, Pirro V, Lombardo T, Di Corcia D, Pellegrino S, Vincenti M. Interpretation of group-level factors from a large population dataset in the determination of ethyl glucuronide in hair. Drug Test Anal 2014; Available from: http://dx.doi.org/10.1002/dta.1697. [Epub ahead of print].

[86] Appenzeller BM, Schuman M, Yegles M, Wennig R. Ethyl glucuronide concentration in hair is not influenced by pigmentation. Alcohol Alcohol 2007;42:326–7.

[87] Morini L, Zucchella A, Polettini A, Politi L, Groppi A. Effect of bleaching on ethyl glucuronide in hair: an in vitro experiment. Forensic Sci Int 2010;198:23–7.

[88] Kerekes I, Yegles M. Coloring, bleaching, and perming: influence on EtG content in hair. Ther Drug Monit 2013;35:527–9.

[89] Gareri J, Rao C, Koren G. Examination of sex differences in fatty acid ethyl ester and ethyl glucuronide hair analysis. Drug Test Anal 2014;6(Suppl. 1):30–6.

[90] Agius R. Utility of coloured hair for the detection of drugs and alcohol. Drug Test Anal 2014;6(Suppl. 1):110–19.

[91] Agius R, Dufaux B, Kahl HG, Nadulski T. Is urine an alternative to cosmetically treated hair for the detection of drugs and alcohol? Drug Test Anal 2014;6(Suppl. 1):120–2.

[92] Agius R, Nadulski T, Kahl HG, Dufaux B. Ethyl glucuronide in hair—A highly effective test for the monitoring of alcohol consumption. Forensic Sci Int 2012;218:10–14.

[93] Dufaux B, Agiusm R, Nadulski T, Kahl HG. Comparison of urine and hair testing for drugs of abuse in the control of abstinence in driver's license re-granting. Drug Test Anal 2012;4:415–19.

[94] Ettlinger J, Kirchen L, Yegles M. Influence of thermal hair straightening on ethyl glucuronide content in hair. Drug Test Anal 2014;6(Suppl. 1):74–7.

[95] Sporkert F, Kharbouche H, Augsburger MP, Klemm C, Baumgartner MR. Positive EtG findings in hair as a result of a cosmetic treatment. Forensic Sci Int 2012;218:97–100.

[96] Arndt T, Schröfel S, Stemmerich K. Ethyl glucuronide identified in commercial hair tonics. Forensic Sci Int 2013;231:195–8.

[97] Sporkert F, Sizun A, Giroud C. Ethyl glucuronide in plant extracts. Toxicol Anal Clin 2014;26:S11.

[98] Binz TM, Baumgartner MR, Kraemer T. The influence of cleansing shampoos on ethyl glucuronide concentrations in hair analyzed with an optimized and validated LC-MS/MS method. Toxicol Anal Clin 2014;26:S34.

[99] Ammann D, Becker R, Kohl A, Hänisch J, Nehls I. Degradation of the ethyl glucuronide content in hair by hydrogen peroxide and a non-destructive assay for oxidative hair treatment using infra-red spectroscopy. Forensic Sci Int 2014;244:30–5.

[100] Laposata M. Fatty acid ethyl esters: nonoxidative ethanol metabolites with emerging biological and clinical significance. Lipids 1999;34:281–5.

[101] Doyle KM, Cluette-Brown JE, Dube DM, Bernhardt TG, Morse CR, Laposata M. Fatty acid ethyl esters in the blood as markers of ethanol intake. JAMA 1996;276:1152–6.

[102] Laposata M. Fatty acid ethyl esters: ethanol metabolites which mediate ethanol induced organ damage and serve as markers of ethanol intake. Prog Lipid Res 1998;37:307–16.

[103] Dan L, Laposata M. Ethyl palmitate and ethyl oleate are the predominant fatty acid ethyl esters in the blood after ethanol ingestion and their synthesis is differentially influenced by the extracellular concentrations of their corresponding fatty acids. Alcohol Clin Exp Res 1997;21:286–92.

[104] Borucki K, Schreiner R, Dierkes J, Jachau K, Krause D, Westphal S, et al. Determination of recent ethanol intake with new markers: comparison of fatty acid ethyl esters in serum and of ethyl glucuronide and the ratio of 5-hydroxytryptophol to 5-hydroxy indole acetic acid in urine. Alcohol Clin Exp Res 2005;29:781–7.

[105] Borucki K, Dierkes J, Wartberg J, Westphal S, Genz A, Luley C. In heavy drinkers, fatty acid ethyl esters remain elevated for up to 99 hours. Alcohol Clin Exp Res 2007;31:423–7.

[106] Doyle KM, Bird DA, Al-Salihi S. Fatty acid ethyl esters are present in human serum after ethanol ingestion. J Lipid Res 1994;35:428–37.

[107] Auwärter V, Sporkert F, Hartwig S, Pragst F, Vater H, Diefenbacher A. Fatty acid ethyl esters in hair as markers of alcohol consumption. Segmental hair analysis of alcoholics, social drinkers, and teetotalers. Clin Chem 2001;47:2114–23.

[108] Hartwig S, Schwarz M, Nadulski T, Kienast T, Pragst F. Nachweis von chronischem Alkoholmissbrauch—Eignung von Fettsäureäthylestern im Sebum. Rechtsmedizin 2010;20:251–7.

[109] Pichini S, Pellegrini M, Gareri J, Koren G, Garcia-Algar O, Vall O, et al. Liquid chromatography-tandem mass spectrometry for fatty acid ethyl esters in meconium: assessment of prenatal exposure to alcohol in two European cohorts. J Pharm Biomed Anal 2008;48:927–33.

[110] Kwak HS, Kang YS, Han KO, Moon JT, Chung YC, Choi JS, et al. Quantitation of fatty acid ethyl esters in human meconium by an improved liquid chromatography/tandem mass spectrometry. J Chromatogr B Analyt Technol Biomed Life Sci 2010;878:1871–4.

[111] Himes SK, Concheiro M, Scheidweiler KB, Huestis MA. Validation of a novel method to identify in utero ethanol exposure: simultaneous meconium extraction of fatty acid ethyl esters, ethyl glucuronide, and ethyl sulfate followed by LC-MS/MS quantification. Anal Bioanal Chem 2014;406:1945–55.

[112] Pragst F, Auwaerter V, Sporkert F, Spiegel K. Analysis of fatty acid ethyl esters in hair as possible markers of chronically elevated alcohol consumption by headspace solid-phase microextraction (HS-SPME) and gas chromatography-mass spectrometry (GC-MS). Forensic Sci Int 2001;121:76–88.

[113] Hastedt M, Herre S, Pragst F, Rothe M, Hartwig S. Workplace alcohol testing program by combined use of ethyl glucuronide and fatty acid ethyl esters in hair. Alcohol Alcohol 2012;47:127–32.

[114] Albermann ME, Madea B, Musshoff F. A SPME-GC/MS procedure for the determination of fatty acid ethyl esters in hair for confirmation of abstinence test results. J Chromatogr Sci 2014;52:955–60.

[115] Süsse S, Selavka CM, Mieczkowski T, Pragst F. Fatty acid ethyl ester concentrations in hair and self-reported alcohol consumption in 644 cases from different origin. Forensic Sci Int 2010;196:111–17.

[116] De Giovanni N, Donadio G, Chiarotti M. The reliability of fatty acid ethyl esters (FAEE) as biological markers for the diagnosis of alcohol abuse. J Anal Toxicol 2007;31:93–7.

[117] Bertol E, Bravo ED, Vaiano F, Mari F, Favretto D. Fatty acid ethyl esters in hair: correlation with self-reported ethanol intake in 160 subjects and influence of estroprogestin therapy. Drug Test Anal 2014;6:930–5.

[118] Zimmermann CM, Jackson GP. Gas chromatography tandem mass spectrometry for biomarkers of alcohol abuse in human hair. Ther Drug Monit 2010;32:216–23.

[119] Bossers LC, Paul R, Berry AJ, Kingston R, Middendorp C, Guwy AJ. An evaluation of washing and extraction techniques in the analysis of ethyl glucuronide and fatty acid ethyl esters from hair samples. J Chromatogr B Analyt Technol Biomed Life Sci 2014;953–954:115–19.

[120] Peters FT, Drummer OH, Musshoff F. Validation of new methods. Forensic Sci Int 2007;165:216–24.

[121] Politi L, Mari F, Furlanetto S, Del Bravo E, Bertol E. Determination of fatty acid ethyl esters in hair by GC-MS and application in a population of cocaine users. J Pharm Biomed Anal 2011;54:1192–5.

[122] Albermann ME, Musshoff F, Madea B. Comparison of ethyl glucuronide (EtG) and fatty acid ethyl esters (FAEEs) concentrations in hair for testing abstinence. Anal Bioanal Chem 2011;400:175–81.

[123] Kulaga V, Pragst F, Fulga N, Koren G. Hair analysis of fatty acid ethyl esters in the detection of excessive drinking in the context of fetal alcohol spectrum disorders. Ther Drug Monit 2009;31:261–6.

[124] Jones MK, Jones BM. Ethanol metabolism in women taking oral contraceptives. Alcohol Clin Exp Res 1984;8:24–8.

[125] Suesse S, Blueml M, Pragst F. Effect of the analyzed hair length on fatty acid ethyl ester (FAEE) concentrations in hair—Congruence of cut-offs for 0–3 and 0–6 cm hair segment. Forensic Sci Int 2014;249:1–5.

[126] Kulaga V, Velazquez-Armenta Y, Aleksa K, Vergee Z, Koren G. The effect of hair pigment on the incorporation of fatty acid ethyl esters (FAEE). Alcohol Alcohol 2009;44 (3):287–92. Available from: http://dx.doi.org/10.1093/alcalc/agn114.

[127] Hartwig S, Auwärter V, Pragst F. Effect of hair care and hair cosmetics on the concentrations of fatty acid ethyl esters in hair as markers of chronically elevated alcohol consumption. Forensic Sci Int 2003;131:90–7.

[128] De Giovanni N, Donadio G, Chiarotti M. Ethanol contamination leads to Fatty acid ethyl esters in hair samples. J Anal Toxicol 2008;32:156–9.

[129] Gareri J, Appenzeller B, Walasek P, Koren G. Impact of hair-care products on FAEE hair concentrations in substance abuse monitoring. Anal Bioanal Chem 2011;400:183–8.

[130] Hartwig S, Auwärter V, Pragst F. Fatty Acid ethyl esters in scalp, pubic, axillary, beard and body hair as markers for alcohol misuse. Alcohol Alcohol 2003;38:163–7.

[131] Kintz P, Nicholson D. Testing for ethanol markers in hair: discrepancies after simultaneous quantification of ethyl glucuronide and fatty acid ethyl esters. Forensic Sci Int 2014;243C:44–6.

[132] Pragst F. Interpretation problems in a forensic case of abstinence determination using alcohol markers in hair. Forensic Sci Int 2012;217:e4–7.

[133] Kintz P, Nicholson D. Interpretation of a highly positive ethyl glucuronide result together with negative fatty acid ethyl esters result in hair and negative blood results. Forensic Toxicol 2014;32:176–9.

[134] Bendroth P, Kronstrand R, Helander A, Greby J, Stephanson N, Krantz P. Comparison of ethyl glucuronide in hair with phosphatidylethanol in whole blood as post-mortem markers of alcohol abuse. Forensic Sci Int 2008;176:76–81.

[135] Bianchi V, Arfini C, Premaschi S, Raspagni A, Secco S, Vidali M. A comparison between serum carbohydrate-deficient transferrin and hair ethyl glucuronide in detecting chronic alcohol consumption. Toxicol Anal Clin 2014;26:S11.

[136] Muskovich M, Haag-Dawoud M. Alcohol consumption among drivers subject to the Swiss license restriction of zero tolerance when driving. Traffic Inj Prev 2012;13:537–43.

[137] Marques PR, Tippetts AS, Yegles M. Ethyl glucuronide in hair is a top predictor of impaired driving recidivism alcohol dependence, and a key marker of the highest BAC interlock tests. Traffic Inj Prev 2014;15:361–9.

[138] European Monitoring Centre for Drugs and Drug Addiction. Legal status of drug testing in the workplace, <http://www.emcdda.europa.eu/html.cfm/index16901EN.html>; 2014 [accessed 09.05.14].

[139] Morini L, Marchei E, Tarani L, Trivelli M, Rapisardi G, Elicio MR, et al. Testing ethyl-glucuronide in maternal hair and nails for the assessment of fetal exposure to alcohol: comparison with meconium testing. Ther Drug Monit 2013;35:402–7.

[140] Kulaga V, Shor S, Koren G. Correlation between drugs of abuse and alcohol by hair analysis: parents at risk for having children with fetal alcohol spectrum disorder. Alcohol 2010;44:615–21.

[141] Kulaga V, Gareri J, Fulga N, Koren G. Agreement between the fatty acid ethyl ester hair test for alcohol and social workers' reports. Ther Drug Monit 2010;32:294–9.

[142] Caprara DL, Klein J, Koren G. Diagnosis of fetal alcohol spectrum disorder (FASD): fatty acid ethyl esters and neonatal hair analysis. Ann Ist Super Sanita 2006;42:39–45.

[143] Morini L, Marchei E, Vagnarelli F, Garcia Algar O, Groppi A, Mastrobattista L, et al. Ethyl glucuronide and ethyl sulfate in meconium and hair-potential biomarkers of intra-uterine exposure to ethanol. Forensic Sci Int 2010;196:74–7.

[144] Laslett AM, Room R, Dietze P, Ferris J. Alcohol's involvement in recurrent child abuse and neglect cases. Addiction 2012;107:1786–93.

[145] High Court of Justice London, Family Division. LB Richmond v B & W & B & CB [2010] EWHC 2903 (Fam), <http://www.familylawweek.co.uk/site.aspx?i = ed71271>; 2010 [accessed 14.09.14].

[146] Rudolph J. Haaranalyse als Beweismittel in familiengerichtlichen Verfahren—richterliche Erwägungen. Praxis der Rechtspsychologie 2010;20:123–30.

[147] Allen JP, Wurst FM, Thon N, Litten RZ. Assessing the drinking status of liver transplant patients with alcoholic liver disease. Liver Transpl 2013;19:369–76.

[148] Sterneck M, Yegles M, von Rothkirch G, Staufer K, Vettorazzi E, Schulz KH, et al. Determination of ethyl glucuronide in hair improves evaluation of long-term alcohol abstention in liver transplant candidates. Liver Int 2013;34:469–76. Available from: http://dx.doi.org/10.1111/liv.12243.

[149] Haller DL, Acosta MC, Lewis D, Miles DR, Schiano T, Shapiro PA, et al. Hair analysis versus conventional methods of drug testing in substance abusers seeking organ transplantation. Am J Transplant 2010;10:1305–11.

[150] Haller DL, Schiano T, Lewis D. Is there a better way to monitor abstinence among substance abusers awaiting transplantation? Curr Opin Organ Transplant 2012;17:180–7.

[151] Musshoff F, Brockmann C, Madea B, Rosendahl W, Piombino-Mascali D. Ethyl glucuronide findings in hair samples from the mummies of the Capuchin Catacombs of Palermo. Forensic Sci Int 2013;232:213–17.

[152] Politi L, Morini L, Mari F, Groppi A, Bertol E. Ethyl glucuronide and ethyl sulfate in autopsy samples 27 years after death. Int J Legal Med 2008;122:507–9.

[153] Kłys M, Woźniak K, Rojek S, Rzepecka-Woźniak E, Kowalski P. Ethanol-related death of a child: an unusual case report. Forensic Sci Int 2008;179:e1–4.

[154] Kłys M, Scisłowski M, Rojek S, Kołodziej J. A fatal clomipramine intoxication case of a chronic alcoholic patient: application of postmortem hair analysis method of clomipramine and ethyl glucuronide using LC/APCI/MS. Leg Med (Tokyo) 2005;7:319–25.

[155] Grant BF, Harford TC. Concurrent and simultaneous use of alcohol with cocaine: results of national survey. Drug Alcohol Depend 1990;25:97–104.

[156] Farré M, de la Torre R, Llorente M, Lamas X, Ugena B, Segura J, et al. Alcohol and cocaine interactions in humans. J Pharmacol Exp Ther 1993;266:1364–73.

[157] McCance-Katz EF, Price LH, McDougle CJ, Kosten TR, Black JE, Jatlow PI. Concurrent cocaine-ethanol ingestion in humans: pharmacology, physiology, behavior, and the role of cocaethylene. Psychopharmacology (Berl) 1993;111:39–46.

[158] Martin G, Macdonald S, Pakula B, Roth EA. A comparison of motivations for use among users of crack cocaine and cocaine powder in a sample of simultaneous cocaine and alcohol users. Addict Behav 2014;39:699–702.

[159] Perez-Reyes M, Jeffcoat AR. Ethanol/cocaine interaction: cocaine and cocaethylene plasma concentrations and their relationship to subjective and cardiovascular effects. Life Sci 1992;51:553–63.

[160] Perez-Reyes M, Jeffcoat AR, Myers M, Sihler K, Cook CE. Comparison in humans of the potency and pharmacokinetics of intravenously injected cocaethylene and cocaine. Psychopharmacology (Berl) 1994;116:428–32.

[161] McCance EF, Price LH, Kosten TR, Jatlow PI. Cocaethylene: pharmacology, physiology and behavioral effects in humans. J Pharmacol Exp Ther 1995;274:215–23.

[162] McCance-Katz EF, Kosten TR, Jatlow P. Concurrent use of cocaine and alcohol is more potent and potentially more toxic than use of either alone—a multiple-dose study. Biol Psychiatry 1998;44:250–9.

[163] Brzezinski MR, Abraham TL, Stone CL, Dean RA, Bosron WF. Purification and characterization of a human liver cocaine carboxylesterase that catalyzes the production of benzoylecgonine and the formation of cocaethylene from alcohol and cocaine. Biochem Pharmacol 1994;48:1747–55.

[164] Harris DS, Everhart ET, Mendelson J, Jones RT. The pharmacology of cocaethylene in humans following cocaine and ethanol administration. Drug Alcohol Depend 2003;72:169–82.

[165] Herbst ED, Harris DS, Everhart ET, Mendelson J, Jacob P, Jones RT. Cocaethylene formation following ethanol and cocaine administration by different routes. Exp Clin Psychopharmacol 2011;19:95–104.

[166] López-Guarnido O, Álvarez I, Gil F, Rodrigo L, Cataño HC, Bermejo AM, et al. Hair testing for cocaine and metabolites by GC/MS: criteria to quantitatively assess cocaine use. J Appl Toxicol 2013;33:838–44.

[167] Rubio NC, Hastedt M, Gonzalez J, Pragst F. Possibilities for discrimination between chewing of coca leaves and abuse of cocaine by hair analysis including hygrine, cuscohygrine, cinnamoylcocaine and cocaine metabolite/cocaine ratios. Int J Legal Med 2014;129:69–84.

[168] Politi L, Zucchella A, Morini L, Stramesi C, Polettini A. Markers of chronic alcohol use in hair: comparison of ethyl glucuronide and cocaethylene in cocaine users. Forensic Sci Int 2007;172(1):23–7.

[169] Natekar A, Motok I, Walasek P, Rao C, Clare-Fasullo G, Koren G. Cocaethylene as a biomarker to predict heavy alcohol exposure among cocaine users. J Popul Ther Clin Pharmacol 2012;19:e466–72.

[170] Natekar A, Koren G. Interpretation of combined hair fatty acid ethyl esters, cocaine and cocaethylene. Ther Drug Monit 2011;33:284.

[171] Casale FJ, Boudreau DK, Jones LM. Tropane ethyl esters in illicit cocaine: isolation, detection, and determination of new manufacturing by-products from the clandestine purification of crude cocaine base with ethanol. J Forensic Sci 2008;53:661–7.

[172] Wilson AS, Brown EL, Villa C, Lynnerup N, Healey A, Ceruti MC, et al. Archaeological, radiological, and biological evidence offer insight into Inca child sacrifice. Proc Natl Acad Sci USA 2013;110:13322–7.

[173] Radun I, Ohisalo J, Rajalin S, Radun JE, Wahde M, Lajunen T. Alcohol ignition interlocks in all new vehicles: a broader perspective. Traffic Inj Prev 2014;15:335–42.

[174] Leffingwell TR, Cooney NJ, Murphy JG, Luczak S, Rosen G, Dougherty DM, et al. Continuous objective monitoring of alcohol use: twenty-first century measurement using transdermal sensors. Alcohol Clin Exp Res 2013;37:16–22.

[175] Yaldiz F, Daglioglu N, Hilal A, Keten A, Gülmen MK. Determination of ethyl glucuronide in human hair by hydrophilic interaction liquid chromatography-tandem mass spectrometry. J Forensic Leg Med 2013;20:799–802.

Chapter 5

Clinical Applications of Hair Analysis

Pascal Kintz[1], Evan Russell[2], Marta Baber[3] and Simona Pichini[4]

[1]Institute of Legal Medicine, University of Strasbourg, France and X-Pertise Consulting, Oberhausbergen, France, [2]Department of Physiology and Pharmacology, University of Western Ontario, London, Ontario, Canada, [3]Department of Pharmacology and Toxicology, Faculty of Medicine, University of Toronto, Ontario, Canada, [4]Department of Therapeutic Research and Medicine Evaluation, Istituto Superiore di Sanità, Rome, Italy

5.1 INTRODUCTION

The potential for clinical applications of hair analysis is substantial and has been reported with increasing frequency over the last decade, largely due to several advantages over routinely available biomedical testing methodologies employing body fluids such as urine or serum. In particular:

1. drugs and metabolites remain sequestered in the hair shaft with relatively little observed degradation over time, providing a window of detection which is much wider (of the order of weeks to several months) than that of serum or urine, in which drug levels decrease rapidly over a relatively short period of time (hours to days),
2. hair collection is simple and noninvasive with low potential for evasion or manipulation of results from hair testing,
3. there is little to no biosafety risk (i.e., disease transmission) in the handling of samples,
4. hair samples are highly stable and can be stored indefinitely at room temperature, allowing for easy future confirmation of initial results when necessary.

As opposed to a point measure of drug exposure such as those determined through serum or urine analysis, hair analysis represents the average area-under-the-concentration curve for a given exposure over a long period of time and can therefore allow an estimation of the average intensity of exposure over several months [1,2].

As described earlier in this book, since hair grows at a relatively consistent average rate of 1.0–1.3 cm/month, it is possible to establish patterns of chronic exposure over defined, clinically relevant time periods. There is potential for such information to provide invaluable diagnostic information where long-term exposure history is a required component of the medical assessment; such as thoroughly assessing substance abuse histories in psychiatric and addiction patients, identifying preconceptional risks such as excessive mercury intake or frequent ethanol consumption (i.e., risk for fetal alcohol spectrum disorder, FASD), or determining the potential role of chronic exposures in acute poisoning situations (i.e., "acute-on-chronic") [3]. The prospect of using hair for the purposes of therapeutic drug monitoring (TDM) carries not only some promise, but also challenges with regard to interpretation of dose and time relationships. Whereas some authors presented data indicating a linear relationship between drug dose and amount of parent drug and/or metabolites found in hair [4,5], some others showed the lack of dose-concentration association, advocating intersubject variability of drug incorporation in hair, drug diffusion along the hair shaft with time, and finally incorporation of drug into hair by multiple mechanisms (i.e., through sweat or sebum) [6,7]. Nonetheless, it seems apparent that hair analysis can distinguish categorized patterns of drug consumption and has shown some success when evaluated for potential applications in TDM [8]. While the clinical potential for hair analysis begins with the well-established methods reported identifying exposure to xenobiotics, a wider scope of application has emerged with the assessment of endogenous molecules in hair to identify endocrinological disturbances.

The objective of this chapter is to present selected clinical applications of hair analysis published in the last decade in order to highlight the tremendous potential of this matrix. Due to limited space we have chosen not to review some long-used clinical methods of measuring heavy metals (mostly lead, arsenic, and mercury) in the diagnosis of chronic poisoning. Rather, we have selected to focus on relatively new clinical methods which have not been reported extensively before.

5.2 HAIR ANALYSIS IN MATERNAL, NEONATAL, AND CHILD HEALTH

Worldwide, approximately 10–20% of pregnant women and children experience some form of mental disorder, with neuropsychiatric conditions being the leading cause of disability in young people [9]. Addiction, often perceived as a linked but separate entity to mental illness, is in fact a form of mental illness in and of itself, long-accepted as a disease by health professionals based on objective biomedical criteria [10]. It is estimated that the lifetime prevalence of a substance use disorders is approximately 17% in the general population, with a lifetime prevalence of approximately 50% in individuals diagnosed with bipolar disorder or schizophrenia [11]. The continued use of drugs and/or ethanol into the late stages of pregnancy in spite of known adverse impacts on the health of the neonate is a strong indicator of addiction. As such, evidence of late gestational drug and/or ethanol use obtained via toxicological analysis can serve as a surrogate marker of unresolved mental health issues in the mother as well as developmental risk to the neonate.

A maternal hair sample at birth can be segmented into three equal segments thus providing a comparative representation of drug or ethanol use behaviors over each trimester of a pregnancy. Detection windows associated with certain lengths of hair are estimated by applying a specific growth rate, the international consensus being 1 cm/month [12]. Specific hair growth rates among individuals are variable as is the case with any physiological parameter, but still provide a relatively reliable estimation of chronological exposure history.

The use of neonatal or maternal hair testing in the context of prenatal substance abuse can serve a number of clinical purposes.

5.2.1 Fetal Alcohol Spectrum Disorder

By far the most clinically relevant assessment of gestational exposure is the identification of prenatal ethanol exposure. Ethanol is the most teratogenic drug of abuse currently known and the only drug of abuse associated with a diagnosable, irreversible syndrome. FASD is an umbrella term used to describe a wide range of physical, behavioral, and cognitive harms resulting from prenatal alcohol exposure. The most severe manifestation of this disorder is known as fetal alcohol syndrome (FAS), which is characterized by a triad of primary disabilities, including facial dysmorphology, growth retardation, and CNS neurodevelopmental abnormalities [13]. Unless there is the presence of craniofacial dysmorphology, which only occurs in roughly 10% of cases, diagnosis of FASD requires confirmation of maternal alcohol consumption [14,15]. Individuals with more subtle impairments tend to have poorer quality-of-life outcomes mediated by lack of recognition of their neurological impairments [6]. Secondary disabilities associated with FASD, such as dependent living, incarceration, early death, addictions, and early school drop-out, are preventable with early diagnosis and biomedical/social interventions. In North America, the incidence of FASD is approximately 1% of live births, early detection and intervention is crucial to minimize harm and to improve clinical outcomes [16,17].

When evidence of prenatal ethanol exposure is obtained, exposed neonates should be placed in a longitudinal follow-up program to monitor their neurodevelopmental progress and facilitate a full neurobehavioral assessment by the age of 6 years in order to maximize the benefits of early detection [18]. It is critical to note that only approximately 40% of chronically exposed neonates will exhibit diagnosable FASD; children cannot be assumed to be developmentally impaired until they have been rigorously assessed by a qualified physician as mislabeling can lead to negative outcomes as well.

Maternal hair analysis for alcohol use biomarkers provides a unique opportunity to investigate suspicion of gestational ethanol use and identify children at risk of FASD. Maternal hair analysis has been shown to outperform neonatal matrices, such as meconium, cord tissue, and placental tissue, in identifying prenatal exposures [19]. Maternal hair samples collected at or near the time of birth can improve sensitivity in identifying children at risk.

i Primary prevention: FASD is a preventable lifelong neurodevelopmental disorder associated with gestational ethanol use. Preconceptional (i.e., maternal) screening for risky alcohol use behaviors can help physicians identify patients that would benefit from counseling and/or therapy and potentially prevent alcohol-exposed pregnancies. Furthermore, substance-abusing women have a mean occurrence of four pregnancies, therefore identification of a first child exposed to drugs or ethanol can potentially prevent future exposures [20].

ii Secondary prevention: Prenatal alcohol and drug exposure is associated with a number of negative quality-of-life associated outcomes [21]. Identifying exposed neonates can facilitate mobilization of addiction treatment and parenting support to improve early development.

iii Tertiary prevention: Early identification of infants—particularly alcohol-affected infants—can substantially mitigate negative outcomes and improve neurobehavioral functioning [6]. Furthermore, identification of a maternal substance use disorder can allow for clinical interventions with a potential to substantially improve her quality of life through effective treatment of her mental illness [20].

Toxicological testing routinely available in hospitals is primarily conducted via maternal and neonatal urine analysis. While positive findings in urine are clinically valuable in identifying at-risk mothers and neonates, negative findings provide little reassurance of absence of substance/ethanol abuse due to relatively short detection windows; typically on the order of several days for most drugs of abuse and less than 24 h for ethanol [22,23].

In discussing the potential for clinical applications of hair analysis, particularly those involving drugs of abuse, clear understanding of the specific clinical relevance at hand is of utmost importance. Harms associated with gestational drug and ethanol use are substantial, although they are often overstated and subject to intensive stigma [24]. The medical laboratory scientist carries an ethical responsibility to do no harm to the patient while attempting to contribute optimally to their medical care [25]. There are several reported instances where the use of perinatal toxicology has resulted in inappropriate or unethical responses from health-care providers due to a lack of understanding of methodology or an absence of *a priori* planning of care in the case of drug-positive results [26–28]. There are critically important complexities observed in substance-dependent mothers and their substance-exposed neonates, which all medical professionals within the circle of care should be aware of. As with any laboratory screening test, it is critical to objectively identify the specific medically actionable purpose as to why a neonate or its mother is subjected to toxicological assessment. Basic principles of laboratory medicine dictate that a screening or diagnostic test should only be done if treatment options are available to the neonate, and its mother, that confer a direct health benefit [13].

Both ethyl glucuronide (EtG) and fatty acid ethyl esters (FAEE) are validated biomarkers for assessment of chronic excessive alcohol consumption [29]. Hair FAEE appear very effective as a qualitative marker, with concentrations in excess of 0.50 ng/mg in the proximal 3 cm segment and 1.00 ng/mg in the 3–6 cm segment highly suggestive of chronic excessive ethanol intake over the last 3–6 months of pregnancy on a sample collected at or near the time of birth [29]. It is critical to note, however, that the exact threshold of gestation ethanol use required to cause FASD is unknown and likely much lower than the threshold of ethanol intake required to elicit FAEE concentrations above these cutoffs [17]; meaning children of mothers with positive hair FAEE findings are at a high-risk of FASD, but children of mothers with negative/below cutoff FAEE findings cannot be considered devoid of FASD risk if other evidence of gestational exposure exists.

Hair EtG performs better than FAEE as a quantitative biomarker, exhibiting a strong dose-response relationship and a consensus positive cutoff of 30 pg/mg to indicate chronic excessive alcohol consumption [29,30]. The use of EtG in abstinence assessment is much more validated than FAEE, meaning that use of a lower EtG cutoff of 7 pg/mg would be more effective in assessing neonatal risk of FASD than using the chronic excessive consumption cutoffs due to the considerations regarding threshold teratogenic dose mentioned above [31]. It is important to note that use of the "abstinence" cutoff (7 pg/mg) for EtG is likely to be very specific in identifying FASD risk as limited clinical study has demonstrated that consumption of 1–2 standard drinks per day on average is required to achieve hair EtG concentrations in excess of 7 pg/mg; this amount, while considered moderate in most contexts, constitutes a clinically relevant level of consumption in pregnancy with regard to FASD risk.

The propensity for false-negative EtG findings due to cosmetic treatment of hair and a sex bias in EtG production/hair deposition coupled with a risk for false-positive FAEE findings due to use of ethanol-containing hair-care products has led to several proposals for use of both markers, particularly in female populations in order to maximize sensitivity and specificity of ethanol use assessment [32–35]. FAEE are typically tested only in the first 6 cm of hair, representing the second and third trimesters, however EtG perform very well in identifying ethanol intake throughout the entire length of the hair shaft allowing for an assessment of the entire gestational period in individuals with hair length of 9 cm or greater [30,36].

Agius et al. have demonstrated reasonable reliability of EtG analysis performed on the more distal portions of the hair shaft, making segmental EtG analysis in maternal samples over a minimum 9 cm (i.e., 9 months) in length a viable screening option [37]. It would be feasible to conduct maternal assessments even well after birth if concerns arise. Most standards of postnatal infant care include multiple physician visits over the first few months of life. Analysis of maternal hair could be conducted 3 months postpartum and still provide reliable gestational alcohol use information via EtG analysis.

Table 5.1 provides various patterns of alcohol consumption that may be demonstrated by segmental hair analysis and the corresponding clinical conclusions.

5.2.2 Neonatal Abstinence Syndrome

The second diagnosable medical sequel associated with gestational substance use disorders is neonatal abstinence syndrome (NAS). NAS is medically manageable with clinical guidelines readily available in most jurisdictions [38]. NAS symptoms typically peak by 48–72 h after birth with a mean onset time of 33 h, although onset may be delayed as much as several weeks [39]. This could be problematic in infants with undetected chronic opioid or benzodiazepine exposures that are discharged from hospital within 24–36 h postpartum. Launiainen et al. estimated the frequency at which meconium samples test positive for opioids, amphetamines (APs), and cannabis based on suspected or reported drug use and to relate these findings to neonatal outcome [40,41]. The results revealed misuse of medicinal opioids among drug-dependent mothers, suggesting that licit drugs with substance abuse potential should be routinely tested in meconium [40]. Lendoiro et al. showed considerable improvement in sensitivity of maternal methadone and benzodiazepine use by maternal hair analysis over routine self-report with maternal methadone hair concentrations greater than 926.2 pg/mg being highly predictive for NAS [42].

5.2.3 Measuring Drugs of Abuse in Neonatal Hair

5.2.3.1 *Physiology*

Hair follicle development *in utero* begins during the 10th week of gestation with the emergence of cellular buds in the undifferentiated ectoderm that develop into hair follicles over the following 4–6 weeks [43,44]. The 1st anagen phase initiates hair production at approximately the 15th week of gestation with the first transcutaneous emergence of hair occurring about a week later [45]. By the 18–20th week the entire scalp is covered with

TABLE 5.1 Patterns of Maternal Alcohol Consumption Demonstrated by Segmental Hair Analysis and the Corresponding Clinical Indications

	Sample Collected at or Near Time of Birth						
			Sample Collected at 3 Months Postpartum				
Hair Segment	**0–3 cm**		**3–6 cm**		**6–9 cm**		**Clinical Indication**
Alcohol Biomarker	**[FAEE] (ng/mg)**	**[EtG] (pg/mg)**	**[FAEE] (ng/mg)**	**[EtG] (pg/mg)**	**[FAEE] (ng/mg)**	**[EtG] (pg/mg)**	
Hair concentrations	>0.50	>30	>1.0	>30	n/a	>30	High risk of maternal addiction issues, referral for AUD assessment
							Child at high risk for FASD[a]
	>0.50	>7	>1.0	>7	n/a	>7	Risk of maternal addiction issues, referral for AUD assessment
							Child at high risk for FASD[a]
	<0.50	>7	<1.0	>7	n/a	>7	Repeated maternal consumption of alcohol during pregnancy
							Child at risk for FASD[a]
	<0.50	<7	<1.0	<7	n/a	>7	Maternal alcohol consumption decreased/ceased upon pregnancy recognition
							Child at low risk for FASD
	<0.50	<7	<1.0	<7	n/a	<7	No evidence of maternal alcohol consumption.
							No indication of FASD risk

[a]*Children at risk for FASD cannot generally be diagnosed until 4–6 years of age in the absence of cranio-facial malformation. Children identified at risk should be placed into a long-term neurodevelopmental monitoring program under the care of a pediatrician with a full diagnostic assessment at 4–6 years.*

anagen-phase follicles with the first catagen/telogen conversions taking place during the 24–28th weeks of gestation [45]. Neonatal hair present at birth is estimated to reflect drug exposures after the 28th week of gestation, primarily based on the current understanding of *in utero* hair development [44,45]. Drugs are incorporated into fetal hair through two main mechanisms. Drugs that are present in the fetal circulation are carried by the capillary blood supply to the hair follicle and deposited into the growing hair shaft [46]. Alternatively, drugs may also be deposited through the soaking of the hair in drug-contaminated amniotic fluid [46].

Neonatal hair is available for collection for several months after birth; however, due to high interindividual variability in the rate of hair growth over the first year of life and the telogen/catagen:anagen phase ratios on the scalp, it is unclear as to when prenatal exposure history can be excluded on interpreting infant hair toxicology findings. It is appropriate to consider the possibility of prenatal exposure in any drug-positive hair sample on a child less than 1 year of age. Environmental exposures and exposures through breast milk are two potential postnatal sources of drugs detected in infants.

5.2.3.2 Dose-Response

Forman et al. studied the dose-response relationship between cocaine consumption and benzoylecgonine (BZ) and cocaine accumulation in the hair of guinea pig pups exposed to cocaine *in utero* [47]. At doses of 10, 15, and 20 mg/kg/day, there was a strong linear correlation between cocaine consumption and concentration of BZ found in the hair of pups. Similarly, at doses of 15 and 20 mg/kg/day, there was a significant linear correlation between cocaine consumption and accumulation of cocaine hair. Goodwin et al. assessed the dose-response relationship between maternal dose of buprenorphine and concentrations of buprenorphine and norbuprenorphine in neonatal hair [48]. Due to the small sample size ($n = 4$) a quantitative relationship could not be established, however, it was found that all neonatal hair samples tested positive for the presence of buprenorphine and norbuprenorphine among mothers who were on buprenorphine maintenance during third trimester. Results also revealed a positive linear correlation between buprenorphine concentration in unwashed maternal hair and neonatal hair.

5.2.3.3 Extraction

Drug extraction from neonatal hair typically requires a sample of 10–20 mg; however, several drugs of abuse have been successfully extracted using 0.5 mg of hair [49]. Nakamura et al. uses heptane to extract methamphetamine (MA) and AP [50], whereas Moller et al., Garcia-Bournissen et al., and Klein et al. use methanol as an extraction solvent to derive various drugs abuse from neonatal hair [22,49,51].

A number of authors have reviewed the literature on recently published methods to test for drugs of abuse in neonatal hair [52–54]. This review will address papers that have not been covered by these authors. Wada et al. performed neonatal hair analysis in a low birth weight neonate whose mother was suspected of using MA during her pregnancy [55]. A semi-micro-HPLC-peroxyoxalate chemiluminescence method was performed to test for the presence of MA and AP in the root and tip of neonatal hair according to a previously described method [50]. A significantly higher concentration of MA was found in the root of the hair compared to the tip of the hair. The authors inferred from these results that MP was primarily deposited through the fetal circulation rather than through exposure to amniotic fluid, since there was a nonuniform MP distribution. AP was not detected in neonatal hair samples. Klein et al. conducted a method to analyze various drugs of abuse in neonatal hair using radioimmunoassay (RIA) and enzyme-linked immunosorbent assay (ELISA) techniques followed by gas-chromatography mass spectrometry (GC-MS) to confirm positive results. Sample mass is a major consideration when developing an appropriately sensitive method for neonatal hair analysis, as large amounts of hair may be unavailable for collection. Furthermore, cutting of large quantities of hair, such as the typically required 20 mg (up to 100 mg) required for many published adult hair methods, may cause considerable distress to the parents due to unattractive alterations in the neonate's appearance. Several recent hair toxicology methods have been published requiring between 2.5 mg and 10 mg of hair using HPLC-HRMS (High Performance Liquid Chromatography–High Resolution Mass Spectrometry) or liquid chromatography–mass spectrometry (LC-MS/MS) for analysis of opioids (including fentanyl and norfentanyl), APs, benzodiazepines, ketamine, and norketamine that would be highly applicable to the smaller sample sizes required in neonatal analysis [56–58].

5.2.4 Social Pediatric Risks

5.2.4.1 Neonatal Considerations

Different forms of poor clinical outcome associated with prenatal drug exposure are well documented in the literature; however, it is difficult in most cases to establish a causal relationship due to a number of confounding factors. For instance, while *in utero* exposures to illicit drugs have been linked to poor neurodevelopmental outcomes; the high prevalence of unemployment, domestic abuse and depression among pregnant women with substance abuse problems, and early infant exposure to suboptimal nutrition, poor maternal attachment, and domestic violence may equally contribute to these negative outcomes [59–61]. It has been well documented that pregnant women involved in substance abuse during pregnancy are more likely to have a chaotic lifestyle, poor general health, improper nutrition, a history of mental illness, limited social support, and exposure to domestic violence [60,62,63].

Taylor et al. outlined a number of high risk factors that may indicate prenatal drug use: lack of prenatal care, physical evidence of withdrawal or substance abuse, history of drug abuse or treatment, scent of alcohol or chemicals, and inapposite behavior are factors that may warrant immediate testing for substance abuse in pregnancy [63]. There are also a number of intermediate risk factors that might be suggestive of drug use in pregnancy. These factors may need to be evaluated according to their clinical context prior to testing. It is critically important to understand, in these cases, that both the mothers and neonates being tested are highly vulnerable individuals in need of a relatively complex continuum of supportive care [20]. In these cases of toxicological investigation, no specific diagnosis is being made but rather maternal addiction risk and early infant developmental risk is being identified. While most drugs are not causally implicated in acute neonatal adverse outcomes, the fact remains that substance-exposed infants have a three-fold higher rate of Neonatal Intensity Care Unit (NICU) admissions, making the detection of drugs a biomarker for associated risks [63]. Social support, addiction assessment, and addiction treatment services should be made readily accessible to mothers exhibiting gestational substance abuse as effective, advocative intervention can have a substantial beneficial impact on the child's early development and the overall quality of life of the mother–child dyad.

5.2.4.2 Pediatric Hair Analysis and Passive Exposure Assessment

One of the major benefits of hair analysis is that it not only provides information about active drug use in adult users, but because the hair is located external to the body, it can absorb valuable information about environmental drug exposure. This is of particular benefit when assessing social pediatric concerns that may emerge due to habitual drug use by caregivers.

In situations where psychosocial risk to a child is suspected due to frequent caregiver drug use, the analysis of child hair can provide valuable information regarding the drug use behavior occurring in the home. The presence of drugs in a child's hair indicates that caregiver drug use is occurring within the context of caring for the child. It is common, for example, in child-welfare investigations for drug-using parents to claim drug use as recreational and occurs only outside of the home, when they are not involved in child care. Analysis of children's hair can provide evidence supporting or disproving such assertions.

Children are much more sensitive to picking up environmental exposure to drugs than adults are, likely because they are so frequently handled. Minute drug residues present on a caregiver's hands and/or clothing after use will be transferred to a child's hair if they are handled shortly after drug use has occurred. Moreover, routine hygiene will remove externally deposited drugs over time, therefore, when a child's hair tests positive, this indicates a high risk that this child is repeatedly exposed to drugs passively in their environment [64].

The fundamental conclusions that can be drawn from a positive hair test in a child are as follows:

i This child has a caregiver who is a regular user of the drug in question.
ii This child's home environment may be contaminated with drug residues or drug smoke.
iii This child may be at risk for drug ingestion or inhalation.

It is important to recognize that while drug ingestion may be identified as a risk, it is much less probable than scenarios (i) and (ii) listed above. The presence of BZ (cocaine is converted to BZ by the body after administration of the drug) is generally an indicator of active cocaine use in adult hair samples; however, it is often present in child hair samples in the absence of any drug intake [65]. In children, especially infants and toddlers, frequent handling by regular cocaine users can result in the transfer of BZ present in caregiver sweat to the child's hair. This means that while the presence of BZ in a child's hair sample may indicate a risk of systemic (i.e., internal) cocaine exposure, it is likely only present as a result of external passive exposure to cocaine. In approximately 90% of cases where a parent's hair test result is positive for cocaine, the children in the home test positive for cocaine as well [65].

Due to a number of factors, it is common for older children in families with cocaine-using caregivers to exhibit lower results than their younger siblings [66]. First, in the case of passive cocaine smoke exposure, very young children have higher respiratory rates than older children and adults, making passive inhalation more significant with decreasing age. Second, older children are handled less, and therefore have a lower rate of cocaine transfer via hand-to-hair contact with drug using caregivers. Third, older, school-age children tend to spend less time in the home and therefore have a lower average duration of exposure to the cocaine-contaminated environment.

It is important to bolster one's understanding of the child's level of risk by incorporating information from the home inspection into the assessment of positive hair test results. For example, does the home smell like smoke? If the answer is yes, this significantly raises the risk of inhalational exposure. If caregivers are smoking their cigarettes inside, they are more likely to be smoking their other drugs (e.g., cocaine, marijuana) inside as well. If the home does not smell smoked in, then caregiver-handling (soon after drug use) is the more plausible explanation for the positive results.

5.3 CLINICAL APPLICATIONS OF HAIR CORTISOL ANALYSIS

During the last decade, the use of hair cortisol as a biomarker of stress as well as for Cushing's disease has emerged, and will be described here. Unlike alcohol and drugs of abuse which are xenobiotics, in the case of cortisol one deals with a naturally occurring hormone.

5.3.1 Measuring Chronic Stress

Biomarkers of *acute* stress have been well established and primarily assess catecholamine release. [67]. In contrast, finding a "gold standard" biomarker for *chronic* stress has proved to be challenging given its complex etiology and the highly individual manifestations.

Glucocorticoids are commonly used as biomarkers of stress. In humans, nonhuman primates, and many larger mammals cortisol is the most common glucocorticoid, while in other vertebrates including rodents, corticosterone is the primary stress hormone. As there are only very few studies on corticosterone in hair, cortisol will be focused on as the primary biomarker of stress.

During times when an organism undergoes physiologic duress, cortisol acts to mobilize energy stores and to modulate the immune system. Cortisol is a steroid hormone produced by the cortices of the adrenal glands in response to stress (be it physiological or psychological). The signal for cortisol production arises in the hypothalamus that releases corticotropin-releasing hormone (CRH) onto the anterior pituitary, which in turn releases adrenocorticotropin hormone (ACTH) into circulation to act upon the adrenal glands [68]. Once released, 90% of cortisol circulates in the blood in an inactive protein-bound fraction, and the remaining 10% of free cortisol is physiologically active. This system is subject to negative feedback; when sufficiently high, cortisol levels act to inhibit the system both at the level of the anterior pituitary and the hypothalamus [69].

Cortisol is a derivative of a sequence of biochemical reactions that modify cholesterol. Cortisol is converted into its inactivate metabolite, cortisone, via 11-beta hydroxysteroid dehydrogenase 2 (11β-HSD2). Cortisone can then be conjugated in the liver, making it soluble and capable of excretion via the kidneys. It should be noted that 11-beta hydroxysteroid dehydrogenase 1 (11β-HSD1) opposes the action of 11β-HSD2, and therefore a significant portion of cortisone produced is converted back to cortisol.

The ability of hair to effectively detect changes in cortisol concentrations has been convincingly demonstrated in several proof-of-concept human studies. One such study by Thomson et al. involved patients with Cushing's synsdrome [70]. In this study, hair samples were obtained from patients at the time of first presentation in the clinic. When analyzed month by month, a steady increase in cortisol concentration was observed up until the point of presentation, consistent with clinical symptoms of increasing cortisol exposure. Following a successful surgical intervention to correct the condition another hair sample was obtained, and the cortisol content of those samples was significantly reduced. This was corroborated by Manenschijn and colleagues, who took hair samples during the clinical course of a patient with Cushing's disease (hypercortisolism caused by a pituitary ACTH producing adenoma). Hair cortisol concentrations were elevated initially, and showed a marked decline following a corrective surgery [70]. In aggregate, these studies support the notion that cortisol in hair provides a reflection of long-term systemic cortisol exposure.

5.3.2 Stress and Acute Myocardial Infarction

Recently, Pereg et al. investigated the role of chronic stress as measured by hair cortisol, in the development of an acute myocardial infarction (AMI) [71]. As chronic psychosocial stressors (e.g., financial concerns, marital stress, job stress) are frequently listed as risk factors for AMIs, the authors hypothesized that hair cortisol analysis could potentially be a useful tool to quantify these stressors. Hair samples representing the past 3 months of cortisol production were obtained from patients within 2 days of admission to a hospital for chest pain. The study group consisted of 56 patients who had a confirmed AMI, with a control group consisting of 56 patients in whom chest pain was attributed to other causes. Median cortisol concentration of the AMI group was significantly higher than that of the control group (295.3 vs. 224.9 ng/g; $P = 0.006$). In logistic regression, accounting for age, lipid status, smoking, and other predictors, hair cortisol was the strongest predictor of AMI, followed by BMI. It should be emphasized that the hair cortisol measurement reflected the 3 months *before* the heart attack, and not the stress caused by the heart attack. Thus, this suggests that chronic stress plays a causative role in the pathophysiology of AMI. This type of information cannot usually be obtained using other matrices, except as part of a prospective study, as illustrated by the predictive effect of urinary cortisol excretion on cardiovascular morbidity and mortality [72].

While further study is needed, this work outlines a strong potential to apply hair cortisol analysis clinically in identifying patients at elevated risk of acute myocardial infarction.

5.3.3 Analysis of Cortisol in Hair

Overall, the methods used for measurement of cortisol in hair are very similar, with some variations in procedures among laboratories. Briefly, to extract cortisol from hair, the sample is carefully sectioned into segment lengths that will approximate the time period of interest (e.g., the most proximal 3 cm for the last 3 months of cortisol production). Then, the hair is finely minced with scissors or ground with a ball mill, and incubated in a solvent such as methanol. The resulting solution is evaporated to dryness, and then reconstituted in a solution such as phosphate buffered saline [73]. Following the extraction, ELISA, RIA, or LC-MS/MS have all been used for cortisol quantification [74]. Presently, there is significant interassay variability in the commercially available immunoassays. The immediate implication is that researchers in this field must try to perform all tests of a particular

TABLE 5.2 A Comparison of Properties of the Various Matrices for Cortisol Measurement

Property	Serum	Saliva	Urine	Hair
Subjective level of invasiveness associated with sample collection	High	Low	Moderate	Low
Cortisol affected by stress of sampling procedure?	Possibly	Possibly	Possibly	No
Storage requirements	Spinning and refrigeration followed by freezing	Refrigeration or freezing	Refrigeration or freezing	Room temperature; stable for years
Time periods of cortisol production represented	Single point measure	Single point measure	12–24 h; integral of exposure	Months to years; integral of exposure
Affected by changes in cortisol binding globulin?	Yes; total cortisol measured	No; only free cortisol measured	No; only free cortisol measured	No; only free cortisol measured
Clinically relevant reference ranges established?	Yes	Yes	Yes	No

protocol using the same batch of cortisol immunoassay, using internal positive controls as standards, and preferably using assays that have low interassay variability (Table 5.2).

To date, the majority of studies have investigated cortisol responses using samples of serum, saliva, or urine. The most commonly used assays to detect cortisol in these samples are immunoassays, LC-MS/MS and ELISA [75]. A recent international interlaboratory comparison study revealed strong correlations among the commonly used immunoassays and two LC-MS/MS methods [75].

5.4 FUTURE CLINICAL APPLICATIONS OF HAIR TESTING

One of the most difficult challenges in clinical medicine is in ascertaining patient adherence to their medications, and lack of therapeutic response to medications is often due to patients' not taking their medications. The same way hair analysis is used in adherence for treatment of addiction, it can be very effective in establishing adherence to therapy of chronic conditions such as psychiatric diseases, HIV, and cardiovascular conditions to mention a few [76–78]. While presently hair analysis is perceived as an expensive method, unavailable in most clinical settings, the tremendous cost of nonadherence in terms of disease exacerbation, morbidity, and mortality may change this approach over the next decades.

REFERENCES

[1] Pragst F, Balikova MA. State of the art in hair analysis for detection of drug and alcohol abuse. Clin Chim Acta 2006;370(1–2):17–49.

[2] Kharbouche H, Steiner N, Morelato M, et al. Influence of ethanol dose and pigmentation on the incorporation of ethyl glucuronide into rat hair. Alcohol 2010;44(6):507–14.

[3] Taguchi N, Mian M, Shouldice M, Karaskov T, Gareri J, Nulman I, et al. Chronic cocaine exposure in a toddler revealed by hair test. Clin Pediatr (Phila) 2007;46(3):272–5.

[4] Mizuno A, et al. Analysis of nicotine content of hair for assessing individual cigarette smoking behaviour. Ther Drug Monit 1993;15:99.

[5] Welp EA, et al. Amount of self-reported illicit drug use compared to quantitative hair test results in community-recruited young drug users in Amsterdam. Addiction 2003;98:987.

[6] Kintz P, Bundeli P, Brenneisen R, Ludes B. Dose-concentration relationships in hair from subjects in a controlled heroin-maintenance program. J Anal Toxicol 1998;22:231.

[7] Wennig R. Potential problems with the interpretation of hair analysis results. Forensic Sci Int 2000;10:5.

[8] Pepin G, Gaillard Y. Concordance between self-reported drug use and findings in hair about cocaine and heroin. Forensic Sci Int 1997;17:37.

[9] CDC. Organic Solvent Neurotoxicity, <http://www.cdc.gov/niosh/docs/87-104/#appnote2>; 1987 [accessed 21.04.14].

[10] Hyman SE. Addiction: a disease of learning and memory. Am J Psychiatry 2005;162(8):1414–22.

[11] Regier DA, Farmer ME, Rae DS, et al. Comorbidity of mental disorders with alcohol and other drug abuse. Results from the Epidemiologic Catchment Area (ECA) study. JAMA 1990;264(19):2511–18.

[12] Cooper GA, Kronstrand R, Kintz P. Society of hair testing guidelines for drug testing in hair. Forensic Sci Int 2012;218(1–3):20–4.

[13] Hermeren G. Neonatal screening: ethical aspects. Acta Paediatr Suppl 1999;88(432):99–103.

[14] Sokol RJ, Delaney-Black V, Nordstrom B. Fetal alcohol spectrum disorder. JAMA 2003;290(22):2996–9.

[15] Sampson PD, Streissguth AP, Bookstein FL, et al. Incidence of fetal alcohol syndrome and prevalence of alcohol-related neurodevelopmental disorder. Teratology 1997;56(5):317–26.

[16] Abel EL. An update on incidence of FAS: FAS is not an equal opportunity birth defect. Neurotoxicol Teratol 1995;17(4):437–43.

[17] Koren G, Nulman I. The Motherisk guide to diagnosing fetal alcohol spectrum disorder. Toronto, Canada: The Hospital for Sick Children; 2002.

[18] Zelner I, Shor S, Lynn H, et al. Clinical use of meconium fatty acid ethyl esters for identifying children at risk for alcohol-related disabilities: the first reported case. J Popul Ther Clin Pharmacol 2012;19(1):e26–31.

[19] Concheiro M, González-Colmenero E, Lendoiro E, Concheiro-Guisán A, de Castro A, Cruz-Landeira A, et al. Alternative matrices for cocaine, heroin, and methadone in utero drug exposure detection. Ther Drug Monit 2013;35(4):502–9.

[20] Pepler DJ, Moore TE, Motz M, Leslie M. Breaking the cycle: a chance for new beginnings. 1995–2000 evaluation report. Toronto, Canada: Breaking the Cycle; 2002.

[21] Eyler FD, Behnke M. Early development of infants exposed to drugs prenatally. Clin Perinatol 1999;26(1):107–50.

[22] Moller M, Karaskov T, Koren G. Opioid detection in maternal and neonatal hair and meconium: characterization of an at-risk population and implications to fetal toxicology. Ther Drug Monit 2010;32(3):318–23.

[23] Jones AW. Evidence-based survey of the elimination rates of ethanol from blood with applications in forensic casework. Forensic Sci Int 2010;200(1–3):1–20.

[24] Bandstra ES, Morrow CE, Mansoor E, Accornero VH. Prenatal drug exposure: infant and toddler outcomes. J Addict Dis 2010;29(2):245–58.

[25] McQueen MJ. Ethics and laboratory medicine. Clin Chem 1990;36(8 Pt 1):1404–7.

[26] Chavkin W, Breitbart V. Reproductive health and blurred professional boundaries. Womens Health Issues 1996;6(2):89–96.

[27] Ferguson v. City of Charleston, S.C. Wests Fed Rep 1999; p. 186469–89.

[28] CBC News. Montreal hospital regrets drug-testing error after baby seized, <http://www.cbc.ca/news/canada/montreal/montreal-hospital-regrets-drug-testing-error-after-baby-seized-1.1339238>; 2013 [accessed 21.04.14].

[29] Kintz P. Consensus of the Society of Hair Testing on hair testing for chronic excessive alcohol consumption 2011. Forensic Sci Int 2012;218(1–3):2.

[30] Appenzeller BM, Agirman R, Neuberg P, Yegles M, Wennig R. Segmental determination of ethyl glucuronide in hair: a pilot study. Forensic Sci Int 2007;173(2–3):87–92.

[31] Morini L, Politi L, Polettini A. Ethyl glucuronide in hair. A sensitive and specific marker of chronic heavy drinking. Addiction 2009;104(6):915–20.

[32] Albermann ME, Madea B, Musshoff F. A SPME-GC/MS procedure for the determination of fatty acid ethyl esters in hair for confirmation of abstinence test results. J Chromatogr Sci 2013.

[33] Kerekes I, Yegles M. Coloring, bleaching, and perming: influence on EtG content in hair. Ther Drug Monit 2013;35(4):527–9.

[34] Jones M, Jones J, Lewis D, et al. Correlation of the alcohol biomarker ethyl glucuronide in fingernails and hair to reported alcohol consumed. Alcohol Clin Exp Res 2011;35 (Suppl. 1):16A.

[35] Gareri J, Rao C, Koren G. Examination of sex differences in fatty acid ethyl ester and ethyl glucuronide hair analysis. Drug Test Anal 2014;6(Suppl. 1):30–6.

[36] Agius R, Nadulski T, Kahl HG, Dufaux B. Ethyl glucuronide in hair—A highly effective test for the monitoring of alcohol consumption. Forensic Sci Int 2012;218(1–3):10–14.

[37] Agius R, Ferreira LM, Yegles M. Can ethyl glucuronide in hair be determined only in 3 cm hair strands?. Forensic Sci Int 2012;218(1–3):3–9.

[38] Dow K, Ordean A, Murphy-Oikonen J, et al. Neonatal abstinence syndrome clinical practice guidelines for Ontario. J Popul Ther Clin Pharmacol 2012;19(3):e488–506.

[39] Oei J, Lui K. Management of the newborn infant affected by maternal opiates and other drugs of dependency. J Paediatr Child Health 2007;43(1–2):9–18.

[40] Launiainen T, Nupponen I, Halmesmaki E, Ojanpera I. Meconium drug testing reveals maternal misuse of medicinal opioids among addicted mothers. Drug Test Anal 2013;5 (7):529–33.

[41] Ristimaa J, Gergov M, Pelander A, Halmesmaki E, Ojanpera I. Broad-spectrum drug screening of meconium by liquid chromatography with tandem mass spectrometry and time-of-flight mass spectrometry. Anal Bioanal Chem 2010;398(2):925–35.

[42] Lendoiro E, Gonzalez-Colmenero E, Concheiro-Guisan A, et al. Maternal hair analysis for the detection of illicit drugs, medicines, and alcohol exposure during pregnancy. Ther Drug Monit 2013;35(3):296–304.

[43] Furdon SA, Clark DA. Scalp hair characteristics in the newborn infant. Adv Neonatal Care 2003;3(6):286–96.

[44] Hollbrook K. Structural and biochemical organogenesis of skin and cutaneous appendages in the fetus and newborn. In: Polin R, Fox W, editors. Fetal and neonatal physiology. Philadelphia, PA: WB Saunders; 1998. p. 729–51.

[45] Gareri J, Koren G. Prenatal hair development: implications for drug exposure determination. Forensic Sci Int 2010;196(1–3):27–31.

[46] Bailey B, Klein J, Koren G. Noninvasive methods for drug measurement in pediatrics. Pediatr Clin North Am 1997;44(1):15–26.

[47] Forman R, Schneiderman J, Klein J, Graham K, Greenwald M, Koren G. Accumulation of cocaine in maternal and fetal hair; the dose response curve. Life Sci 1992;50 (18):1333–41.

[48] Goodwin RS, Wilkins DG, Averin O, et al. Buprenorphine and norbuprenorphine in hair of pregnant women and their infants after controlled buprenorphine administration. Clin Chem 2007;53(12):2136–43.

[49] Garcia-Bournissen F, Rokach B, Karaskov T, Gareri J, Koren G. Detection of stimulant drugs of abuse in maternal and neonatal hair. Forensic Sci Med Pathol 2007;3(2):115–18.

[50] Nakamura S, Wada M, Crabtree BL, et al. A sensitive semi-micro column HPLC method with peroxyoxalate chemiluminescence detection and column switching for determination of MDMA-related compounds in hair. Anal Bioanal Chem 2007;387(6):1983–90.

[51] Klein J, Karaskov T, Koren G. Clinical applications of hair testing for drugs of abuse--the Canadian experience. Forensic Sci Int 2000;107(1–3):281–8.

[52] Gray T, Huestis M. Bioanalytical procedures for monitoring in utero drug exposure. Anal Bioanal Chem 2007;388(7):1455–65.

[53] Lozano J, Garcia-Algar O, Vall O, de la Torre R, Scaravelli G, Pichini S. Biological matrices for the evaluation of in utero exposure to drugs of abuse. Ther Drug Monit 2007;29(6):711–34.

[54] Joya X, Friguls B, Ortigosa S, et al. Determination of maternal-fetal biomarkers of prenatal exposure to ethanol: a review. J Pharm Biomed Anal 2012;69209–22.

[55] Wada M, Sugimoto Y, Ikeda R, Isono K, Kuroda N, Nakashima K. Determination of methamphetamine in neonatal hair and meconium samples: estimation of fetal drug abuse during pregnancy. Forensic Toxicol 2012;30(1):80–3.

[56] Favretto D, Vogliardi S, Stocchero G, Nalesso A, Tucci M, Ferrara SD. High performance liquid chromatography-high resolution mass spectrometry and micropulverized extraction for the quantification of amphetamines, cocaine, opioids, benzodiazepines, antidepressants and hallucinogens in 2.5 mg hair samples. J Chromatogr A 2011;1218(38):6583–95.

[57] Kim J, Ji D, Kang S, et al. Simultaneous determination of 18 abused opioids and metabolites in human hair using LC-MS/MS and illegal opioids abuse proven by hair analysis. J Pharm Biomed Anal 2014;89:99–105.

[58] Chang YJ, Chao MR, Chen SC, Chen CH, Chang YZ. A high-throughput method based on microwave-assisted extraction and liquid chromatography-tandem mass spectrometry for simultaneous analysis of amphetamines, ketamine, opiates, and their metabolites in hair. Anal Bioanal Chem 2014;406(9–10):2445–55.

[59] Thompson BL, Levitt P, Stanwood GD. Prenatal exposure to drugs: effects on brain development and implications for policy and education. Nat Rev Neurosci 2009;10(4):303–12.

[60] Taylor L, Hutchinson D, Rapee R, Burns L, Stephens C, Haber PS. Clinical features and correlates of outcomes for high-risk, marginalized mothers and newborn infants engaged with a specialist perinatal and family drug health service. Obstet Gynecol Int 2012;2012:867265.

[61] Hans SL. Demographic and psychosocial characteristics of substance-abusing pregnant women. Clin Perinatol 1999;26(1):55–74.

[62] Johnson HL, Glassman MB, Fiks KB, Rosen TS. Resilient children: individual differences in developmental outcome of children born to drug abusers. J Genet Psychol 1990;151 (4):523–39.

[63] Taylor P, Bailey D, Green SR, McCully C. Substance abuse during pregnancy: guidelines for screening. Seattle, USA: Washington State Department of Health; 2012.

[64] Cairns T, Hill V, Schaffer M, Thistle W. Removing and identifying drug contamination in the analysis of human hair. Forensic Sci Int 2004;145(2–3):97–108.

[65] Smith FP, Kidwell DA, Cook F. Cocaine in children's hair when they live with drug dependent adults. Presented at the SOFT Conference on Drug Testing in Hair, Tampa, FL, Oct 29-30;1994.

[66] Mieczkowski T. Distinguishing passive contamination from active cocaine consumption: assessing the occupational exposure of narcotics officers to cocaine. Forensic Sci Int 1997;84(1–3):87–111.

[67] Goldstein DS. Clinical assessment of sympathetic responses to stress. Ann. N.Y. Acad. Sci. 1995;771:570–93.Miller GE, Chen E, Zhou ES. If it goes up, must it come down? Chronic stress and the hypothalamic-pituitary-adrenocortical axis in humans. Psychol Bull 2007;133:25–45.

[68] Henley DE, Russell GM, Douthewaite JA, Wood SA, Buchanan F, Gibson R, et al. Hypothalamic-pituitary-adrenal axis activation in obstructive sleep apnea: the effect of continuous positive airway pressure therapy. J Clin Endocrinol Metab 2009;94 (11):4234–42.

[69] Thomson S, Koren G, Fraser LA, Rieder M, Friedman TC, Van Uum SH. Hair analysis provides a historical record of cortisol levels in Cushing's syndrome. Exp Clin Endocrinol Diabetes 2010;118:133–8.

[70] Manenschijn L, Koper JW, Lamberts SW, van Rossum EF. Evaluation of a method to measure long term cortisol levels. Steroids 2011;76:1032–6.

[71] Pereg D, Gow R, Mosseri M, Lishner M, Rieder M, Van US, et al. Hair cortisol and the risk for acute myocardial infarction in adult men. Stress 2011;14:73–81.

[72] Vogelzangs N, Beekman AT, Milaneschi Y, Bandinelli S, Ferrucci L, Penninx BW. Urinary cortisol and six-year risk of all-cause and cardiovascular mortality. J Clin Endocrinol Metab 2010;95:4959–64.

[73] Gow R, Thomson S, Rieder M, Van US, Koren G. An assessment of cortisol analysis in hair and its clinical applications. Forensic Sci Int 2010;196:32–7.

[74] Gatti R, Antonelli G, Prearo M, Spinella P, Cappellin E, De Palo EF. Cortisol assays and diagnostic laboratory procedures in human biological fluids. Clin Biochem 2009;42:1205–17.

[75] Russel E, Kirschbaum C, Laudenslager ML, Stalder T, de Rijke Y, van Rossum EC, et al. Toward standardization of hair cortisol measurement: results of the first international inter-laboratory round robin. Ther Drug Monit 2015;37(1):71–5.

[76] Baxi SM, Liu A, Bacchetti P, Mutua G, Sanders EJ, Kibengo FM, et al. Comparing the novel method of assessing PrEP adherence/exposure using hair samples to other pharmacologic and traditional measures. J Acquir Immune Defic Syndr 2015;68(1):13–20.

[77] Liu AY, Yang Q, Huang Y, Bacchetti P, Anderson PL, Jin C, et al. Strong relationship between oral dose and tenofovir hair levels in a randomized trial: hair as a potential adherence measure for pre-exposure prophylaxis (PrEP). PLoS One 2014;9(1):e83736. http://dx.doi.org/10.1371/journal.pone.0083736. eCollection 2014.

[78] Gandhi M, Yang Q, Bacchetti P, Huang Y. Short communication: a low-cost method for analyzing nevirapine levels in hair as a marker of adherence in resource-limited settings. AIDS Res Hum Retroviruses 2014;30(1):25–8.

Chapter 6

Experiences in Child Hair Analysis

Pascal Kintz
X-Pertise Consulting, Oberhausbergen, France and Institute of Legal Medicine, University of Strasbourg, Strasbourg, France

6.1 INTRODUCTION

Although the detection of drugs in a child's hair unambiguously shows drug handling in the environment of the child, it is difficult to distinguish between systemic incorporation into hair after ingestion or inhalation and external deposition into hair from smoke, dust, or contaminated surfaces. However, the interpretation of hair results with respect to systemic or only external exposure is particularly important in case of children for a realistic assessment of the toxic health risk.

Except the lower amount of biological material in children versus adults, there is no specific analytical problem when processing samples from children. Obviously, the same procedure can be used. However, the issue is the interpretation of the findings, with respect to the different pharmacological parameters.

To perform successful toxicological examinations, the analyst must follow some important rules: (1) to obtain the corresponding biological specimens (blood, urine, and hair) in appropriate time; (2) to use sophisticated analytical techniques (GC-LC/MS-MS, headspace/GC/MS, and accurate mass spectrometry); and (3) to take care of the interpretation of the findings.

For all compounds involved in child poisoning, the detection times in blood and urine depend mainly on the dose and sensitivity of the method used. Prohibiting immunoassays and using only hyphenated techniques, substances can be found in blood for 6 h to 3–4 days and in urine for 12 h to up to 10 days [1,2]. After 48–72 hours the exposure occurred, sampling blood or urine has low interest due to drug clearance from the body.

To address a response to this important caveat, hair was suggested as a valuable specimen. While there are a lot of papers focused on the

Hair Analysis in Clinical and Forensic Toxicology.

identification of drugs (mainly drugs of abuse) in hair following chronic use, those dealing with a controlled single dose are very scarce [3].

When using hair analysis as a matrix during investigative analysis, the question of importance is to know whether the analytical procedure was sensitive enough to identify traces of drugs; this is particularly important when the urine sample(s) of the subject was positive and the hair sample(s) was negative. It has been accepted in the forensic community that a negative hair result cannot exclude the administration of a particular drug, or one of its precursors and the negative findings should not overrule a positive urine result. Nevertheless, the negative hair findings can, on occasion, cast doubt on the positive urine analysis, resulting in substantial legal debate and various consequences for the subject.

The concept of minimal detectable dosage in hair is of interest to document the negative findings, but limited data are currently available in the scientific literature. Such data include cocaine, codeine, ketamine, some benzodiazepines, and some unusual compounds [3].

Until laboratories will have sensitive methodologies to detect a single use of drug, care should be taken to compare urine and hair findings.

This is even more complicated at the scale of life, when dealing with young children. Immature enzymatic processes in children will influence the circulating drug blood concentrations and therefore the level of impairment. In some cases, what can be interpreted as a normal concentration for a healthy adult can be fatal for children. Moreover, the physiology of hair is different according to the age of the subject.

6.2 AGE AS A FACTOR OF INFLUENCE OF DRUG DISTRIBUTION

Changes in the rate but not the extent of drug absorption are usually observed with age. Factors that affect drug absorption (gastric pH and emptying, intestinal motility, blood flow) change with age. Gastric acid secretion does not approach adult levels until the age of 3 and gastric emptying and peristalsis is slow during the first months of life. Higher gastric pH, delayed gastric emptying, and decreased intestinal motility and blood flow are observed in elderly individuals [4].

Children are not "little adults" but rather immature individuals whose bodies and organ functions are in a continuing state of development. More particularly, the newborn infant has to adapt very rapidly to a new environment by going through a series of rapid and continuous anatomical and physiological changes. It is not surprising, therefore, that the pharmacokinetics and toxicity of most drugs vary considerably throughout the pediatric age range and may differ profoundly from findings in adults [5].

There are differences between adults and children in terms of drug distribution; body composition and protein binding are responsible for many of these differences. Water-soluble drugs will have a greater volume of distribution in the neonate. Sometimes, the neonate may require a greater loading dose per kilogram compared to the older child to have a similar effect. Muscle and fat content as a proportion of total body mass is smaller in neonates compared to older children. Therefore, anesthetic drugs that redistribute to muscle and fat would be expected to have a prolonged clinical effect. Protein binding is altered during (approximately) the first 6 months of life. The effect of reduced quantity and quality of protein binding has a marked effect on the concentration of "free" active drug as well as its ability to cross membranes. It is of particular importance in those drugs that are highly protein bound, such as phenytoin, diazepam, bupivacaine, barbiturates, many antibiotics, and theophylline. Other factors that alter distribution include regional blood flow and the maturation of the blood–brain barrier.

The metabolism of many drugs is dependent on the liver and its blood flow. The hepatic blood flow is reduced in the neonate and increases as a proportion of the cardiac output as the infant matures. The complex enzyme systems involved in drug metabolism matures at differing rates in the pediatric population. Many drugs undergo Phase I metabolism, and are metabolized by enzymes of the cytochrome p450 system. The important gene families of these iso-enzymes are CYP1, CYP2, and CYP3. The enzymes of these families develop at very different rates. The development is also variable between individuals which emphasizes the need to titrate drugs to effect or to concentration levels if available. As an example of this variability, the enzyme CYP2D6 is responsible for transforming codeine into its active form of morphine. The activity of this enzyme is very low in neonates and can take more than 5 years to develop adult levels. In contrast, CYP3A4 has an important role in the metabolism of many drugs, for example, midazolam, diazepam, paracetamol; however it matures rapidly to adult levels in the first 6–12 months of life.

The renal efficiency in neonates is considerably reduced compared to the adult. This is due to a combination of factors: incomplete glomerular development with immature glomerular filtration and tubular function, low renal perfusion pressure, and inadequate osmotic load to produce full countercurrent effects [6].

6.3 DIFFERENCES IN CHILDREN'S HAIR VERSUS ADULTS

In adults, hair shaft begins in cells located in a germination center, called the matrix, located at the base of the follicle. Hair does not grow continually, but in cycles, alternating between periods of growth and quiescence. A follicle that is actively producing hair is said to be in the anagen phase. Hair is produced during 4–8 years for head hair (<6 months for nonhead hair) at a rate of approximately 0.22–0.52 mm/day or 0.6–1.42 cm/month for head hair (growth rate depending of hair type and anatomical location). After this period, the follicle enters a relatively short transition period of about 2 weeks, known as the catagen phase, during which cell division stops and the follicle begins to degenerate. Following the transition phase, the hair follicle enters a resting or quiescent period, known as the telogen phase (10 weeks), in which the hair shaft stops growing completely and begins to shut down. Factors such as race, disease states, nutritional deficiencies, and age are known to influence both the rate of growth and the length of the quiescent period. On the scalp of an adult, approximately 85% of the hair is in the growing phase and the remaining 15% is in a resting stage [7].

An extensive literature search, in libraries (University of Strasbourg, medicine and science), database, Google, and on PubMed, with the key words "hair," "anatomy," "neonate," "newborn," "growth," and "physiology," was unable to produce recent citations about the anatomy of hair in newborns and very young children. It was necessary to go back to 1927 [8] to identify the following anatomical features. *In utero*, the activation of the hair follicles of the fetus starts at about the 4th month of pregnancy to produce the lanugo, which is very thin, curly, and nonpigmented. The lanugo will be replaced by a down, short, nonpigmented, thin hair (between 5 and 40 μm), with a slow growing rate and located between 0.5 and 1.5 mm under the skin. Starting the 8th month of pregnancy, the final hair will appear. It is long, pigmented, thick (more than 80 μm), with a rapid growing rate, and located between 2.5 and 5 mm under the skin. After delivery, to avoid becoming bald, hair growing is asynchronous (variation in the anagen/catagen phases) during the first 3–4 months. One will observe hair loss during the first 6 months after delivery, followed by a slow growing rate during the next 6 months. After 1 year, the normal rate of hair growth (1 cm/month such as in adults) starts. The growth rate is first 0.2 mm/day, and then it increases to 0.3–0.5 mm/day, to finally be stabilized at about 0.35 mm/day.

As a consequence, it is very difficult to put any window of detection when testing for drugs in young children. It is even more complicated as it has been demonstrated that drugs can be incorporated during pregnancy in the hair of the fetus, which will contribute to the positive findings after delivery [9–12]. It has been demonstrated that the drugs are transferred transplacentally, with accumulation in fetal hair. Hair measurement for drugs of abuse or pharmaceuticals in neonates is a useful screening method to detect intrauterine exposure to the drug. It seems that there is a good correlation between the amount of drug that was consumed by the mother during pregnancy and the hair findings of the newborn at the time of delivery. Several weeks or months after delivery, identification of a drug in hair can indicate: (1) *in utero* exposure or (2) exposure after delivery or (3) a mix of both situations. No paper has been published about the disappearance of drug after discontinuation of use or exposure (in case of *in utero* exposure).

In adults, it can take 3–6 months to have a negative hair result after heroin or cocaine abstinence [13,14]. Other drugs have not yet been studied. The time course after delivery of disappearance of a drug incorporated during pregnancy has never been reported. It has been proposed by Kintz [15] to consider 100% *in utero* contribution to the final interpretation when the ratio concentration of the proximal segment to the concentration of the distal segment is lower than 0.5. This can be applied only when the child is under 1 year old and the hair shaft length is at least 4 cm (to achieve suitable segmentation). These restrictive criteria (age of the child, length of the hair strand) were included in his statement as they correspond to the hair nonhomogenous growing phases and variability of growth rates in children aged less than 1 year. A minimal hair length (4 cm) is needed to achieve segmental analyses with good representation of the various period of growth.

It must be considered that the amount of hair from children, available for analysis, can be low, particularly when several drugs have to be tested. This has consequences on the limit of quantitation (LOQ) and the identification of the metabolite(s), which is/are generally at a lower concentration when compared to the parent drug. It must be also noted that hair from children is finer and more porous in comparison with adult (risk of higher contamination by sweat versus adults).

6.4 CASE REPORTS

From our experience, children are exposed to drugs of abuse or pharmaceuticals in five major situations:

- *in utero* exposure
- by accidental ingestion of residue(s)
- by administration to obtain sedation (to have them quiet or to sexually assault them)
- by administration to kill them or to sedate them before killing them
- being passively in contact with drug abuser(s)

It seems that the third situation is underreported and only discovered by the authorities when the child is experiencing an overdose or even when death has occurred.

Although numerous controlled studies have been performed in adults, those dealing with children are very scarce, except for antiepileptic drugs. The following examples are a selection from my daily practice and are presented to demonstrate how interpretation was difficult. A full range of literature is available in a recent contribution [16].

The increasing number of people consuming drugs leads necessarily to an increase of pregnant women under the effect of these drugs. Because of the immediate and long-term problems, newborns born to women exposed to drugs during pregnancy should be identified soon after birth so that appropriate intervention and follow up can be done.

The currently used methods to verify drug abuse are the following:

- maternal self-reported drug history (but the information is generally unreliable),
- maternal urinalysis (risk of false-negative results due to the short elimination half-life of the drugs and positive results only reflects exposure during the preceding 1–3 days), and
- analysis of amniotic fluid, urine, or meconium of the baby at the time of delivery (qualitative test at the moment of delivery, risk of false-negative results due to abstinence during the preceding 1 or 3 days).

In case of newborn's hair analysis, the window of detection (from weeks to months) is enhanced and may provide information concerning the severity and pattern of the mother's drug use. Numerous papers have been published on this topic [9–12] and it is not unusual to identify a drug or a mixture of drugs in the hair of a newborn at concentrations close to what can be detected in chronic adult abuser.

Niaprazine, under the trade name Nopron, is largely used in France as a hypnotic agent for occasional insomnia of children. This compound is available without medical prescription. Three children (2 girls and 1 boy) were repetitively sedated and assaulted by their father-in-law for several years. Niaprazine's liquid formulation represents a good potential access to surreptitiously administer it in beverages. According to the request of the judge in charge of this case, hairs of victims were collected, segmented, and screened for sedatives by LC-MS/MS. Niaprazine was detected in the range of 21–382, <LOQ to 315, 2,642, and 3,431 pg/mg for the three children, respectively. These concentrations could not be compared with previous results due to lack of literature. In particular, it was not possible to put any quantitative interpretation on the dosage that was administered to the children. It is however obvious that repetitive administration has occurred but it is not possible to determine the number of exposures. Given the length of the hair, exposure to niaprazine should have occurred at least during the previous months. The surreptitious administration of niaprazine to obtain sedation was considered as a drug-facilitated crime, even in an intrafamilial situation. According to the French law, the drug can be considered as a chemical weapon [17].

Under the trade name Subutex®, buprenorphine is largely used for the substitution management of opiate-dependent individuals, but can also be easily found on the black market. A 14-year-old boy was found dead at the home of a well-known sex offender of minors. According to a report of the police, this was not the first time the boy was seen around this apartment. A blister of Subutex® was discovered on the scene. At the autopsy, no particular morphological changes were noted, except for pulmonary and visceral congestion. There was no evidence of violence and no needle mark was found by the pathologist. Toxicological analyses, as achieved by LC/MS, demonstrated both recent and repetitive buprenorphine exposure in combination with nordiazepam. Buprenorphine concentrations were 1.1 ng/mL and 23 pg/mg in blood and hair, respectively. Noruprenorphine concentration was 0.2 ng/mL in blood and it was not detected in hair. Blood concentrations were considered as corresponding to a therapeutic treatment in the case of a heroin addict. The death of the boy was attributed to accidental asphyxia, in a facilitated repetitive sexual abuse situation, due to the combination of buprenorphine and benzodiazepines, even at therapeutic concentrations. To make the boy vulnerable to sexual activity, he was administered a mixture of buprenorphine and benzodiazepine. The aim of these drugs was to induce sedation and to lower inhibitions. There was no intent to poison the boy, even in the case of repetitive administrations. The use of buprenorphine as a sedative drug was not challenged by the perpetrator who was charged for accidental homicide [18].

Trimeprazine or alimemazine is largely used as antipruritic agent but also for insomnia, cough, and oral premedication in pediatric day surgery. This case involved repetitive sedation linked to the use of trimeprazine as a drug-facilitated crime and subsequent impairment of two children. They were living with their mother-in-law. She was unable to take care of them due to low interest. Drowsiness, ataxia, sedation, muscular weakness, and marked somnolence were noted in both kids at school and during the weekend. These symptoms were present at least for 3 months. Due to the long delay between the alleged crime and clinical examination, collection of blood or urine was of little value. A strand of hair from each child was sampled about 2 months after the first suspicion of administration and was cut into small segments. In the hair of the two subjects, trimeprazine was detected at concentrations in the range of 23–339 pg/mg. Segmental analyses are presented in Table 6.1. These concentrations could not be compared with previous results due to lack of literature. In particular, it was not possible to put any quantitative interpretation on the dosage that was administered to the children. Given the length of the boy's hair, exposure to trimeprazine should have occurred at

TABLE 6.1 Trimeprazine Concentrations after Segmental Analyses of the Hair of Both Children

Segment	Boy	Girl
Segment 1	126 pg/mg (0–2.5 cm)	339 pg/mg (0–2 cm)
Segment 2	127 pg/mg (2.5–5 cm)	199 pg/mg (2–4 cm)
Segment 3		23 pg/mg (4–6 cm)
Segment 4		Not detected (6–8 cm)

least during the previous 5 months. This is confirmed by the analysis of the girl's hair. According to the police investigations, the window of potential drug exposure was about 5 months, during the time the children were under the responsibility of their mother-in-law. All symptoms of impairment disappeared after the children were placed in another family. The mother-in-law that was the perpetrator in both cases did not challenge the use of trimeprazine as a sedative drug [19].

The scientific literature dealing with carbamazepine (an antiepileptic drug but also an antidepressant with slight sedative properties) detection in hair is well documented. In various papers, the concentrations of carbamazepine in the hair of adult patients under daily therapy are >10 ng/mg. At this time, nothing has been published about the detection of carbamazepine in the hair of children. We have been recently involved in a case dealing with child custody, where the final outcome was difficult to establish. The following concentrations were measured by LC-MS/MS in the hair of a 21-month-old girl: 154 (0–1 cm), 198 (1–2 cm), 247 (2–3 cm), and 368 pg/mg (3–4 cm) after decontamination. Obviously, the concentrations measured in the hair are much lower than those observed in patients under daily treatment. In that sense, the frequency of exposures appears as infrequent (low level of exposure), with marked decrease in the more recent period. However, the girl was never prescribed carbamazepine and the mother, who was under carbamazepine therapy, denied any administration. As a consequence, contamination was considered as an issue and interpretation of the results was a challenge that deserves particular attention. It was asked by the Judge if this could result from a single exposure and at which period. There are many differences between the hair from children and those from adults: the hair from children is thinner and more porous, the ratio of anagen/catagen phases is not maintained, and the growth rate can be different, at some periods, from the usual 1 cm/month. At least, three possible interpretations of the measured carbamazepine concentrations were addressed: (1) decrease in administration in the more recent period; (2) increase of body weight due to growing, so the same dosage will result in lower concentrations in hair; and (3) sweat contamination from the mother at the time the girl is with her in bed, the older hair being longer in contact with the bedding. In this case, it was impossible to conclude that the child was deliberately administered carbamazepine. The results of the analysis of hair could indicate that she was in an environment where carbamazepine was being used and where the drug was not being handled and stored with appropriate care. In view of these results we concluded that a single determination should not be used firmly to discriminate long-term exposure to a drug [20].

A 9-year-old boy was admitted to the Emergency Unit for coma and seizures after a stay with his mother. Blood glucose was 0.32 g/L. Blood screening for general unknowns revealed the presence of glibenclamide at 16 ng/mL. A strand of hair (blonde, 4 cm), collected at the time of the event tested positive for glibenclamide at 41 pg/mg. Glibenclamide is a potent, second generation oral sulfonylurea antidiabetic agent widely used to lower glucose levels in patients with type II noninsulin-dependent diabetes mellitus. It acts mainly by stimulating endogenous insulin release from beta cells of pancreas. The mother did not challenge the administration of glibenclamide.

Munchausen by proxy syndrome (MBPS) is a relatively rare form of child abuse that involves the exaggeration or fabrication of illnesses or symptoms by a primary caretaker. Also known as "medical child abuse," MBPS was named after Baron von Munchausen, an eighteenth-century German dignitary known for making up stories about his travels and experiences in order to get attention. "By proxy" indicates that a parent or other adult is fabricating or exaggerating symptoms in a child, not in himself or herself. In MBPS, an individual—usually a parent or caregiver—causes or fabricates symptoms in a child. The adult deliberately misleads others (particularly medical professionals), and may go as far as to actually cause symptoms in the child through poisoning, medication, or even suffocation. In most cases, the mother is responsible for causing the illness or symptoms. Typically, the cause is a need for attention and sympathy from doctors, nurses, and other professionals. Some experts believe that it is not just the attention that is gained from the "illness" of the child that drives this behavior, but also the satisfaction in deceiving individuals who they consider to be more important and powerful than themselves. Because the parent or caregiver appears to be so caring and attentive, often no one suspects any wrongdoing. Diagnosis is made extremely difficult due to the ability of the parent or caregiver to manipulate doctors and induce symptoms in their child. Often, the perpetrator is familiar with the medical profession and knowledgeable about how to induce illness or impairment in the child. Medical personnel often overlook the possibility of MBPS because it goes against the belief that parents and caregivers would never deliberately hurt their child. Most victims of MBPS are preschoolers (although there have been cases in kids up to 16 years old), and there are equal numbers of boys and girls. Often, hospitalization is required. And because they may be deemed a "medical mystery," hospital stays tend to be longer than usual. Whatever the cause, the child's symptoms decrease or completely disappear when the perpetrator is not present. According to experts, common conditions and symptoms that are created or fabricated by parents or caregivers with MBPS can include: failure to thrive, allergies, asthma, vomiting, diarrhea, seizures, and infections. The long-term prognosis for these children depends on the degree of damage created by the illness or impairment and the amount of time it takes to recognize and diagnose MBPS. Some extreme cases have been reported in which children developed destructive skeletal changes, limps, mental retardation, brain damage, and blindness from symptoms caused by the parent or caregiver. If the child lives to be old enough to comprehend what is happening, the psychological damage can be significant. The child may come to feel that he or she will only be loved when ill and may, therefore, help the parent try to deceive doctors, using self-abuse to avoid being abandoned. And so, some victims of MBPS are at risk of repeating the cycle of abuse. If MBPS is suspected, health care providers are required by law to report their concerns. However, after a parent or caregiver is charged, the child's symptoms may increase as the person who is accused attempts to prove the presence of the illness. If the parent or caregiver repeatedly denies the charges, the child would likely be removed from the home and legal action would be taken on the child's behalf. In some cases, the parent or caregiver may deny the charges and move to another location, only to continue the behavior. Even if the child is returned to the perpetrator's custody while protective services are involved, the child may continue to be a victim of abuse while the perpetrator avoids treatment and interventions [21].

With the exception of the treatment of withdrawal syndrome, methadone cannot be prescribed to children under 15 years old. Low dosages (1 mg/kg) can be lethal to children, hence the presence of methadone in a household (for use by parents, grandparents, etc.) with little or no respect for the safe storage of the drug, poorly closed caps (even those with child-resistant caps), poor domestic hygienic conditions (e.g., contamination of dishes and utensils). In cases of accidental or intentional administration, if there is a lack of immediate care in case of respiratory depression and/or intense sedation being observed can prove fatal. During the last years [22], we have been asked to test for methadone and EDDP, its major metabolite, in hair from children that were admitted to hospital unconscious and where methadone had already been identified in a body fluid (4 cases) or where the children were deceased and evidence of methadone overdosage have already been established (2 cases).

Case 1: the boy was taken to hospital unconscious where a urine sample was obtained and confirmed the presence of methadone. The police requested testing of a hair sample for any evidence of previous methadone administration. A strand of hair that was 4 cm in length and blonde in color was collected.

Case 2: parents were suspected of administering methadone to a 16-month-old child. A strand of hair that was 5 cm in length and blonde in color was collected.

Case 3: this case involved a suggestion that methadone was administered to a child over period of time. A strand of hair that was 4.5 cm in length and blonde in color was collected. The hair tested positive for both heroin and cocaine.

Case 4: the child was admitted to hospital after allegedly drinking some of her mother's methadone. The investigators wanted to know if the child had been exposed to the drug on other occasions. A strand of hair that was 15 cm in length and brown in color was collected.

Case 5: a 2-year-old child died of methadone overdose. A history of use was required to assist the police in assessing if it was a one-off dose or had it been given on more than one occasion. A strand of hair that was 4 cm in length and light brown in color was collected.

Case 6: the child (3 years old) was found dead at home. Both parents were reported to have been under methadone therapy. Analysis of cardiac blood tested positive for methadone at 0.24 mg/L. The police have requested the testing of a hair sample to assess if there was any historical evidence that the mother or the father may have administered the drug to the child on more than one occasion. A strand of hair that was 4.5 cm in length and brown in color was collected.

After decontamination with dichloromethane and segmentation, the hair was cut into small pieces, incubated overnight at 40°C, liquid–liquid extracted, and analyzed with LC-MS/MS, using two transitions per compound. The LOQ for both methadone and EDDP was 10 pg/mg. Individual data from each child are presented in Table 6.2. In all cases the methadone concentration was lower than 1 ng/mg. Although 20 mg of hair per segment were used, it must be considered that the amount available for analysis can be lower, particularly when other drugs of abuse have to be tested. This has consequences on the LOQ and the identification of the metabolite EDDP, which is always at a lower concentration when compared to the parent drug. It must be also noted that hair from children is finer and more porous in comparison with adult. The relatively homogenous concentrations of methadone along the hair lock in each specific case were surprising. In the past [23], we have considered such a situation as indicative of external contamination. Thus the presence of homogenous consecutive concentrations after segmental analysis in hair samples obtained from children known to have had methadone in their body close to the time of sampling may be considered as indicative of potential contamination from an individual's body fluids or tissues. The routine decontamination procedure of the laboratory involves two consecutive washes with 5 mL of dichloromethane for 5 min at room temperature, when about 200 mg of hair are processed. This procedure has been used for 20 years and is efficient for various compounds, including drugs of abuse, pharmaceuticals, and doping. In cases involving dirty specimens (blood stains, vomit stains, ground, etc.), the specimens are prewashed with warm water until a clear effluent is obtained, in addition to the 2 × 5 mL dichloromethane washes. In cases where there is some suspicion of external contamination, the second dichloromethane wash is analyzed and the ratio of total concentration in hair (in ng) to concentration in wash is established. When this ratio is higher than 10, this indicates drug exposure as opposed to

TABLE 6.2 Concentrations of Methadone and EDDP Measured in the Six Cases

Case	Segment (cm)	Methadone (ng/mg)	EDDP (ng/mg)
1	0–1	0.05	> LOQ
	1–2	0.07	< LOQ
	2–3	0.07	< LOQ
	3–4	0.08	< LOQ
2	0–1	0.13	0.02
	1–2	0.14	0.02
	2–3	0.15	0.02
	3–5	0.15	0.02
3	0–1.5	0.08	0.02
	1.5–3	0.07	0.01
	3–4.5	0.09	0.03
4	0–2	0.13	0.03
	2–4	0.07	0.01
	4–6	0.07	0.01
	6–8	0.07	0.01
5	0–2	0.53	< LOQ
	2–4	0.58	< LOQ
6	0–1	0.44	0.04
	1–2	0.63	0.05
	2–3	0.77	0.05
	3–4.5	0.65	0.06

external contamination. In the six cases, the last dichloromethane wash was negative for the target drugs. As hair damage or degradation promotes both contamination and incorporation via drug containing body fluids, a visual evaluation of the specimens was undertaken and this did not identify actual degradation. The differentiation between drug use and external contamination has been frequently referred to as one of the limitations of drug testing in hair. The detection of relevant metabolite(s) has been proposed to minimize the possibility of external contamination causing a misinterpretation. Difficulty arises when a metabolite is not detected either due to the absence of specific metabolite or to low doses of the drug being used. Moreover, the presence of a metabolite cannot be considered as an absolute discrimination tool, as it can also be present in the biological tissues, as it is the case with EDDP.

In providing an interpretation of these findings, it can be proposed that potential explanations are as follows:

- at the time of the incident the children were exposed to methadone (as the urine or blood were positive for methadone);
- the children were living in a household where methadone was being used;
- the results of hair analysis may reflect the ingestion of methadone by the children on a fairly regular basis, although the frequency of the ingestion cannot be established;
- the possibility that a significant proportion of the methadone and EDDP present in hair samples was the result of external contamination (sweat, body fluids during the course of the postmortem examination) cannot be excluded.

Hair analysis for drugs and drugs of abuse is increasingly applied in child protective cases. To estimate the potential risk to a child living in a household where drugs are consumed, not only the hair of the parents can be analyzed but also the hair of the child.

Contamination is still an issue in interpreting the analytical data, particularly in children. For example, the differentiation between an active cannabis consumption and a passive drug exposure (i.e., through sidestream marijuana smoke, or transfer from contaminated hands) remains an unsolved problem in hair analysis. The presence in hair for Δ^9-tetrahydrocannabinolic acid A (THCA-A; a nonpsychoactive precursor of Δ^9-tetrahydrocannabinol (THC) and the main cannabinoid in a fresh plant) has been proposed as a specific marker of external contamination. As THCA-A is not incorporated into the hair matrix through the bloodstream in relevant amounts it may act as a marker for external contamination. However, the evidence of active consumption of cannabis (inhaled and/or oral) currently remains based on the detection of 11-nor-9-carboxy-Δ^9-tetrahydrocannabinol acid (THC-COOH) in hair. Nevertheless, the missing detection of this metabolite does not exclude an active consumption, as the low pg/mg range of THC-COOH concentrations in hair of cannabis users is unable to detect, despite using extremely sensitive analytical methods. Recently, Moosmann et al. [24] analyzed 41 hair samples of children and 35 hair samples of drug consuming parents. In all but one of the samples, the concentration of THCA-A was higher than the concentration of THC and in 14 cases no THC could be detected despite the presence of THCA-A, suggesting that in almost all cases a significant external contamination had occurred. The authors concluded that the major part of the cannabinoids detected in the hair samples from these children arose from an external contamination through "passive" transfer by, for example, contaminated fingers or surfaces and not from inhalation or deposition from sidestream smoke.

Children living in homes with drug-addicted parents are in a steady danger of poisoning and may suffer from neglect, maltreatment, and lagging behind in development. Hair analysis could be a suitable way to examine this endangering exposure to drugs. Pragst et al. [25] demonstrated that investigation of children's hair proved to be a useful way to detect endangering drug use in their environment and lead to a more thorough inspection and measures to improve their situation in many of the cases. Within the families (children—aged 1–14 years—living with parents, substituted by methadone and/or suspected for abuse of illegal drugs), hair samples of children and parents provided often the same drug pattern. External deposition from smoke and by contact with contaminated surfaces or parent's hands and systemic deposition after passive smoking, administration, or oral intake by hand-to-mouth transfer were discussed as alternative incorporation mechanisms into hair.

Several authors [26–28] have identified drugs in the hair of children from environmental contamination. It has been proposed that the identification of specific metabolites can help in interpretation. For example, despite an inability to definitively rule out external contamination, the presence of norcocaine in hair is strongly associated with elevated cocaine levels and performs as a highly specific surrogate marker for frequent/intensive cocaine use and highly sensitive marker for intensive/daily use of cocaine. Hair analysis offered a more sensitive tool for identifying drug environmental exposure in children than urine or oral fluid testing. A negative urine or hair test does not exclude the possibility of drug exposure, but hair testing provided the greatest sensitivity for identifying drug-exposed children. Children living in homes where drugs are being manufactured can have drug identified in their hair regardless of hair color. This testing can aid in illuminating the child's presence in an at-risk environment and a family in need of services. Measurement of drug concentrations in the hair can allow estimation of the degree of environmental drug exposure in young children. Infants seem to have a disproportionately increased risk for systemic exposure, compared with older children.

The lack of published data on blood, urine, and hair concentrations in children makes interpretation of analytical results problematic; one cannot simply apply the same interpretation (for adults) to children's samples testing results. Significantly, less data are available concerning controlled laboratory studies on drug intake and subsequent pharmacological effects.

6.5 CONCLUSION AND PERSPECTIVES

It appears that the value of biological analysis for the identification of drug in the specific context of drug exposure in children is steadily gaining recognition. Despite late sampling or even lack of collection of traditional biological fluids, such as blood and/or urine, results for hair testing allow to document exposure to a drug. Although there are still controversies on how to interpret the results, pure analytical work has reached a sort of plateau, having solved almost all the analytical problems. In children, hair testing should be used to complement conventional blood and urine analysis as it increases the window of detection and permits differentiation, by segmentation, of long-term therapeutic use from a single exposure. However, the potential contamination by external sources of drugs remains the critical point when dealing with children. In young children, the identification of drugs in hair does not always mean administration. It can be the result of administration or accidental intake, but also of environmental contamination and *in utero* exposure. Incorporation mechanisms and concentration of drugs in children's hair depend strongly on conditions of living and drug-use habits of the adults.

In case of late crime declaration, positive hair findings are of paramount importance for a victim, in order to start under suitable conditions a psychological follow-up. It can also help in the discrimination of false report of assault, for example, in case of revenge. These cases are often sensitive with little other forensic evidence. The concentration in hair is not always a measure of the degree of danger and a child with lower hair values may be more endangered than another with higher values. Therefore, hair results should not be used in isolation but should be seen as a part of the whole picture and be corroborated by further investigations before decisions are taken in favor of the child [29].

Hair testing is a useful tool to investigate the prevalence of unsuspected chronic exposure to drugs of abuse in pediatric populations. The results of several study cohorts demonstrated a significant prevalence of unsuspected pediatric exposure to drugs of abuse, which mainly involved cocaine, for example, in Barcelona, Spain [30].

Several other pediatric applications of hair testing have been described in the literature, including determination of exposure to pollutants [31], evaluation of stress via cortisol measurement [32], verification of long-term compliance to prescribed medication in individuals displaying a nonnegligible tendency to refuse drugs and to lie on the adherence to therapy as a specific symptom of the disease [33], or investigation of the potential effects on neurodevelopment and behavioral disorders in children exposed to elements and heavy metals [34].

Obviously, there is no analytical limitation when testing hair from children for drugs of abuse, pharmaceuticals, or pollutants. Often, hair from children is used as blank hair during validation of methods or routine QC samples. When using very sensitive procedures, traces of signal can be obtained due to environmental contamination (for drugs of abuse and pharmaceuticals), food intake (pollutants, hormones, pesticides), or endogenous presence (ethyl glucuronide). Pirro et al. [35] have tentatively estimated basal ethyl glucuronide, the phase II metabolite of ethanol concentrations in hair around 0.8 ± 0.4 pg/mg. Although ethanol markers have been mostly tested in adults to document abstinence or chronic excessive drinking behavior, it is possible to test for them in children to demonstrate that the donor has been occasionally imbibed alcohol prior to collection [36].

Overall, the interpretation of the analytical results remains difficult, because children are not "little adults."

REFERENCES

[1] Wells D. Drug administration and sexual assault: sex in a glass. Sci Justice 2001;41:197–9.

[2] Verstraete A. Fenêtre de détection des xénobiotiques dans le sang, les urines, la salive et les cheveux. Ann Toxicol Anal 2002;14:390–4.

[3] Kintz P. Value of the concept of minimal detectable dosage in human hair. Forensic Sci Int 2012;218:28–30.

[4] Jenkins AJ. Pharmacokinetics: drug absorption, distribution, and elimination. In: Karch SB, editor. Drug abuse handbook. Boca Raton, FL: CRC Press; 2007. p. 167.

[5] Rylance G. Clinical pharmacology. Drugs in children. Br Med J 1981;282:50–1.

[6] Buatois S, Le Merdy M, Labat L, Schermann JM, Decleves X. Principales modifications pharmacocinétiques chez l'enfant. Toxicol Anal Clin 2014;26:156–64.

[7] Saitoh M, Uzaka M, Sakamoto M. Rate of hair growth. In: Montagna W, Dobson RL, editors. Advances in biology of skin. Oxford: Pergamon Press; 1969. p. 183–201.

[8] Pinkus F. Entwicklungsgeschichte der Haut, Handbuch der Haut u. Geschlechts-krankheiten 1/1; 1927.

[9] Concheiro M, Gonzalez-Colmenero E, Lendoiro E, Concheiro-Guisan A, de Castro A, Cruz-Landeira A, et al. Alternative matrices for cocaine, heroin, and methadone in utero drug exposure detection. Ther Drug Monit 2013;35:502–9.

[10] Su PH, Chang YZ, Chen JY. Infant with in utero ketamine exposure: quantitative measurement of residual dosage in hair. Pediatr Neonatol 2010;51:279–84.

[11] Gareri J, Koren G. Prenatal hair development: implications for drug exposure determination. Forensic Sci Int 2010;196:27–31.

[12] Lozano J, Garcia-Algar O, Vall O, de la Torre R, Scaravelli G, Pichini S. Biological matrices for the evaluation of in utero exposure to drugs of abuse. Ther Drug Monit 2007;29:711–34.

[13] Garcia-Bournissen F, Moller M, Nesterenko M, Karaskov T, Koren G. Pharmacokinetics of disappearance of cocaine from hair after discontinuation of drug use. Forensic Sci Int 2009;189:24–7.

[14] Shen M, Xiang P, Sun Y, Shen B. Disappearance of 6-acetylmorphine, morphine and codeine from human scalp hair after discontinuation of opiate abuse. Forensic Sci Int 2013;227:64–8.

[15] Kintz P. Contribution of in-utero drug exposure when interpreting hair results in young children. Toxicol Anal Clin 2014;26:S4–5.

[16] Kintz P. The specific problem for children and old people in drug-facilitated crime cases. In: Kintz P, editor. Toxicological aspects of drug-facilitated crimes. London: Academic Press; 2014. p. 255–81.

[17] Villain M, Vallet E, Cirimele V, Kintz P. Mise en évidence d'une soumission chimique à le niaprazine chez des enfants par analyse des cheveux en CL-SM/SM. Ann Toxicol Anal 2008;20:85–8.

[18] Kintz P, Villain M, Tracqui A, Cirimele V, Ludes B. Buprenorphine as a drug-facilitated sexual abuse. A fatal case involving a 14-year old boy. J Anal Toxicol 2003;27:527–9.

[19] Kintz P, Villain M, Cirimele V. Determination of trimeprazine-facilitated sedation in children by hair analysis. J Anal Toxicol 2006;30:400–2.

[20] Kintz P. Interpretation of hair findings in children: about a case involving carbamazepine. Drug Test Anal 2014;6:2–4.

[21] MBPS, <http://kidshealth.org/parent/general/sick/munchausen.html>; [accessed 16.07.13].

[22] Kintz P, Evans J, Villain M, Cirimele V. Interpretation of hair findings in children after methadone poisoning. Forensic Sci Int 2010;196:51–4.

[23] Kintz P. Segmental hair analysis can demonstrate external contamination in postmortem cases. Forensic Sci Int 2012;215:73–6.

[24] Moosmann B, Roth N, Hastedt M, Jacobsen-Bauer A, Pragst F, Auwärter V. Cannabinoids findings in children hair—what do they really tell us? An assessment in the light of three different analytical methods with focus on interpretation of Δ9-tetrahydrocannabinolic acid A concentrations. Drug Test Anal 2014; http://dx.doi.org/10.1002/dta.1692.

[25] Pragst F, Broecker S, Hastedt M, Herre S, Andresen-Streichert H, Sachs H, et al. Methadone and illegal drugs from children with parents in maintenance treatment or suspected for drug abuse in a German community. Ther Drug Monit 2013;35:737–52.

[26] Smith FP, Kidwell DA. Cocaine in hair, saliva, skin swabs, and urine of cocaine users' children. Forensic Sci Int 1996;83:179–89.

[27] Joya X, Papaseit E, Civit E, Pellegrini M, Vall O, Garcia-Algar O, et al. Unsuspected exposure to cocaine in preschool children from a Mediterranean city detected by hair analysis. Ther Drug Monit 2009;31:391–5.

[28] Bassindale T. Quantitative analysis of methamphetamine in hair of children removed from clandestine laboratories—Evidence of passive exposure? Forensic Sci Int 2012;219:179–82.

[29] Eysseric H, Allibe N, Kintz P, Bartoli M, Michard-Lenoir AP, Stranke F, et al. Amitriptyline poisoning of a baby: how informative can be hair analysis? Toxicol Anal Clin 2014;26:S6–7.

[30] Pichini S, Garcia-Algar O, Alvarez AT, Mercadel M, Mortali C, Gottardi M, et al. Pediatric exposure to drugs of abuse by hair testing: monitoring 15 years of evolution in Spain. Int J Environ Res Public Health 2014;11:8267–75.

[31] Tzatzarakis MN, Barbounis EG, Kavvalakis MP, Vakonaki E, Renieri E, Vardavas AI, et al. Rapid method for the simultaneous determination of DDTs and PCBs in hair of children by headspace solid phase microextraction and gas chromatography-mass spectrometry (HSSPME/GC-MS). Drug Test Anal 2014;6:85–92.

[32] Groeneveld MG, Vermeer HJ, Linting M, Noppe G, van Rossum EF, van Ijzendoom MH. Children's hair cortisol as a biomarker of stress at school entry. Stress 2013; 16:711–15.

[33] Papaseit E, Marchei E, Mortali C, Aznar G, Garcia-Algar O, Farrè M, et al. Development and validation of a liquid chromatography-tandem mass spectrometry assay for hair analysis of atomoxetine and its metabolites: application in clinical practice. Forensic Sci Int 2012;218:62–7.

[34] Rodriguez-Barranco M, Lacasana M, Aguilar-Garduno C, Alguacil J, Gil F, Gonzalez-Alzaga B, et al. Association of arsenic, cadmium and manganese exposure with neurodevelopment and behavioural disorders in children: a systematic review and meta-analysis. Sci Total Environ 2013;454–455:562–77.

[35] Pirro V, Di Corcia D, Seganti F, Salamone A, Vincenti M. Determination of ethyl glucuronide levels in hair for the assessment of alcohol abstinence. Forensic Sci Int 2013;232:229–36.

[36] Klys M, Wozniak K, Rojek S, Rzepecka-Wozniak E, Kowalski P. Ethanol-related death of a child: an unusual case report. Forensic Sci Int 2008;179:e1–4.

Chapter 7

Hair Analysis for the Biomonitoring of Human Exposure to Organic Pollutants

Brice M.R. Appenzeller
Laboratory of Analytical Human Biomonitoring - Luxembourg Institute of Health, Luxembourg

7.1 INTRODUCTION

7.1.1 A Recently Investigated Field

Although the first studies reporting the use of hair for the analysis of metals and drugs of abuse date back to the fifties–seventies and seventies–eighties, respectively, the use of this matrix for the detection of organic pollutants (OPs) with a view to assess human exposure is much more recent (Figure 7.1). As a result, the number of publications dealing with OP analysis in human hair is limited to about 70 studies, more than a half of which over the past 5 years.

The will to analyze a compound (pollutant) in a matrix is due to a combination of several factors: the presence of the compound in human surroundings leading to exposure (due to agricultural use, industrial release, food contamination, occupation, etc.), the scientific interest for this compound with regard to public health issues (suspected adverse effects), and of course the technical feasibility. As long as analytical sensitivity was not sufficient to properly detect OPs in hair, results obtained from hair analysis were mostly negative [1,2]. This has driven to the erroneous idea that most compounds were not incorporated in hair and that no reliable results could be obtained from this matrix, at least with regard to the assessment of human exposure to pollutants. As a result, although interest in the biomonitoring of human exposure to several OPs was already evident, technical constraints limited the use of hair which was not considered a relevant matrix for human biomonitoring, and biological fluids were therefore still preferred for the assessment of exposure.

Hair Analysis in Clinical and Forensic Toxicology.

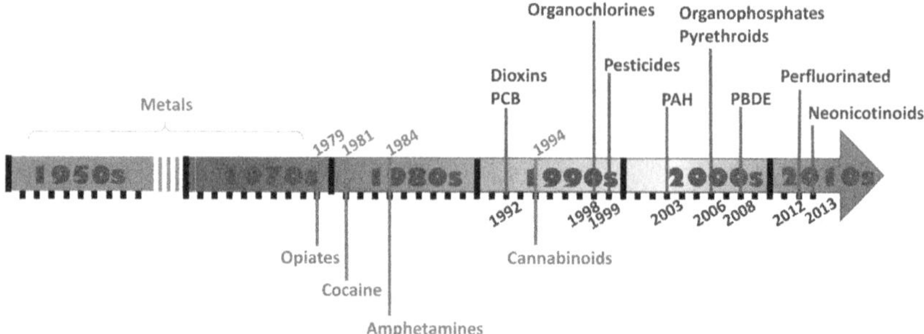

FIGURE 7.1 Chronology of publications reporting the first detection in human hair of different chemical families.

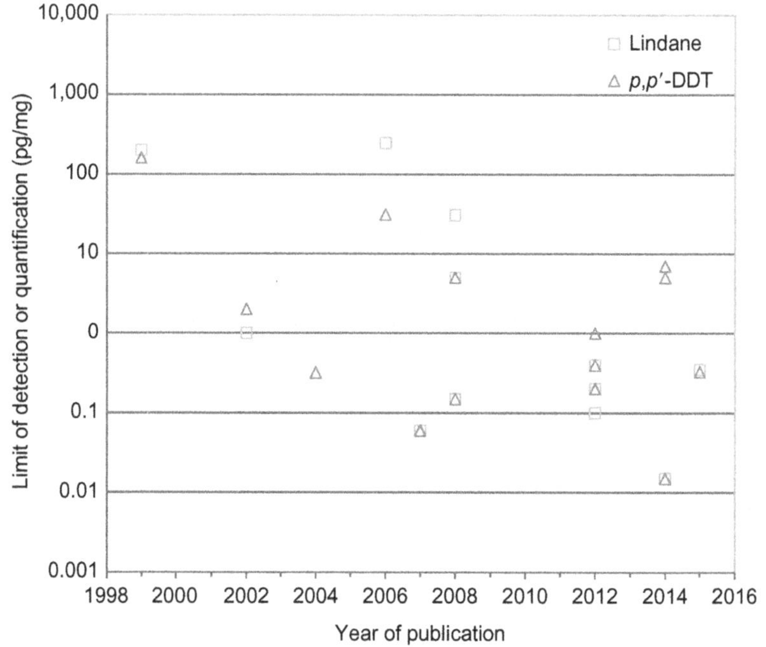

FIGURE 7.2 Limit of detection reported for lindane and *p,p′*-DDT analysis in human hair in studies published since 1999.

Progress in analytical techniques observed over the last 10–15 years yet enabled to reach the sensitivity level required to detect OP in hair. As presented in Figure 7.2 with the examples of lindane (γ-hexachlorobenzene) and *p,p′*-DDT (dichlorodiphenyltrichloroethane), the limit of detection reported in the literature for these organochlorine pesticides was decreased by approximately thousand since 1998–1999 when they were first tested in human hair.

7.1.2 From Feasibility to Field Studies

Over the decade from 1991 to 2001 only about 10 works were published which actually consisted of feasibility studies including a limited number of subjects [1,3–9], with the exception of Ohgami et al. [10] who analyzed polychlorinated biphenyls (PCBs) and polychlorinated quaterphenyls (PCQs) in hair and blood collected from 27 patients with acute PCBs poisoning and 22 controls, and Neuber et al. [2] who analyzed DDT and lindane in hair samples collected from 193 preschool children but obtained low level of positive detection (below 10%). The studies published over this period only focused on chlorinated persistent organic pollutants (POPs), namely dioxins, PCBs, and organochlorine pesticides (Figure 7.3).

The next decade has been marked by an increase in the number of publications on OP analysis in hair, generally including tens to hundreds of subjects, and particularly by investigations on new chemical classes such as pesticides (organophosphates [11,12], pyrethroids, and others [11]), as well as polycyclic aromatic hydrocarbons (PAHs) [13,14] and polybrominated diphenyl ethers (PBDEs) [15,16], therefore widening the possibilities of hair analysis. Over this period, most of the studies reported that hair analysis was able to show differences in OP concentration in hair for groups with supposed different levels of exposure. These works therefore demonstrated the reliability of the results obtained from hair analysis and its relevance for human biomonitoring. Pioneering studies thereby highlighted high levels of

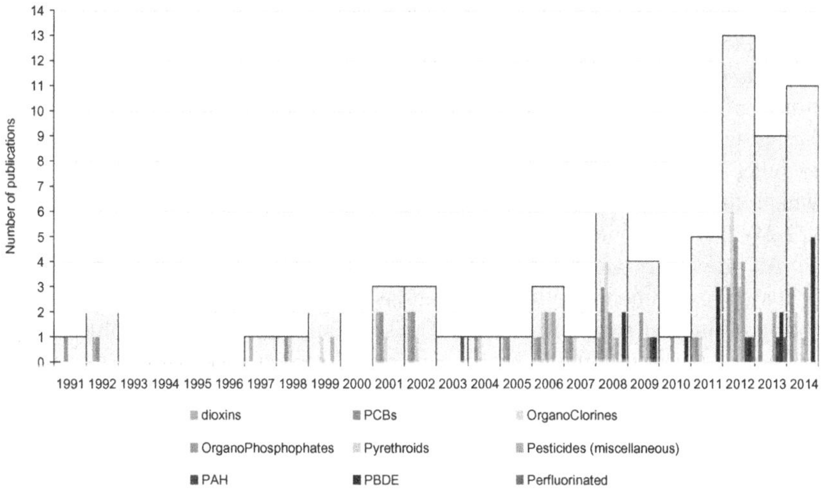

FIGURE 7.3 Number of publications per year presenting experimental results on OPs in human hair. Colored thin bars present the number of publications for each class of OP. Large gray bars present the total number of publications on OP for each year. As some studies present results for several OP classes, gray bars may be inferior to the sum of colored bars. (For interpretation of the references to color in this figure legend, the reader is referred to the online version of this book.)

exposure to pollutants of subpopulations due to their occupation (e.g., greenhouse workers [12], waste incineration workers [17], e-waste recycling workers [18]) or to the contamination of their residence area [15,19,20]. Significantly higher exposure to PAHs was also demonstrated for smokers compared to nonsmokers [13,14].

Once the issue of technical feasibility was addressed and reliability of the results was demonstrated, the interest in hair for the biomonitoring of exposure dramatically increased. As a consequence, more than 50% of the publications dealing with OP analysis in hair were published between 2011 and 2014, which definitely demonstrates that similar to the fields of forensic and clinical toxicology, hair is becoming a reference matrix for the biomonitoring of human exposure to OP as well.

7.2 THE CHALLENGE OF ASSESSING ENVIRONMENTAL EXPOSURE

7.2.1 Facing Low Concentration Levels

When assessing environmental exposure by the quantification of chemicals in a biological matrix, analytical sensitivity is a fundamental parameter since contrary to cases of acute exposure (e.g., intoxication), levels of concentration resulting from environmental exposure are relatively low. For instance, serum concentration reported in cases of acute poisoning with pesticides ranged from 20 μg/L (for bifenthrin) up to 1.5 and 6.5 mg/L (for carbofuran and endosulfan, respectively) [21]. In comparison, levels of pesticide concentration detected in serum collected from raw population is generally about 10 ng/L or lower [22,23], which is between one thousand and one million times inferior. Sensitivity is particularly relevant in the case of hair analysis in that the weight of material used is often limited to 50–200 mg compared to other matrices such as urine, blood/serum, or meconium in the case of which the amount used for the detection of OPs is commonly 2–5 mL [22,24–27] and 10 mL or more for breast milk [28,29].

As previously mentioned, the lack of sufficiently sensitive methods has probably been among the main limitations to the use of hair for the biomonitoring of human exposure to OPs. Consequently, no reference values have yet been provided in public health authorities' reports regarding OPs concentration in hair, and urine and blood remain the reference matrices [30]. As quoted above, the lack of sufficiently sensitive analytical methods has also probably contributed to supporting erroneous statements concerning hair analysis, such as the common idea that hair analysis could not be a reliable indicator of environmental exposure or internal body burden and should only be viewed as a supportive tool [31]. Although similar doubts had first called into question the use of hair analysis for clinical or forensic purposes [32], the current widespread use of this matrix in this field has definitely demonstrated its relevance and proved wrong the initial criticisms [33–35]. The rise in the use of hair for the biomonitoring of human exposure to OPs that we are currently noticing has been delayed because their concentration is significantly lower (typically picograms per milligram of hair down to below picograms per gram for PCBs and dioxins) than those of medical drugs and drugs of abuse, that are mostly present at concentration levels of nanograms per milligram (Figure 7.4).

7.2.2 Is the Method Sensitive Enough?

The difficulty to evaluate whether a method is sensitive enough to demonstrate human exposure has been highlighted through several studies. Although the ultimate proof of method remains positive detection in field samples, results below the limit of detection may be interpreted in different ways. On the one hand, negative results may indicate that the method is not sensitive enough; on the other hand, it may lead to think that examined people are not exposed to the target compounds or are exposed at levels that cannot be detected with the methods used. However, the question that remains is: what is the expected level of exposure? For drugs, the concentration in a biological matrix corresponds to a "typical" intake (to reach the therapeutic zone for medical drugs or the amount corresponding to a classical dose for drugs of abuse) [33]. Even if the concentration can vary, on the whole, there is limited need to reach limits of detection far below the expected concentration. In the field of environmental/occupational exposure, there is no such "therapeutic zone" below which sensitivity would not be relevant, in that the level of exposure can vary in several orders of magnitude between people (Figure 7.4). Moreover, history demonstrated that levels of exposure previously believed safe may actually be harmful, as shown for PCBs and dioxins, for which the current "no observable adverse effect level" (NOAEL) is nearly one million-fold lower than it initially was [36].

In the absence of other indications, one possible way to assess whether a method is sensitive enough is to compare its sensitivity with previously published values, although this approach is of course limited to chemicals that have previously been investigated. Moreover, as presented here for lindane (Figure 7.5), published values may concern different countries or regions with different levels of exposure of the populations, which limits direct transposition. This was demonstrated for DDT and dichlorodiphenyldichloroethylene (DDE), which are detected in human tissues (e.g., adipose tissue, breast milk, serum) at highly varying concentration levels depending on the country where samples were collected [37]. In that regard, the possibility to obtain values from different areas corresponding to populations with different lifestyles (diet, industrialization, agriculture, etc.) can gain lot of knowledge.

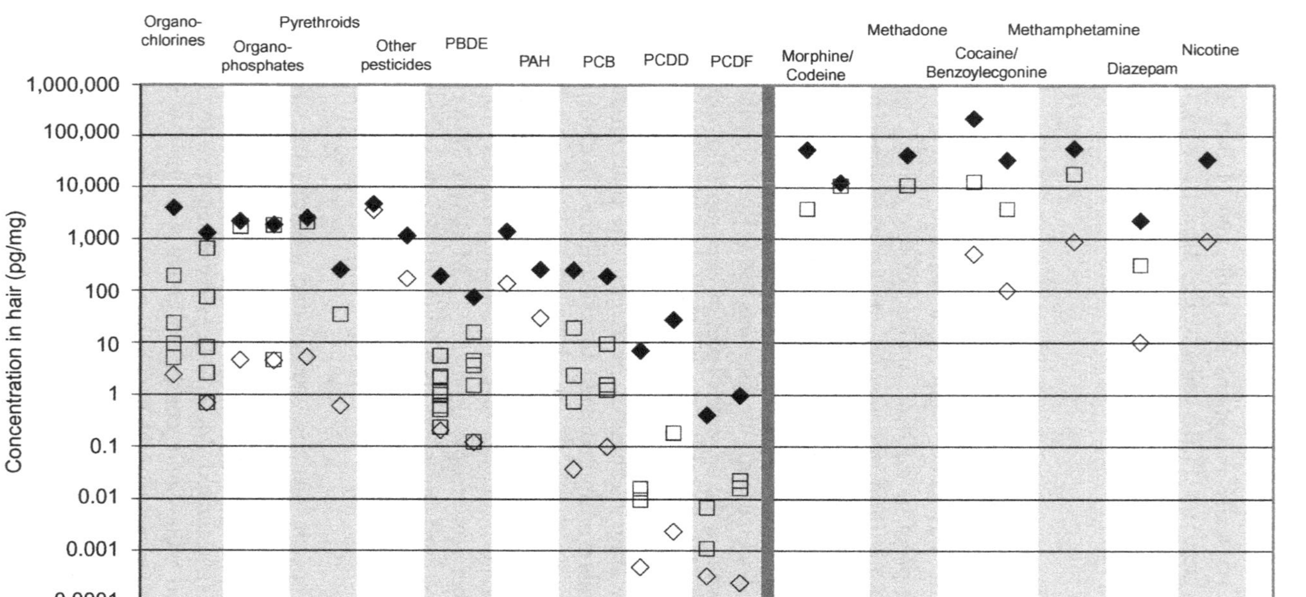

FIGURE 7.4 Levels of concentration in hair reported for OPs (the two most representative compounds from each class) and for some common drugs. *Adapted from Appenzeller and Tsatsakis, [37a], and Pragst and Balikova [33].*

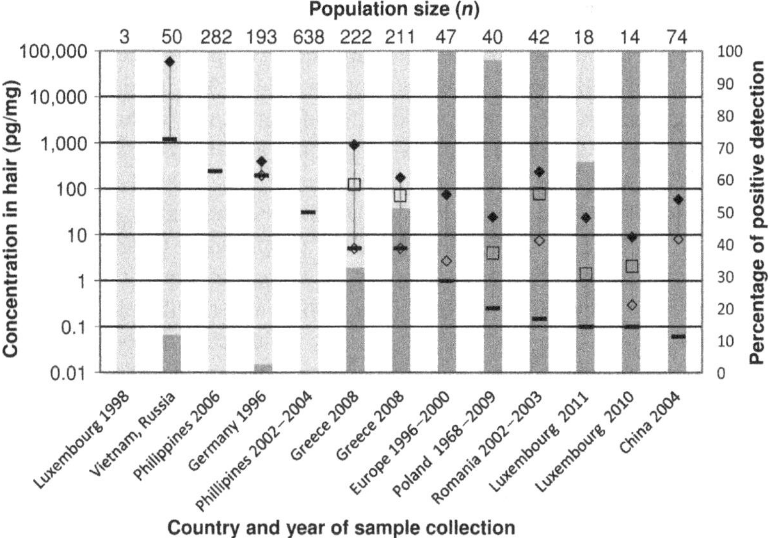

FIGURE 7.5 Percentage of positive detection (bars), levels of concentration ("♦" highest value, "□" median, and "◇" lowest value), and limit of detection/quantification reported for lindane in human hair for different countries [1,2,11,12,20,39–46]. *Adapted from Appenzeller and Tsatsakis, 2012.*

7.2.3 Sensitivity and Rate Positive Detection

The example of lindane (γ-hexachlorohexane), an organochlorine pesticide categorized as POPs, which is presented in Figure 7.5, is a good illustration of the aforementioned issue. Since the first reports describing its detection in hair were published in 1998–1999 [1,2], the sensitivity of the methods has been increased by 100 to 1,000 (Figure 7.2), mainly due to technical progresses in general, and more recently to the use of solid phase microextraction associated with gas chromatography tandem mass spectrometry (GC-MS/MS) [38]. In the process, the rate of positive detection has increased from values close to zero to 100% in most of the recent publications. A good illustration of the gain in analytical sensitivity can be observed through studies focusing on Luxembourg population [1,39,40]: the first analysis of lindane in hair conducted in 1998 only provided negative results (below limit of detection) whereas the studies carried out 14 years later on similar subjects reported 66–100% positive detection. This difference is of course very likely due to the increase in method sensitivity which enabled to reach high rate of positive detection despite the relatively low level of exposure of the Luxembourg population in comparison with other countries (Figure 7.5).

7.3 SAMPLE PRETREATMENT

Several OPs may be detected in environmental compartments (particularly air and dust) at various levels of concentration [47–52]. Their presence in human surroundings may therefore lead to human exposure through inhalation, ingestion, or dermal contact, but may also lead to their deposition on hair surface. Although chemicals incorporated by means of biological mechanisms, likely to be located in the whole hair structure, can be interpreted as representative of the internal dose people have undergone during the period of hair growth, on the contrary, chemicals externally deposited on hair surface only represent recent contamination and are therefore less biologically relevant. Differentiating chemicals biologically incorporated into hair from those deposited on hair surface due to contaminated air or dust thus appears to be crucial for correct interpretation of the results.

7.3.1 Decontamination Procedures

In forensic toxicology, distinguishing external deposition from internally incorporated drugs is mainly used for the identification of adulterated specimens (e.g., by voluntary or involuntary application on hair) and to differentiate consumers from people exposed to indirect contamination (typically smoke of cannabis cigarettes) [53,54]. Decontamination before drug analysis has therefore been extensively documented and several washing procedures with different degree of complexity have been proposed [33]. Hair decontamination before drug testing is generally performed with organic solvents (e.g., dichloromethane, acetone, methanol, acetonitrile) and/or with aqueous solutions of detergents (e.g., sodium dodecyl sulfate) or buffers (e.g., phosphate buffer) [33,55–59]. Unlike for drug testing, the decontamination with organic solvents before environmental pollutants analysis was less used [6,13,20,40,60] and more gentle decontamination solvents like water [14,46,61–63] and water with shampoo [15,52,64–68] were generally preferred. In some studies, hair was even analyzed without being decontaminated, the authors considering that external deposition also represents chemicals to which the subjects have been exposed, and should not be removed before analysis [69]. Nevertheless, no standardized "universal" procedure for the decontamination of hair before analysis has been set up to date and in most studies, the efficiency of the washing procedure was even not assessed.

The few studies which investigated the effect of the washing procedure used to clean hair before analysis yet highlighted its relevance. Comparing unwashed hair and hair washed with shampoo just before analysis, Schramm et al. [3] observed a significant but not uniform decrease in the concentration of polychlorinated dibenzodioxin (PCDD)/Fs congeners. Octa-CDD/Fs were reduced by 50% but the hexachlorinated congeners were removed by a factor of nearly 100. These results led to the conclusion that the PCDD/F burden could not be mainly attributed to particle deposition on hair surface. Later on, Nakao et al. [64] demonstrated that washing hair with common surfactant decreased the levels of PCDDs and PDCFs in hair samples by 50% and 64% respectively. They also observed that washing once more had no further effect on the elimination of either chemical and that both the unwashed and washed samples contained similar composition of PCDDs and polychlorinated dibenzofurans (PCDFs). They concluded that PCDD/Fs were mainly deposited on hair surface via atmospheric transfer and are completely removed by the first wash. The residual amounts of these compounds were thought to be contained in the inner part of the hair. Altshul et al. [65] reported similar observation when comparing the concentration of PCBs and organochlorine pesticides in hair washed with hot water only and hair washed with shampoo once and twice. In fact, washing with shampoo once decreased the levels of PCBs, pesticides, and lipids by 25–33% on average. For the less-chlorinated congeners (PCB-8 and PCB-18), the decrease was even larger (48% and 62%). The study also demonstrated that most of the decrease occurred after the first shampoo washing: 82% of the total loss for $\sum$PCBs and 88% for *p,p′*-DDE. More recently, Ostrea et al. [69a] reported that washing hair with shampoo prior to analysis had no influence on the concentration of propoxur, but significantly decreased the concentration of bioallethrin.

The effect of washing hair with organic solvents was tested by Toriba et al. [13] who analyzed PAHs in hair. Three different solvents were tested (methanol, *n*-hexane, and dichloromethane), each within a cycle of three successive washings. Although different results were obtained for both the different molecules analyzed and the different solvents, the results were in line with the studies that used shampoo, in that a part of the target molecules was removed by washing (most likely the external chemicals) and a remaining part seemed to be unaffected by washing (most likely the inner chemicals). Although the most significant part of the removable chemicals was removed during the first washing, chemicals were sometimes also removed during the second and the third washing. This might be explained by the fact that washing was carried out without agitation and was therefore less efficient, contrary to studies using shampoo.

7.3.2 Artificial Contamination

Although the efficiency of different decontamination procedures can be assessed on native hair, interpretation of the results remains limited by the fact that the respective part of external contamination and biologically incorporated compounds is unknown. For this purpose, producing artificially contaminated samples can be useful for testing the ability of the washing to remove externally deposited chemicals. The most classical contamination procedure is the utilization of a soaking aqueous solution containing the compounds of interest [57,59,70,71]. Even if this artificial contamination is effective, it is neither representative of the external deposition of compounds on hair surface by biological fluids nor of the environmental contamination occurring mainly through airborne dust particle. Moreover, an important limitation associated with hair dipping in aqueous solutions is the possible irreversible penetration of the compounds within the hair shaft [59]. In the last decade, more "realistic" artificial contamination procedures were proposed in the field of clinical and forensic toxicology: hair incubation with spiked blood to demonstrate postmortem external contamination [72], hair briefly dipped into authentic urine from a methamphetamine user to investigate the contamination/decontamination of pubic hair [73], hair coating with drugs followed by sweat conditioning using synthetic sweat to study drug contamination/decontamination in the analysis of human hair [59], and subjects' hair own contamination using their hands powdered with a drug mixture [74]. Although such procedures might be relevant to clinical or forensic contexts, they are not representative of hair surface contamination by OPs present in air or dust and specific approaches have to be developed for the latter purpose.

The importance of the procedure of artificial contamination of the hair sample applied to assess decontamination efficiency was recently investigated [75].

As atmospheric OPs generally present low volatility and are rather adsorbed on airborne particles, hair surface contamination may occur as contaminated particle deposit and/or dry transfer from particle to hair. Artificial contamination of hair by dipping into aqueous solution containing analytes may therefore not be the most representative artificial contamination procedure.

In order to simulate realistic contamination of hair surface by OPs, Duca et al. [75] recently tested different procedures: hair contact under agitation with solid particles (silica or cellulose) previously contaminated with a pesticide cocktail, and the more classical dipping into pesticide aqueous solution. With a view to cover a wide range of physicochemical properties, the pesticide cocktail consisted of several compounds from different chemical classes. After artificial contamination, different washing solvents (organic and aqueous) were tested for the decontamination of hair samples. The results demonstrated that the efficiency of the washing was depending on both the pesticide and the contamination procedure. In particular, the transfer of pesticides from aqueous solution to hair was about 10 times higher than from solid particles. In addition, the decontamination was less efficient after wet contamination (solution) than after dry contamination from particles. Since no unique solvent (organic or aqueous) was able to remove all the pesticides deposited on hair surface, the best procedure was found to combine successive washing with sodium dodecyl sulfate and methanol subsequently. This procedure allowed the removal of about 100% of the pesticides deposited

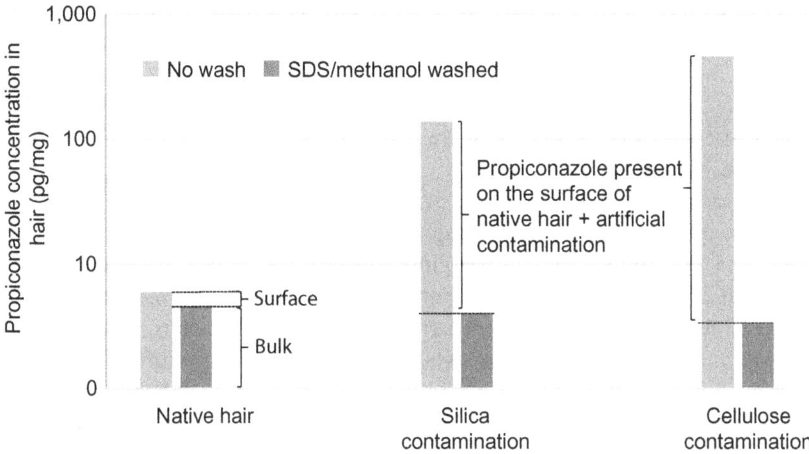

FIGURE 7.6 Propiconazole detected before and after decontamination with SDS/methanol, in native hair and in hair after contact with contaminated silica and cellulose particles. *Adapted from Duca et al. [75].*

from contaminated cellulose particles. In addition, the same procedure applied to native hair samples (without artificial contamination) before multiclass pesticide analysis removed between 5% and 100% of the pesticides initially present in hair. Figure 7.6 presents the example of propiconazole that was detailed in Duca et al. [75].

The difference observed in native hair analyzed directly or after decontamination with sodium dodecyl sulfate (SDS)/methanol demonstrates that a part of the propiconazole initially present in hair has been removed by washing. The part of propiconazole that was removed was assumed to be located on hair surface. On the contrary, the part that was not removed by washing was likely to be located in the bulk structure of hair, corresponding therefore to biological incorporation. After artificial deposition on hair surface by contact with contaminated particles (silica or cellulose), the amount of pesticide detected was significantly increased, but returned to the same value as native hair after washing. These observations tended to support that the part of chemicals that was not removed by washing was actually located in the bulk structure of the hair and that the part removed by washing in native hair was rather located on the surface.

7.3.3 The Specificity of OPs

As long as hair scales are synthetized along with inner compartments (cortex and medulla), hair surface must also contain chemicals that have been incorporated from blood into living cells, even though no external deposition occurred. In that sense, one cannot exclude that any decontamination procedure may remove a part of biologically incorporated compounds. On the other hand, although decontamination experiments demonstrate that a part of the target chemicals is removed by washing while another part remains unaffected, it has to be acknowledged that none of these allows conclusive demonstration that externally deposited chemicals were completely removed.

This results in the existence of a zone of uncertainty that can vary between different studies applying different decontamination procedures. How this uncertainty will impact interpretation of results depends on the objective of hair analysis. On the one hand, in studies aiming at assessing subjects' level of exposure from chemical concentration in hair, as with urine or blood, the use of different decontamination procedures in different studies might limit the possibility of interstudies comparison. Although no such experimental work was carried out yet, assessing human's level of exposure to OP from hair analysis is definitely one of next steps in the field. The lack of consensus of decontamination procedure will therefore be a gap to be filled. In that regard, hair testing for OP would be very similar to the fields of forensic toxicology, were procedure normalization has already been pointed out as mandatory in order to set up common cutoffs for chemicals concentration in hair (e.g., Society of Hair Testing, SoHT, cutoffs on ethyl glucuronide in hair for abstinence or chronic alcohol abuse).

On the other hand, the relative uncertainty on the decontamination procedure efficiency has not to be considered a limitation to the use of hair as a biomarker of exposure in case–control studies. As long as all samples are treated the same and washing procedure is repeatable, bias in the decontamination will be similar for all samples. Comparison in the exposure of different subpopulations (e.g., different living areas, occupationally exposed vs controls, age groups) will therefore remain reliable [18,76–78].

In a general manner, if a part of the externally deposited molecules becomes "unremovable," it can be considered as representative of the history of exposure of the subject. Moreover, the removal of a limited part of internal molecules should not dramatically influence the final result, as far as it is repeatable. In this regard, considering the "easily removable chemicals" (ERCs) instead of the strictly externally deposited chemicals would be more relevant.

Chemicals deposited on hair surface are not problematic themselves, keeping in mind that personal exposure may also be assessed using techniques that mainly provide information on pollutants present in air and/or dust (e.g., wearable samplers or wiping hand with swabs) [79–81]. The risk of misinterpretation lies in that chemicals deposited on hair surface are likely to be easily removed during subjects' self-washing, which may induce significant variability in chemical concentration depending on the time elapsed between hair sampling and subjects' last washing if redeposition has occurred. In that regard, hair washing before analysis appears to be a necessary precaution. On the basis of the considerations presented above, some criteria may be proposed to ensure that a washing procedure is suitable to remove the ERCs:

1. The hair total ERC is significantly decreased by the hair washing applied once;
2. A steady state is reached, that is, no more chemicals (or a significantly lower amount) are removed by additional washing procedures;
3. The non-ERC that is extracted after hair destructuration (pulverization or digestion) is not affected by washing procedure.

Another approach to avoid misinterpretation due to external contamination is to analyze chemical metabolites that are produced in the body and are unlikely to be deposited on hair surface from air or dust. In that regard, Schummer et al. [14] investigated human exposure to PAHs by determining the concentration of monohydroxy-PAHs (PAH metabolites) in hair. As expected, the concentration of metabolites was not significantly decreased by washing performed before hair analysis. The approach is of course limited to metabolites which are unlikely to be present in the environment due to their use as active compounds themselves (e.g., dieldrin is both produced from the metabolization of aldrin and directly used as an active compound) or due to environmental degradation (e.g., DDE is produced from both metabolization and environmental degradation of DDT).

REFERENCES

[1] Dauberschmidt C, Wennig R. Organochlorine pollutants in human hair. J Anal Toxicol 1998;22:610–11.

[2] Neuber K, Merkel G, Randow FFE. Indoor air pollution by lindane and DDT indicated by head hair samples of children. Toxicol Lett 1999;107:189–92.

[3] Schramm KW, Kuettner T, Weber S, Lützke K. Dioxin hair analysis as monitoring pool. Chemosphere 1992;24:351–8.

[4] Zupancic-Kralj L, Jan J, Marsel J. Assessment of polychlorobiphenyls in human/poultry fat and hair/plumage from a contaminated area. Chemosphere 1992;25:1861–7.

[5] Luksemburg WJ, Mitzel RS, Hedin JM, Silverbush BB, Wong AS, Zhou HD. Polychlorinated and dibenzofurans (PCDDs/PCDFs) levels in environmental and human hair samples around a pentachlorophenol plan in China. Organohalog Compd 1997;32:38–40.

[6] Cirimele V, Kintz P, Ludes B. Assessment of pesticide exposure by hair analysis. Acta Clin Belg 1999;(S1):59–63.

[7] Tirler W, Voto G, Donega M. PCDD/F, PCB and hexachlorobenzene levels in hair. Organohalog Compd 2001;52:290–2.

[8] Wu WZ, Xu Y, Schramm KW, Kettrup A. Persistence of polychlorinated dibenzo-p-dioxins and dibenzofurans (PCDD/F) in Ya-Er Lake area, China. Environ Int 2001;26:323–6.

[9] Covaci A, Schepens P. Chromatographic aspects of the analysis of selected persistent organochlorine pollutants in human hair. Chromatographia 2001;53:S366–71.

[10] Ohgami T, Nonaka S, Murayama F, Yamashita K, Irifune H, Watanabe M, et al. A comparative study on polychlorinated biphenyls (PCB) and polychlorinated quaterphenyls (PCQ) concentrations in subcutaneous fat tissue blood and hair of patients with Yusho normal control in Nagasaki prefecture. Fukuoka Acta Med 1991;80:307–12.

[11] Posecion Jr. NC, Ostrea Jr. EM, Bielawski Jr. D, Corrion M, Seagraves J, Jin Y. Detection of exposure to environmental pesticides during pregnancy by the analysis of maternal hair using GC-MS. Chromatographia 2006;2006:681–7.

[12] Tsatsakis AM, Tzatzarakis MN, Tutudaki M. Pesticide levels in head hair samples of Cretan population as an indicator of present and past exposure. Forensic Sci Int 2008;176:67–71.

[13] Toriba A, Kuramae Y, Chetiyanukornkul T, Kizu R, Makino T, Nakazawa H, et al. Quantification of polycyclic aromatic hydrocarbons (PAHs) in human hair by HPLC with fluorescence detection: a biological monitoring method to evaluate the exposure to PAHs. Biomed Chromatogr 2003;17:126–32.

[14] Schummer C, Appenzeller BMR, Millet M, Wennig R. Determination of hydroxylated metabolites of polycyclic aromatic hydrocarbons in human hair by gas chromatography-negative chemical ionization mass spectrometry. J Chromatogr A 2009;1216:6012–19.

[15] Zhao G, Wang Z, Dong MH, Rao K, Luo J, Wang D, et al. PBBs, PBDEs, and PCBs levels in hair of residents around e-waste disassembly sites in Zhejiang Province, China, and their potential sources. Sci Total Environ 2008;397:46–57.

[16] Wen S, Yang FX, Gong Y, Zhang XL, Hui Y, Li JG, et al. Elevated levels of urinary 8-hydroxy-2′-deoxyguanosine in male electrical and electronic equipment dismantling workers exposed to high concentrations of polychlorinated dibenzo-p-dioxins and dibenzofurans, polybrominated diphenyl ethers, and polychlorinated biphenyls. Environ Sci Technol 2008;42:4202–7.

[17] Nakao T, Aozasa O, Ohta S, Miyata H. Survey of human exposure to PCDDs, PCDFs, and coplanar PCBs using hair as an indicator. Arch Environ Contam Toxicol 2005;49:124–30.

[18] Ma J, Cheng J, Wang W, Kunisue T, Wu M, Kannan K. Elevated concentrations of polychlorinated dibenzo-p-dioxins and polychlorinated dibenzofurans and polybrominated diphenyl ethers in hair from workers at an electronic waste recycling facility in Eastern China. J Hazard Mater 2011;186:1966–71.

[19] Leung AOW, Chan JKY, Xing GH, Xu Y, Wu SC, Wong CKC, et al. Body burden of polybrominated diphenyl ethers in childbearing-aged women at an intensive electronic-waste recycling site in China. Environ Sci Pollut Res 2010;17:1300–13.

[20] Zhang H, Chai Z, Sun H. Human hair as a potential biomonitor for assessing persistent organic pollutants. Environ Int 2007;33:685–93.

[21] Lacassie E, Marquet P, Gaulier JM, Dreyfuss MF, Lachâtre G. Sensitive and specific multiresidue methods for the determination of pesticides of various classes in clinical and forensic toxicology. Forensic Sci Int 2001;121:116–25.

[22] Barr DB, Barr JR, Maggio VL, Whitehead Jr. RD, Sadowski MA, Whyatt RM, et al. A multianalyte method for the quantification of contemporary pesticides in human serum and plasma using a high-resolution mass spectrometry. J Chromatogr B 2002;778:99–111.

[23] Berman T, Hochner-Celnikier D, Barr DB, Needham LL, Amitai Y, Wormser U, et al. Pesticide exposure among pregnant women in Jerusalem, Israel: results of a pilot study. Environ Int 2011;37:198–203.

[24] Barr DB, Allen R, Olsson AO, Bravo R, Caltabiano LM, Montesano A, et al. Concentrations of selective metabolites of organophosphorus pesticides in the United States population. Environ Res 2005;99:314–26.

[25] Baker SE, Olsson AO, Barr DB. Isotope dilution high-performance liquid chromatography–tandem mass spectrometry method for quantifying urinary metabolites of synthetic pyrethroid insecticides. Arch Environ Contam Toxicol 2004;46:281–8.

[26] Conka K, Drobna B, Petrik KJ. Simple solid-phase extraction method for determination of polychlorinated biphenyls and selected organochlorine pesticides in human serum. J Chromatogr A 2005;1084:33–8.

[27] Zhao G, Xu Y, Li W, Han G, Ling B. Prenatal exposures to persistent organic pollutants as measured in cord blood and meconium from three localities of Zhejiang, China. Sci Total Environ 2007;377:179–91.

[28] Cok I, Yelken C, Durmaz E, Uner M, Sever B, Satir F. Polychlorinated biphenyl and organochlorine pesticide levels in human breast milk from the Mediterranean city Antalya, Turkey. Bull Environ Contam Toxicol 2011;86:423–7.

[29] Damgaard IN, Skakkebaek NE, Toppari J, Virtanen HE, Shen H, Schramm KW, et al. Persistent pesticides in human breast milk and cryptorchidism. Environ Health Perspect 2006;114:1133–8.

[30] CDC CfDCaP. Fourth national report on human exposure to environmental chemicals, Department of Health and Human Services Centers for Disease Control and Prevention; 2009.

[31] Harkins DK, Susten A. Hair analysis: exploring the state of the science. Environ Health Perspect 2003;111:576–8.

[32] Wennig R. Potential problems with the interpretation of hair analysis results. Forensic Sci Int 2000;107:5–12.

[33] Pragst F, Balikova MA. State of the art in hair analysis for detection of drug and alcohol abuse. Clin Chim Acta 2006;370:17–49.

[34] Nakahara Y. Hair analysis for abused and therapeutic drugs. J Chromatogr B 1999;733:161–80.

[35] Boumba VA, Ziavrou KS, Vougiouklakis T. Hair as a biological indicator of drug use, drug abuse or chronic exposure to environmental toxicants. Int J Toxicol 2006;25:143–63.

[36] Solomon GM, Huddle AM. Low levels of persistent organic pollutants raise concerns for future generations. J Epidemiol Community Health 2006;56:826–7.

[37] Jaga K, Dharmani C. Global surveillance of DDT and DDE levels in human tissues. Int J Occup Environ Health 2003;16:7–20.

[37a] Appenzeller BMR, Tsatsakis AM. Hair analysis for biomonitoring of environmental and occupational exposure to organic pollutants: state of the art, critical review and future needs. Toxicol Lett 2012;210:119–40.

[38] Salquebre G, Schummer C, Briand O, Millet M, Appenzeller BMR. Multi-residue pesticide analysis in hair of non-occupationally exposed persons by GC-MS/MS with liquid injection and solid phase microextraction (SPME) sixteenth meeting of the Society of Hair Testing 21–26 March, Chamonix, France; 2011.

[39] Schummer C, Salquebre G, Briand O, Millet M, Appenzeller BMR. Determination of farm workers' exposure to pesticides by hair analysis. Toxicol Lett 2012;210:203–10.

[40] Salquebre G, Schummer C, Millet M, Briand O, Appenzeller BMR. Multi-class pesticide analysis in human hair by gas chromatography tandem (triple quadrupole) mass spectrometry with solid phase microextraction and liquid injection. Anal Chim Acta 2012;710:65–74.

[41] Cuong L, Evgen'ev M, Gumerov F. Determination of pesticides in the hair of Vietnamese by means of supercritical CO2 extraction and GC-MS analysis. J Supercrit Fluids 2012;61:86–91.

[42] Ostrea Jr. EM, Bielawski DM, Posecion Jr. NC, Corrion M, Villanueva-Uy E, Jin Y, et al. A comparison of infant hair, cord blood and meconium analysis to detect fetal exposure to environmental pesticides. Environ Res 2008;106:277–83.

[43] Tsatsakis AM, Tzatzarakis MN, Tutudaki M, Babatsikou F, Alegakis AK, Koutis C. Assessment of levels of organochlorine pesticides and their metabolites in the hair of a Greek rural human population. Hum Exp Toxicol 2008;27:933–40.

[44] Covaci A, Tutudaki M, Tsatsakis AM, Schepens P. Hair analysis: another approach for the assessment of human exposure to selected persistent organochlorine pollutants. Chemosphere 2002;46:413–18.

[45] Wielgomas B, Czarnowski W, Jansen E. Persistent organochlorine contaminants in hair samples of Northern Poland population, 1968–2009. Chemosphere 2012;89:975–81.

[46] Covaci A, Hura C, Gheorghe A, Neels H, Dirtu AC. Organochlorine contaminants in hair of adolescents from Iassy, Romania. Chemosphere 2008;72:16–20.

[47] Schummer C, Mothiron E, Appenzeller BMR, Rizet A-L, Wennig R, Millet M. Temporal variations of concentrations of currently used pesticides in the atmosphere of Strasbourg, France. Environ Pollut 2010;158:576–84.

[48] Schummer C, Tuduri L, Briand O, Appenzeller BMR, Millet M. Application of XAD-2 resin-based passive samplers and SPME-GC-MS/MS analysis for the monitoring of spatial and temporal variations of atmospheric pesticides in Luxembourg. Environ Pollut 2012;170:88–94.

[49] Schummer C, Appenzeller BMR, Millet M. Monitoring of polycyclic aromatic hydrocarbons (PAHs) in the atmosphere of southern Luxembourg using XAD-2 resin-based passive samplers. Environ Sci Pollut Res 2014;21:2098–107.

[50] Raeppel C, Fabritius M, Nief M, Appenzeller BMR, Briand O, Tuduri L, et al. Analysis of airborne pesticides from different chemical classes adsorbed on Radiello® Tenax® passive tubes by thermal-desorption-GC/MS. Environ Sci Pollut Res 2014; http://dx.doi.org/10.1007/s11356-014-3534-z.

[51] Wang W, Wu F, Zheng J, Wong MH. Risk assessments of PAHs and Hg exposure via settled house dust and street dust, linking with their correlations in human hair. J Hazard Mater 2013;263:627–37.

[52] Krol S, Namiesnik J, Zabiegala B. Occurrence and levels of polybrominated diphenyl ethers (PBDEs) in house dust and hair samples from Northern Poland; an assessment of human exposure. Chemosphere 2014;110:91–6.

[53] Tsanaclis L, Wicks JFC. Differentiation between drug use and environmental contamination when testing for drugs in hair. Forensic Sci Int 2008;176:19–22.

[54] Stout PR, Ropero-Miller JD, Baylor MR, Mitchell JM. External contamination of hair with cocaine: evaluation of external cocaine contamination and development of performance-testing materials. J Anal Toxicol 2006;30:490–500.

[55] Kintz P, Mangin P. Opiate concentrations in human head, axillary, and pubic hair. J Forensic Sci 1993;38:657–62.

[56] Montagna M, Stramesi C, Vignali C, Groppi A, Polettini A. Simultaneous hair testing for opiates, cocaine, and metabolites by GC-MS: a survey of applicants for driving licenses with a history of drug use. Forensic Sci Int 2000;107:157–67.

[57] Schaffer MI, Wang WL, Irving J. An evaluation of two wash procedures for the differentiation of external contamination versus ingestion in the analysis of human hair samples for cocaine. J Anal Toxicol 2002;26:485–8.

[58] Balikova MA, Habrdova V. Hair analysis for opiates: evaluation of washing and incubation procedures. J Chromatogr B Analyt Technol Biomed Life Sci 2003;789:93–100.

[59] Cairns T, Hill V, Schaffer M, Thistle W. Removing and identifying drug contamination in the analysis of human hair. Forensic Sci Int 2004;145:97–108.

[60] Margariti MG, Tsatsakis AM. Analysis of dialkyl phosphate metabolites in hair using gas chromatography-mass spectrometry: a biomarker of chronic exposure to organophosphate pesticides. Biomarkers 2009;14:137–47.

[61] Luksemburg WJ, Mitzel RS, Peterson RG, Hedin JM, Maier MM, Schuld M, et al. Polychlorinated dibenzodioxins and dibenzofurans (PCDDs/PCDFs) levels in environmental and human hair samples around an electronic waste processing site in Guiyu, Guangdong province, China. Organohalog Compd 2002;55:347–9.

[62] Zheng J, Chen KH, Luo XJ, Yan X, He CT, Yu YJ, et al. Polybrominated diphenyl ethers (PBDEs) in paired human hair and serum from e-waste recycling workers: source apportionment of hair PBDEs and relationship between hair and serum. Environ Sci Technol 2014;48:791–6.

[63] Tang ZW, Huang QF, Cheng JL, Yang YF, Yang J, Guo W, et al. Polybrominated diphenyl ethers in soils, sediments, and human hair in a plastic waste recycling area: a neglected heavily polluted area. Environ Sci Technol 2014;48:1508–16.

[64] Nakao T, Aozasa O, Ohta S, Miyata H. Assessment of human exposure to PCDDs, PCDFs and co-PCBs using hair as human pollution indicator sample I: development of analytical method for human hair and evaluation for exposure assessment. Chemosphere 2002;48:885–96.

[65] Altshul L, Covaci A, Hauser R. The relationship between levels of PCBs and pesticides in human hair and blood: preliminary results. Environ Health Perspect 2004;112:1193–9.

[66] Tirler W, Voto G, Donega M. PCDD/F and PCB levels in human hair. Organohalog Compd 2006;68:1659–61.

[67] Tadeo JL, Sánchez-Brunete C, Miguel E. Determination of polybrominated diphenyl ethers in human hair by gas chromatography-mass spectrometry. Talanta 2009;78:138–43.

[68] Kang Y, Wang HS, Cheung KC, Wong MH. Polybrominated diphenyl ethers (PBDEs) in indoor dust and human hair. Atmos Environ 2011;45:2386–93.

[69] Ostrea Jr. EM, Bielawski DM, Posecion Jr. NC, Corrion M, Villanueva-Uy E, Bernardo RC, et al. Combined analysis of prenatal (maternal hair and blood) and neonatal (infant hair, cord blood and meconium) matrices to detect fetal exposure to environmental pesticides. Environ Res 2009;109:116–22.

[69a] Ostrea EM, Villanueva-Uy E, Bielawski DM, Posecion NCJ, Corrion ML, Jin Y, et al. Maternal hair - an appropriate matrix for detecting maternal exposure to pesticides during pregnancy. Environ Res 2006;101:312–22.

[70] Blank DL, Kidwell DA. Decontamination procedures for drugs of abuse in hair—are they sufficient? Forensic Sci Int 1995;70:13–38.

[71] Schaffer M, Hill V, Cairns T. Hair analysis for cocaine: the requirement for effective wash procedures and effects of drug concentration and hair porosity in contamination and decontamination. J Anal Toxicol 2005;29:319–26.

[72] Kintz P. Segmental hair analysis can demonstrate external contamination in postmortem cases. Forensic Sci Int 2012;215:73–6.

[73] Lee S, Han E, In S, Choi H, Chung H, Chung KH. Analysis of pubic hair as an alternative specimen to scalp hair: a contamination issue. Forensic Sci Int 2011;206:19–21.

[74] Romano G, Barbera N, Spadaro G, Valenti V. Determination of drugs of abuse in hair: evaluation of external heroin contamination and risk of false positives. Forensic Sci Int 2003;131:98–102.

[75] Duca RD, Hardy E, Salquèbre G, Appenzeller BMR. Hair decontamination procedure prior to multi-class pesticide analysis. Drug Test Anal 2014;6:55–66.

[76] Behrooz RD, Barghi M, Bahramifar N, Esmaili-Sari A. Organochlorine contaminants in the hair of Iranian pregnant women. Chemosphere 2012;86:235–41.

[77] Mercadante R, Polledri E, Giavini E, Menegola E, Bertazzi PA, Fustinoni S. Terbuthylazine in hair as a biomarker of exposure. Toxicol Lett 2012;210:169–73.

[78] Zheng J, Yan X, Chen S-J, Peng XW, Hu GC, Chen KH, et al. Polychlorinated biphenyls in human hair at an e-waste site in China: composition profiles and chiral signatures in comparison to dust. Environ Int 2013;54:128–33.

[79] Scherer G, Frank S, Riedel K, Meger-Kossien I, Renner T. Biomonitoring of exposure to polycyclic aromatic hydrocarbons of nonoccupationally exposed persons. Cancer Epidemiol Biomarkers Prev 2000;9:373–80.

[80] Bouvier G, Blanchard O, Momas I, Seta N. Pesticide exposure of non-occupationally exposed subjects compared to some occupational exposure: a French pilot study. Sci Total Environ 2006;366:74–91.

[81] Besaratinia A, Maas LM, Brouwer EMC, Moonen EJC, De Kok TMCM, Wesseling GJ, et al. A molecular dosimetry approach to assess human exposure to environmental tobacco smoke in pubs. Carcinogenesis 2002;23:1171–6.

Chapter 8

Workplace Drug Testing

Lolita Tsanaclis and John Wicks
Cansford Laboratories, The Cardiff Medicentre; Heath Park, Cardiff, UK and Laboratório, ChromaTox Ltda, São Paulo-SP, Brazil

8.1 INTRODUCTION

Drug testing in the workplace started in the United States in the 1980s following a fatal air crash where crew members tested positive for marijuana and one decade later more than 90% of companies with over 5,000 employees introduced some type of testing program using urine drug testing [1]. Previous studies in the United States showed that workplace drug testing was linked to reduced drug use but any benefit with respect to increased productivity or decreased accidents is more difficult to evaluate [2–5]. Nowadays, with large variations between the different American states, workplace drug testing programs regularly screen millions of employees in aviation, trucking, railroad operation, mass transit, pipeline operation, and other transportation-related industries in preemployment, random, post-accident, reasonable suspicion, return-to-duty, and follow-up drug testing [6,7].

An extensive review to assess the effects of interventions to prevent injuries in construction workers concluded that there is low quality evidence that company-oriented safety interventions such as a multifaceted safety campaign and a multifaceted drug workplace program can reduce nonfatal injuries among construction workers [8]. However, a study performed on work-related fatalities found that about one in five workers killed on the job tested positive for alcohol or other drugs [9].

Substance abuse, particularly of alcohol, is not only a concern in heavy-duty and high-risk activities; it seems to be more prevalent among health workers, particularly physicians than among the general population. Compared to other specialties, anesthesiologists are most vulnerable to controlled substance abuse, basically because of occupational issues of excessive work hours and easy access to drugs. The most abused substances are opioids (fentanyl, in particular), propofol, and inhalation anesthetics [10–12].

It is now recognized that workplace drug testing is a complex interdisciplinary science involving not only the people being tested, but also laboratory personnel, sample collectors, human resources, occupational physicians, risk assessors, compliance officers, lawyers, drug counseling, and policy makers for each separate country [13]. While the majority of employees do not use alcohol and drugs at work, their exposure to a permissive workplace substance use climate is negatively related to perceived safety at work, positively related to work strain, and negatively related to employee morale [14]. However, it is also alleged that on one hand the increase in drug testing in the workplace is likely to increase by employers attempting to protect workers from unhealthy drug effects, improve productivity, and control escalating health care costs, while on the other hand it is argued that such testing could pose a significant threat to workers' privacy, autonomy, and dignity [15,16].

A publication by *Quest* showed positivity rates ranged from 7.8% in 2008 to 5.6% in 2012, from 6.3% to 4.7% and 9.6% to 6.3%, for preemployment and in-employment tests, respectively, over the same period [17]. Studies showed a significant relationship between drug usage and safety performance [18]. Human resource professionals were asked about their organization's drug testing programs and reported that after the implementation of a drug testing program there was an increase in employee productivity and decrease in absenteeism with a net employee turnover [19].

8.2 IDEAL MATRIX IN THE WORKPLACE SCENARIO

The five common samples widely accepted for the detection and monitoring of drug use are blood, sweat, urine, saliva, and hair. They have different characteristics that mean they have different applications and as all have inherent and differing limitations in measuring the timing, duration, frequency, and intensity of drug use. The specimen of choice depends on the context and requirements of the testing [20,21].

In the workplace the choice of the sample to use in drug testing depends on the purpose of the testing. Where it is necessary to ensure that no employee is working under the influence of or has not used drugs in the preceding working hours, drug testing using urine and saliva (oral fluid), including blood samples in case of accidents, are more appropriate. Drug testing using either urine or oral fluid reflects drug use for a relatively short period before sample collection, up to 24 or 72 h depending on the drug. Urine is still the most commonly used matrix in the workplace setting, though oral fluid and hair testing are on the increase [19,22]. Hair as a sample for drug testing has advantages over urine. It is quick and easy to collect, generally not invasive, is easily shipped to the test facility, and can be stored at room temperature. The nature of the specimens allows a long-term historical window of the subject's drug intake (usually around 3 months), which is extremely useful for a hiring decision and certainly preferable to other matrices.

In 2003 the English Cricket board trialed hair testing after the death of Surrey batsman Tom Maynard on a railway line in London in 2012. Maynard, 23, was struck by a Tube train while drunk and high on drugs. The coroner urged cricket and other sports to introduce hair testing to determine long-term drug habits. The Professional Cricketers Association (PCA) and England and Wales Cricket Board instigated a pilot project involving all 18 first-class counties with the aim of discovering how widely cocaine and cannabis, among other drugs, were used by professionals. Cricketers were informed that the results of the pilot test would remain confidential, with counseling and treatment offered initially in the wake of a "positive." This is an example of hair drug testing being used as part of a whole well-being program with the focus on the employee's health (BBC, newspapers).

The current scientific research indicates that drugs deposit in hair by several methods [23,24]. These include transmission from the blood supply, through perspiration (sweat) and skin oil (sebum), and from external environment. The take-up of drugs by hair varies from person to person largely because of differences in the individual's metabolism and the quality of the drugs they use. Drugs and metabolites remain fixed and trapped in the hair indefinitely after they are incorporated. As the hair grows, the drugs maintain their position and grow out with the hair, a feature that extends the detection period. As a result, drug testing with hair samples provides a much longer window of detection, typically 1–3 months.

Abstention from use for 3 days will often produce a negative urine test result and for 3 months can provide a negative hair test on a hair sample of 3 cm. Indeed a major advantage of hair analysis is that it is much more able to monitor drug abstinence. Hence the likelihood of a false-negative test using hair is very much less than with a urine test. A negative hair test is a substantially stronger indicator of a nondrug user than a negative urine test.

Figure 8.1 illustrates the overall periods of detection. While urine or oral fluid drug levels show transient changes in levels of drugs in the body over a

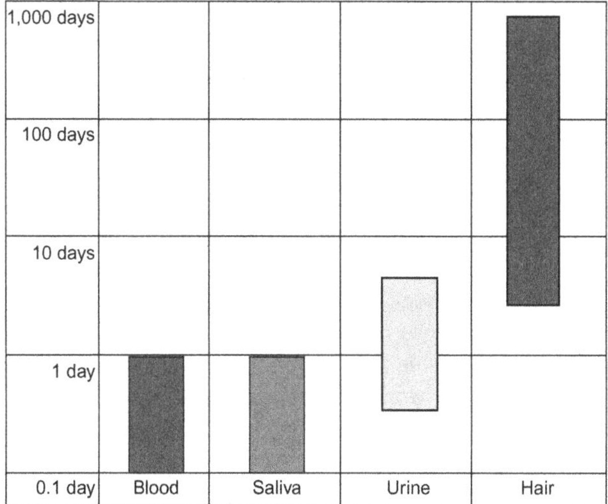

FIGURE 8.1 Typical periods of detection for each matrix. *Adapted from Caplan and Goldberger [25].*

short period (hours), the analysis of hair samples provides an integrated picture of drug use or abstinence over a much more extended time frame (weeks or months).

In the early days of the detection of abuse of drugs, urine was the only choice because the amount of sample that can be obtained and the concentrations of drugs and metabolites found in urine are relatively large. In those early days the technology of detection was relatively less sensitive than that available today. With advances in the technology the smaller quantities of drugs and metabolites in hair samples became detectable and the use and application of such samples increased because of the capability of evaluating the history of drug use [26–28].

The improvement in technology has led to a dramatic increase in the use of hair samples in the detection of drug use in a wide variety of sectors such as clinics, family law firms, the police, and various organizations for workplace testing where the risks associated with drug use are considered too important to use a less sensitive form of testing with its advantages and limitations in relation to the other matrices [21,23,25,29–34].

For example, in the workplace sector when tests are carried out for "preemployment" the approach is different from a "for cause" testing. Some industries need to know whether a candidate is a drug user or not before offering employment. In such cases, preemployment testing in a simple hair test covering a period of 3 months is ideal because the candidate will have to abstain from drugs for 3 months for the test to be negative. Most industries, however, still use urine testing as a requisite for preemployment because of the higher cost associated with a hair test, although one hair sample is equivalent to approximately 18 urine tests.

Other industries require knowing whether the person on duty on that day has drugs in their system or are under the influence of drugs that may impair their performance or put others at risk. These are usually approached by random drug tests. An on-site oral fluid test is ideal in such cases because when drugs are present in oral fluid samples it reflects use in the last 24 h relevant to the assessment of impairment. Urine or oral fluid tests are more appropriate than hair in a "for cause" situation; in a postincident situation testing urine or oral fluid is more suitable to assess impairment. Nevertheless, a hair test will be useful to indicate whether the individual that showed a positive urine or oral fluid test is a regular drug user or just a one-off episode.

Depending on the purpose of the test, the different matrices can be complementary [21,25,35]. Table 8.1 summarizes the appropriateness of the different matrices from the workplace perspective.

The understanding of drug testing in hair has grown greatly with the publication of many comprehensive scientific reviews over the years listing and highlighting the advantages and disadvantages [21,23,25,26,29–31,33,34, 36–38].

TABLE 8.1 Which Matrix from the Workplace Perspective

	Urine	Oral Fluid	Hair
Preemployment	Test will be negative after ~3 to 5 days of abstention from using drugs		Most appropriate
Random drug test	Most appropriate		Useful in conjunction with a positive urine or oral fluid test
Postaccident or incidents	Test will be negative after ~3 days of abstention from using drugs		Useful in conjunction with a positive urine or oral fluid test
Return to work monitoring			Most appropriate

Table 8.1 illustrates the variety of applications of drug testing through the use of different biological matrices in the workplace, highlighting where hair testing gives an advantage.

8.2.1 Preemployment

The risks to an employer of employee drug taking vary from industry to industry. For a manual job in the construction industry it may be sufficient to confirm that the worker has no drugs in their system as they come onto the building site. A short-term urine or oral fluid test at the hiring interview will give assurance that the worker is capable of being drug free. In other situations a lifestyle assessment is of value. So, for example, in the case of a heavy goods driver a hair test would give an indication of whether the driver was in the habit of using stimulants while on the job. He could arrive at interview clear of drugs but his hair sample would reveal usage over the previous 1–3 months. Amphetamine use is a particular risk for long-distance drivers that an employer should know about. A police officer moving to a position where they might be issued with a firearm should be tested more rigorously than just an on-site urine test. In those cases of preemployment testing, where it is important to ensure that the candidate is a nondrug user, a 3 cm hair sample covering a period of 3 months is ideal because the candidate will have to abstain from drugs for at least 3 months for the test to be negative. Most industries, however, still use a urine drug screen test as a requirement for employment. This can only provide a 3- to 5-day window of detection at maximum and if the candidate is prewarned of the test it is probably useless.

There is the perception that a hair test is expensive. However, this perception is illusory, for a sample of hair covering a period of approximately 3 months is equivalent to 18 urine tests in the breadth of coverage. It is for the employer to decide if they really need to know whether the prospective employee is or is not using drugs or whether the drug test is just a deterrent or a regulatory requirement. A hair test will not miss drug taking that a single urine test might, and the candidate could prepare for the test by abstaining for 3–5 days.

8.2.2 Random

Random urine and oral fluid testing is principally a deterrent to drug use.

In industries and institutions that need to know whether a person is responsible and fit to perform specific tasks, it is possible to test urine or oral fluid to ensure that no drugs are present in the person's system that may impair performance or put others at risk. These tests are usually labeled "random." These tests do not accurately reflect or predict fitness to work. They only detect recent past use. They may be the best indicator available but it remains true that no drug test actually measures "fitness to work." The use of saliva (oral fluid) is ideal in such cases. When drugs are present in saliva samples actually reflect the use of drugs in the last 24 h, very relevant in assessing impairment. The main purpose of random urine and oral fluid testing is as a deterrent to drug taking by an employee.

Hair testing for drug use can be applied randomly and as an unannounced test, but this is hardly necessary as a hair test detects historical use of drugs. However, as a deterrent to drug use it must rank higher than urine or oral fluid testing as it will detect use at any time within the time frame determined by the hair sample length. As such it is the ideal sample for a "zero tolerance" policy.

8.2.3 Postaccident or Incident

In the postaccident situation, saliva (oral fluid) testing is best suited to assess very recent use of drugs when a blood test is not possible. However, a hair test may be useful to enable verification that the individual who had a positive urine test or oral fluid test is actually a regular user of drugs or if the positive result reflects only a single episode of use.

8.2.4 Return to Work

In many ways a hair test is the ideal "return to work" drug test. It is the only test that is capable of an overview of what has been taken, or not, over an extended period of time. While regular urine or oral fluid tests can give assurance that the individual is keeping to the clinical contract and is not taking substances both tests are susceptible to evasion. Taking an example from our personal experience of testing for a child custody case, the mother was instructed to attend regularly for random urine tests to demonstrate that she was not using drugs. At the end of the period of treatment and counseling the mother was tested using a hair sample. The hair sample revealed amphetamine at a concentration of 14 ng/mg. The explanation given was that she had a party with her friends every 2 weeks. Because she knew when the urine tests were to be done and she understood how fast amphetamine cleared from her urine it was simple to avoid detection.

8.3 EVIDENCE THAT A HAIR DRUG TEST DETECTS MORE USERS THAN URINE

Baumgartner's paper in 1989 was seminal in demonstrating the increased detection of drug use by hair testing relative to urine testing using radioimmunoassay (RIA) [26]. In 1997 a study conducted by Cook et al. [38] compared drug use assessments using self-reports, urinalysis, and hair testing. It was found that the addition of the chemical testing methods with matrix improved detection over self-reports alone.

This was confirmed at the beginning of the new millennium when urinalysis produced drug use rates that were the lowest of the assessment measures and were typically a fraction of the rates produced by hair analysis, basically as a result of improved technology and laboratory practices [20,25,39–41].

With the dramatic improvement in the sensitivity and specificity of analytical techniques with the introduction of mass spectrometry technologies in the 1990s it became possible to detect many drugs in hair in the picogram per milligram range and below. This, coupled with the extended detection windows, increased the usefulness of hair testing.

More recent studies showed that hair analysis consistently detected more drugs than did urine specimens for job applicants. In a study of a trucking sector in 2010, Mieczkowski [42] found that at preemployment 2% of applicants tested positive by urine testing, while 9% tested positive using hair analysis. In comparison, in-employment testing revealed less than 1% positive by urine testing compared with less than 3% detected by hair testing. In this interesting paper the author showed that the differential is especially pronounced when use of cocaine is examined but is reduced for marijuana, although hair still tends to detect more marijuana use than urine.

The results of a large study comparing the detection rates of illicit drugs in urine and hair samples from individuals trying to regain their revoked driver's license showed that positivity rates for methamphetamines, methylenedioxymethamphetamine (MDMA), cocaine, and monoacetylmorphine were 1.7-, 5.7-, 3.8-, and 9.3-fold higher in hair than in urine. In contrast, the detection rate for benzodiazepines was higher in urine than in hair (oxazepam 0.21% versus 0%, nordiazepam 0.10% versus 0.03%) [43], probably because the levels of benzodiazepines are much lower in hair and more difficult to detect.

8.4 PROCEDURES FOR DRUG TESTING USING HAIR

Drug testing is not restricted to the chemical detection of drugs in hair samples but it is a combination of a sequence of events of policies, processes, and procedures that require strict control of the laboratory chain of custody procedures [44]. Chain of custody of the test starts at the point of the hair sample collection with confirmation that the custody pack has not been tampered with or damaged in transit. If the results may be used in any court or employment tribunal it is vital that the donor be accurately identified by passport or other photo ID like a works ID badge or driving license. As the sample is not an "intimate" sample it is advisable to have a third party present to observe the collection. The orientation of the sample needs to be obvious, perhaps with the cut end marked with a dot or other device. There have been a number of cases where the wrong end has been identified by the laboratory leading to erroneous interpretation of results. This can be easily avoided by proper marking. The key is adequate training of the collection personnel. If the procedure goes wrong at collection then all subsequent steps will be compromised.

Samples and records need to be stored securely within the laboratory. All analytical work needs to be fully traceable, usually by a laboratory information management system (LIMS) using a unique code, and preferably all transfers and input tracked by bar code for each stage of analysis. In this way the progress and status of any sample can be identified at any time. The reports will include the unique chain of custody bar coded sample number, the specific substance(s), and substance grouping, together with the substance level detected.

All records generated from laboratory work are retained in sufficient detail to reconstruct analytical work, inspections, or audits conducted as required and all data are stored securely in order to maintain absolute confidentiality [44–46].

8.4.1 Collection

Sample collection is a critical factor for a successful drug testing with any matrix, requiring appropriate sample collection with secure chain of custody for the results to be valid and legally defensible [45,46].

The collection of head hair is relatively simple as there is no invasion of privacy and is more dignified than the urine collection, where collection is ideally supervised to ensure that no fraud occurred during the collection. There are numerous ways a urine sample can be adulterated that do not apply to hair.

Drinking a lot of water will produce dilute urine and drug concentrations will be reduced. Interferents to the assays may be added at the time of collection. It is possible to substitute clean drug-free urine in the place of the sample, either by hiding the sample or using a prosthetic device such as the Whizzinator.

Hair is preferably obtained from the posterior vertex of the head, an area that has been shown to produce less intraindividual variation in drug levels than other head regions.

While a hair test is more difficult to defeat than urine it is not impossible. Ways that have been used to attempt to beat the hair test include:

1. Shampoos: there are many shampoos available on the internet that claim to remove drugs from hair. None are reliable.
2. Bleaches and hair dyes: it is known that bleaching will reduce drug content of hair in significant amounts. Any cosmetic treatment that potentially damages the cuticle of the hair is likely to lead to the loss of drug from hair. Repeated hair bleaching and dyeing has been the cause of reporting false-negative results. Usually observation of the hair will suggest when this might have happened. However, they are very unlikely to remove all traces of drug except in extreme cases, and the laboratory needs to be alert to the possibility.
3. Substitution of a clean sample has been attempted:
   a. The use of a wig has been tried. On the first occasion a sample taken by a medical doctor was unsuccessful because the laboratory could not dissolve the sample in its extraction protocol. On a later occasion the collector lifted the wig of the donor's head; the donor left. So the collector needs to be alert to the possibility.
   b. The person being tested may be substituted either by another family member or even a twin. Fully identifying the donor is critical.
4. Removal of hair by shaving: if there is no hair then a hair test cannot be taken. Sites other than the head can be used. This usually does suffice, though in rare cases where the donor has been prewarned in some way he/she may remove all their hair. In such cases it might be necessary to warn the donor not to do so. In cases of complete baldness and alopecia nothing can be done and an alternative sample method has to be used.
5. Fraud: a hair sample is a very easy and noninvasive sample to take. And if there are issues or doubts about the chain of custody the whole process can be repeated a week or two later with the second sample covering the same period as the first. It is always advisable to have a witness to the sampling whenever possible. In cases where the result is not as expected or desired accusations of tampering by the collector may be made.
6. The quantity of hair necessary for analysis varies according to the laboratory requirements, but usually it is about 50–100 hair strands. In the workplace setting it is most commonly a 3 cm segmentation of the hair sample for an overview of approximately a 3-month window of detection.

The length of the sample to be analyzed is critical and it will depend on the objective of the test. For example, a 6 cm length of hair might provide a negative result but the analysis of 2 × 3 cm will be above detection, particularly for sporadic drug use [47,48].

The type/site of hair being tested can have a marked influence on the results of the testing.

8.4.2 Growth Rate

Hair grows at a reasonably constant rate of 1 cm/month, with a range of 0.7–1.5 cm/month, but the convention is to use 1 cm/month for time windows' estimates [49–53]. The analysis of adjacent sections of hair can give a historical picture of drug use over a long period time frame. It takes about 5–6 days for hair growing from the root to appear above the scalp and therefore be available for detection. The time window represented by the hair sample is only approximate, not only because of variation of hair growth rate between individuals but also because of collection techniques [50].

8.4.3 Body Hair

When head hair is not available body hair can be used but the time scales are much less clear. The laboratory analysis of body hair is exactly the same whether head hair or body hair is analyzed, however, the time period covered by each hair type will be different because of the biology of the hair. From a drug testing point of view for calculating presumed dates of the hair sample the growth rate of body hair is similar to head hair, but the significant difference is that body hair may represent a longer window of detection than the simple assumption of 1 month per centimeter of growth.

Studies showed that levels detected in body hair are higher than in head hair [37,54] and pubic hair may be contaminated by the individual's own urine sample, depending on the personal hygiene, which might affect the interpretation of the results [55] (Table 8.2).

8.4.4 Window of Detection

Hair will contain a mixture of hairs from previous periods because hair growth occurs in cycles and is composed of three different phases. These phases are called: anagen (the growing phase), catagen (the intermediate phase), and telogen (the shedding phase). Anagen is the active growth phase and lasts between 3 and 7 years. In this phase the hair is produced and makes up to 90% of follicles on a normal human scalp. The catagen phase lasts for between 2 and 4 weeks in the human scalp. Growth of the hair in this phase

TABLE 8.2 Head Hair and Body Hair from the Workplace Perspective

	Head Hair	Pubic, Axilla, Chest, Leg, Arm, Beard, Eyebrow, Sacral
Workplace drug testing	Preferred	Only when head hair is not available
Growth rate	0.7–1.5 cm/month	0.7–1.5 cm/month
Sample adulteration before collection	Wig and hair extensions may be presented	Minimal risk
	Shampoos, hair dyes, and hair preparations to "pass the test" may have been used	
Sample adulteration during and after collection	Minimal risk depending on laboratory chain of custody and collector training	Minimal risk depending on laboratory chain of custody and collector training
Collection	Simple to collect	Simple to collect but with constraint
	African hair and dreadlocks more difficult	May require a medical staff for the collection or same sex collector
	Very thin hair may be collected from various points to produce sufficient sample	
Amount of hair collected	Depending on the laboratory, requirements for sample amount and collector, usually 50 hair strands	Depending on the laboratory requirements for sample amount and collector, usually 50 hair strands
Usual length analyzed	3 cm for 3-month overview	Whole length is analyzed
Window of detection—head hair	1 cm head hair reflects approximately 1 month (0.7–1.5 cm/month)	1 cm head hair reflects approximately one month
		3 cm head hair reflects approximately 3 months
		To cover 1 year, head hair needs to be at least 12 cm long

(*Continued*)

TABLE 8.2 (Continued)

	Head Hair	Pubic, Axilla, Chest, Leg, Arm, Beard, Eyebrow, Sacral
Window of detection—body hair	1 cm head body reflects approximately 1 month (0.7–1.5 cm/month)	Add approximately 3 months to the body hair length due to increased amount of telogen hairs
		1 cm body hair reflects approximately 4 months
		3 cm body hair reflects approximately 6 months
		To cover 1 year, body hair needs to be 6–8 cm long
		If shaved, 1 cm body hair reflects approximately 1 month
Cosmetic effects	Some drugs may be lost due to cosmetic treatment such as bleaching and dyes	Not usually treated cosmetically other than normal hygiene
Contamination	Possible contamination from environment	Possible contamination from urine for pubic hair, but if test is positive it does imply drug use
Race	Dark hair binds more drug than light color hair	Dark hair binds more drug than light color hair

is stopped. About 1% of hair on the head is in the catagen phase. The telogen phase lasts for 3 or 4 months. This hair is not growing. These are the hairs that come out when the hair is shampooed or brushed. It has been estimated that for body hair, 40–60% remains in the resting (telogen) phase in comparison with the 10–15% found in head hair [49,51–53,113]. During the resting phase hair is not growing but does contain traces of drug from previous use.

This means that when people stop using drugs the levels found in the hair drop rapidly to a level of some 10–15% of what they were during active drug use and then the levels decline to zero after a few months if no drug is used [56]. The length of time for body hair is extended. It is more difficult to determine the period of drug exposure when testing body hair in comparison to head hair. Calculation of the window of detection for body hair is based on the length of the tested sample with an average growth rate of 1 cm/month, but in practice the period covered by the sample represents a longer time depending on its length. A 1 cm length of body hair may represent an approximate 4-month period and a 3 cm length of body hair may cover approximately a 6-month period, but to cover a period of 1 year the body hair needs to be at least 6 cm long because the body hair telogen average cycle varies between 2 and 3 months [53]. However, there is potentially greater value in a negative drug result from body hair, like pubic hair, as the result represents a longer time period in comparison to head hair.

8.4.5 Why Hair Decontamination and Analysis of the Wash Residue?

The detection of metabolites is the main approach in hair testing to confirm drug use, and exclude external contamination, as metabolism requires ingestion into the body [37,57]. Decontamination of the hair samples before the analysis is a key step in the preparation of samples before extraction and analysis to remove and minimize the possibility of residue caused by external contamination and other potential assay interferents.

Typical examples are for drugs that are usually smoked or snorted like cocaine or cannabis. It is possible that a nonuser could become contaminated by associating with users. In those cases to confirm ingestion of cocaine the detection of benzoylecgonine is necessary and to confirm cannabis use the detection of 11-nor-delta9-THC-9-carboxylic acid (THC-COOH) or 11-hydroxy-delta9-THC (THC-OH) is necessary to exclude external contamination [37,57–59].

The review of literature contains comparisons of different wash protocols by different authors using a combination of several aqueous buffers and organic solvents, or both aqueous and organic mixtures of different degrees of complication at different incubation length of times and temperatures with debatable effectiveness [60]. Similar diversity in hair wash procedures currently used is illustrated by various authors in Tables 8.3 and 8.4.

The best wash protocol is the one able to solubilize and to remove most of any externally deposited drug with minimum extraction capability at this point, as a small quantity is assumed to leach out during the wash process [87]. The fact is it is unlikely that a perfect wash protocol will be found. Because of the structure of the hair and the binding characteristics of the different drugs, it will be impossible to know for sure whether the wash protocol was able to remove 100% of the drug deposited externally and unable to remove any drug incorporated in the hair shaft.

By measuring the drug component present in the wash residue and comparing it with the amount in the hair samples it is possible to aid the interpretation of results and can lead to successful differentiation of external contamination from drug use in most cases. This was proposed in 2008 by the authors and a further assessment of the criteria was evaluated in 2014 with samples from cocaine users when a conclusive interpretation outcome was achieved in most cases (over 98%) [88,89].

TABLE 8.3 Procedures Using the Immunochemical Methods in Conjunction with Confirmations by Chromatography

Reference	Drug Groups	Amount of Hair	Decontamination	Extraction	Immunoassay Screening	Confirmation
[61]	Amphetamines, cocaine, methadone, opiates	10 mg	2 × dichlormethane/1 min	Methanol incubation for 18 h at 36°C > ELISA > SPE extraction	ELISA	HS-SPME and GC/MS
[62]	Amphetamines, cocaine, methadone, opiates	30 mg	1 mL SLV-VMA-T	0.4 mL VMA-T reagent for 1 h at 100°C	Immunometric DRI reagents CD × 90 analyzer	
	Amphetamines, cocaine, methadone, opiates		15 mL water/2 min > 10 mL acetone/2 min > 10 mL hexane/2 min	Acidic methanol incubation for 15 h		LC-MS/MS-ESI
	THC			Digestion with KOH > SPE		GC-MS; SIM
[63]	Amphetamines, cocaine, opiates	10 mg	1 mL methanol/1 min and acetone/1 min	0.5 mL buffer pH 4.2 and incubate/2 h at 75°C	ELISA	
[64]	Cocaine, benzoilecgonina, cocaethylene, and norcocaine	11–13 mg	2 mL isopropanol/37°C/15 min >2 × 2 mL phosphate buffer-BSA 0.1% pH 6/37°C/30 min >2 × 2 mL phosphate buffer-BSA 0.1% pH 6/37°C/60 min	Enzyme digestion > SPE	RIA	LC-MS/MS

[65]	THC and THC-COOH	20 mg	2 mL methanol/5 min	Incubate with 3 mL methanol/70–75°C/2 h	ELISA
			2 mL KH_2PO_4 1 M/75°C/30 min >1 mL water >1 mL methanol >1 mL water	1 mL NaOH 1 M/75°C/30 min > SPE > elute THC with ethyl hexane-ethyl acetate (1:1) and THC-COOH with hexane—ethyl acetate with 1% de Glacial acetic acid (9:1)	GC-MS/MS
[66]	Methamphetamine	1 mg (ELISA)	Sodium Dodecyl sulfate 0.1/5 > water > methanol/3 min	Dry pulverization:	ELISA
				100 μL PBS 10 mM > filtration	
		2 mg (LC-MS/MS)		Dry pulverization:	LC-MS/MS-ESI
				LC-MS/MS: 100 μL TFA 0.1 M in water-acetonitrile (9:1) > filtration	
[67]	Amphetamines, cocaine, methadone, opiates	33 mg	1 mL SLV-VMA-T	0.4 mL VMA-T reagent for 1 h at 100°C	Immunometric DRI reagents CD × 90 analyzer
					ELISA
				Methanolic ultrasonication > SPE for GC-MS or direct injection for LC-MS	GC-MS and LC-MS
[68]	Amphetamines, cocaine, methadone, opiates, THC	50 mg	5 mL dichlormethane	Methanol/16 h at 40°C	ELISA (OneStep)
					GC-MS

(Continued)

TABLE 8.3 (Continued)

Reference	Drug Groups	Amount of Hair	Decontamination	Extraction	Immunoassay Screening
					Confirmation
[69]	Amphetamines, cocaine, methadone, opiates	30 mg	1 mL SLV-VMA-T	VMA-T reagent	Immunometric DRI reagents Hitachi Analyzer
		50 mg	2 × 2 mL dichlormethane	Hydrolyses: 1 mL 0.1 M HCl/1 h at 100°C	GC-MS
				SPE Extraction	
[70]	Amphetamines, cocaine, methadone, opiates	10 mg	2 × methanol	Methanol sonication > Alkaline digestion > SPE	ELISA
					GC-MS

TABLE 8.4 Procedures Using Confirmations by Chromatography

Reference	Drug Groups	Amount of Hair	Decontamination	Extraction	Confirmation
[71]	Amphetamines, cocaine, methadone, opiates	50 mg	3 × dichlormethane/2 min	Hydrolyses with 1 mL 0.1 M HCl/1 h at 100°C and automated SPE extraction	GC/MS
[72]	Unknown screening (STA)	—	1 × Water/1 min > 2 × acetone/2 min	0.5 mL methanol/acetonitrile/2 mM ammonium formate/18 h at 37°C	LC-QTOF-MS
[73]	Opiates, amphetamines, cocaine, methadone, THC	50 mg	2 × 2 mL dichlormethane/3 min	Methanol incubation for 15 h at 55°C; direct injection	LC-MS/MS-ESI; SRM
[74]	Amphetamines, cocaine, cannabinoids, opiates, etc.	20 mg	Shampoo and water > acetone	Sonication with 4 mL methanol/8 h a 50°C; direct injection	LC-TOF-MS
				Acid extraction: 2 mL 0.1 M HCl/18 h at 50°C; SPE	
				Alkaline digestion: 2 mL 0.35 M NaOH/30 min at 100°C; SPE	
[75]	Cannabinoids	10 mg	2 mL petroleum ether > 2 mL water > 2 mL dichlormethane	HF-LPME	GC-MSMS
[76]	Amphetamines, benzodiazepines, antidepressives, and halucinogenics	10 mg	10 mL sodium dodecyl sulfate 10%/3 min > 2 × 10 mL water/3 min > 10 mL acetone/3 min	Dry pulverization:	LC-HRMS (Orbitrap) ESI
				Ammonium acetate 5 mM pH 5 and acetonitrile with 0.1%	

(Continued)

TABLE 8.4 (Continued)

Reference	Drug Groups	Amount of Hair	Decontamination	Extraction	Confirmation
				formic acid > sonicate 1 h and incubate for 18 h at 37°C	
				Dry and pulverization:	
				Methanol and TFA (70:30) > sonicate 1 h and incubate for 18 h at 45°C	
				Dry and pulverization:	
				Methanol and TFA (90:10) > sonicate 1 h and incubate for 18 h at 45°C	
				Dry and pulverization:	
				water-acetonitrile-TFA 1 M (80:10:10) > shake for 10 min; filtration	
[77]	28 compounds	50 mg	10 mL water > 10 mL dichlormethane	Phosphate buffer/18 h at 45°C > SPE	LC-MS/MS
[78]	Amphetamines, methamphetamines, MDA, MDMA, and ketamine	10 mg	10 mL water > 2 × 10 mL acetone	Dry pulverization:	GC-MS, EI, SIM
				sonicate with 1 mL methanol/50°C/1 h; filtration	

[79]	35 compounds	50 mg	3 × 2 mL dichlormethane/2 min, dry, add PI, and incubate with 2 mL acetonitrile at 50°C/12 h	LLE with 4 mL hexane-ethylacetate (55:45) > SPE	LC-MS/MS, MRM
[80]	Cocaine and opiates	50 mg	Diluted soap solution	Pulverization and extraction using matrix solid-phase dispersion (MSPD) and SPE	GC/MS
[81]	96 compounds	10 mg	Isopropanol > 2 × water	Dry and pulverization:	LC-MS/MS, ESI and ESI-MRM
				200 μL extraction buffer > incubate at 37°C/18 h; filtration	
[82]	Benzodiazepines	20 mg	2 × dichlormethane > methanol	Liquid–liquid extraction:	LC-MS/MS, ESI, MRM
				phosphate buffer pH 8.4 > sonicate/1 h > 4 mL dichloromethane-diethyl ether (90:10)	
				Direct injection:	
				sonication with 4 mL methanol/8 h a 50°C;	
[67]	Amphetamines, cocaine, methadone, opiates	33 mg	1 mL SLV-VMA-T	0.4 mL VMA-T reagent for 1 h at 100°C	Immunometric DRI reagents CD × 90 analyzer
					ELISA
				Methanolic ultrasonication > SPE for GC-MS or direct injection for LC-MS	GC-MS and LC-MS

(*Continued*)

TABLE 8.4 (Continued)

Reference	Drug Groups	Amount of Hair	Decontamination	Extraction	Confirmation
[83]	Amphetamines, cocaine, methadone, opiates, and others	150–250 mg	5 mL ethanol	Alkaline digestion >SPE	LC-TOF-MS
[84]	Amphetamines, cocaine, methadone, opiates	20 mg	1 × water > 2 × acetone	Methanol/acetonitrile/2 mM ammonium formate/18 h at 37°C	LC-QTOF-MS
[85]	THC, THC-COOH, cannabinol (CBN), cannabidiol (CBD)	50 mg	4 mL water/4 min >4 mL acetone/4 min >4 mL ether/4 min	Incubate sample with 2 mL methanol/4 h	LC-MS/MS, ESI negative for THC-COOH
					ESI positive for THC, CBN, and CBD
[86]	Benzodiazepines	30 mg	Water > acetone > hexane	1.5 mL methanol/90 min > 1 mL mobile phase and methanol (1:1) > shake 90 min	LC-MS/MS, ESI, MRM
			Dry pulverization		

In our experience the issue of the interpretation regarding external contamination is rare except in those individuals dealing in drugs and heavy users. The likelihood of a hair sample from a noncocaine user to contain significant levels of cocaine deposited as external contamination is possible but the levels would be much lower than those of a cocaine user. This was confirmed in a study of the 47 hair samples from crack cocaine users that showed levels of cocaine at a maximum ratio of 9% [90]. These values confirm that drug users are likely to have a degree of drug deposited externally on their own hair sample from handling of the drugs they consume or from their own smoke.

8.4.6 Extraction

Hair samples are usually digested or disintegrated in order to release the analytes before analysis. There are many extraction methods in the literature describing distinct approaches used to extract hair samples for the screening and confirmation of drugs. Most common are variations of methanolic, acid or alkaline, extraction or enzymatic digestion examples are shown in Tables 8.3 and 8.4, which demonstrate this diversity [61–69,71–86,91].

Some laboratories prepare hair samples to be screened by immunochemical methods [61–70] (Table 8.3). Methods involving gas chromatography with mass spectrometry (GC-MS) usually do require sample cleanup, generally by traditional solid-phase extraction (SPE), but alternative procedures have been described such as SPE-microextraction like hollow fiber liquid-phase microextraction (HF-LPME), headspace solid-phase microextraction (HS-SPME), or matrix solid-phase dispersion (MSPD) [71,75,80,91].

However, there are many publications in the last few years where the extraction method for chromatographic methods used is a simple methanolic extraction with direct injection, with no sample cleanup with the traditional SPE [67,73,76,78,81,82,84]. Domínguez-Romero and coworkers in 2011 evaluated three different methods of extraction and found that direct methanolic extraction was the most efficient method for extraction when compared with acidic or alkaline extraction followed by SPE cleanup procedure [74].

Favretto and coworkers evaluated four extraction procedures of a combination of pulverization, sonication, and incubation at different temperatures with cleanup before analysis by liquid chromatography–high-resolution mass spectrometry (HPLC-HMRS) [76].

Pulverization is a very convenient method of extraction that has the potential for producing improved recovery [66,67,76,78,81]. However, these publications do not evaluate the effect of high throughput of uncleaned samples on the delicate instrumentation in the long term.

Direct injections are quick and provide high extraction recovery but may shorten the instrument life. It is a balance between simplicity of the extraction process and speed of the analysis with robustness of the method. The best method is the one that works well for the samples a laboratory needs to analyze.

8.4.7 Immunochemical Screens

The practice of the immunoassay is still used by many laboratories as a quick and cost-effective method to screen out negative samples, very practical in the workplace setting where most samples are negative. However, any presumptively positive results when using an immunoassay require confirmation by GC-MS/MS or LC-MS/MS to confirm the presence of drugs or not.

RIA and enzyme-linked immunosorbent assay (ELISA) are the two methods traditionally used for the screening of drugs in samples of hair because they provide highly sensitive and semiquantitative analysis of a large number of samples quickly [64]. RIA, while very sensitive, has been supplanted by enzyme immunoassay because it does not require the associated handling difficulties associated with radioactive materials [92]. ELISA kits are commercially available from several suppliers, some of which are specifically adapted for hair samples. This method has been most commonly used by different laboratories as a screening technique of hair samples because of its high sensitivity and amenability to automation [25,61,63,65–68,70].

Alternative immunochemical methods have been evaluated for drug detection in hair samples using commercially available reagents that digest hair samples and enable the screening using biochemistry analyzers utilized in the screening of urine samples [62,67,69]. This is a practical approach for a high throughput laboratory; an example is the immunometric test VMA-T coupled to Diagnostic Reagents Inc (DRI) enzyme immunoassay. The study carried out by Baumgartner et al. using the immunometric test VMA-T procedure and the cutoffs recommended by the Society of Hair Testing (SoHT) found an overall acceptable performance capable of discriminating negative and presumptively positive samples, but the drawback was that for the confirmation of cocaine a new hair extract using a different protocol needed to be utilized [62].

Musshoff et al. compared immunometric test VMA-T coupled to DRI with ELISA, using lower cutoffs specifically designed for the purpose to control abstinence, and obtained interesting results when compared to GC/MS and LC/MS [67]. The authors confirmed the good discrimination between positive and negative samples by both methods, immunometric test VMA-T coupled to DRI and ELISA, using the cutoffs suggested by SoHT. At lower cutoffs ELISA was not satisfactory to screen for Δ^9-tetrahydrocannabinol and DRI enzyme immunoassay was only useful for morphine and cocaine.

8.4.8 Confirmations

With the advent of the tandem mass spectrometry (MS/MS), which is a combination of the two techniques, chromatographic and spectrometric, such as gas chromatography–mass spectrometry (GC-MS/MS) or liquid chromatography–mass spectrometry (LC-MS/MS), laboratories switched to using solely this technique often called the "molecular fingerprint" that reliably confirms and identifies the compound(s) chemically [71,73,74,76,78,79,81,82,84–86].

In 2010 Wada and coworkers published a comprehensive review of the methods used in the analysis of drugs in hair [93]. More recently, Vincenti and coworkers reviewed the improvements in the field of LC-MS/MS instrumentation available for the detection of smaller quantities of drugs in hair [94].

LC-MS/MS is gradually replacing GC-MS in both screening and confirmation procedures, and is increasingly acknowledged as the technique of choice [95]. This trend is also being applied to hair analysis because it provides increasing performance and decreasing costs of LC-MS/MS instrumentation, as well as the absence of derivatization steps for the simultaneous screening, separation, and detection of a wide variety of substances of toxicological interest in a single run of only few minutes [94].

8.5 ACCREDITATION

Regulated procedural guidelines regarding substance abuse testing in hair are key factors to improving the service provided by drug testing laboratories. The European Workplace Drug Testing Society (EWDTS) issued guidelines to laboratories in order to promote the provision of reliable results in workplace drug testing. The guidelines should help to harmonize laboratory practices and also provide a framework that can be used for accreditation to ISO 17025 for the purpose of drug testing. In the United States, SAMHSA fulfills this role [45,46,58].

Accreditation of laboratories testing drugs in hair samples adds value to the practice by improving compliance with standards and guidelines for the performance of laboratory analysis.

Different countries have different accreditation bodies and individual countries have their own accreditation body that follows international standards that usually certify laboratory competency such as ISO/IEC 17025. In the United Kingdom it is the United Kingdom Accreditation Services (UKAS); in the United States, the ANSI-ASQ National Accreditation Board (ANAB); in Germany, the Trägergemeinschaft für Akkreditierung German Association for Accreditation GmbH (TGA); in France, the Comité Français d'Accréditation (COFRAC); in Portugal, the Instituto Português de Acreditação (IPAC); in Chile, the Instituto Nacional de Normalización (INN); in Brazil, the Instituto Nacional de Metrologia, Normalização e Qualidade Industrial (Inmetro), to cite a few examples.

For a laboratory to get accreditation it is required that they participate in proficiency testing (PT) schemes, which are probably the most important element of the accreditation process, as they provide a direct assessment of the accuracy of the laboratory results. SoHT started distribution of proficiency testing for hair samples in 1995 [96,97], but other commercial PT schemes are available in the market.

8.6 INTERPRETATION AND REPORTING

From the analytical point of view, the uncertainties of the final analytical measurement of hair samples can be greater than those in urine or oral fluid and variable between different laboratories. This is because the analysis of hair samples may be more cumbersome to analyze than urine or oral fluid. For example, the hair needs to be decontaminated and digested before analysis and losses during these processes may increase variability of the results and are likely to affect the detection of drugs and metabolites. Awareness of the analytical pitfalls of hair testing is key to understanding its results [98]. For example, during the analysis of cocaine and 6-acetylmorphine, hydrolysis can occur at extreme pH conditions during extraction with the generation of benzoylecgonine and morphine respectively, which can mislead the interpretation.

As hair covers a longer period of use, it is reasonable that a hair sample is positive and a urine test is negative. The reverse is also possible; a non-drug user might have used cocaine a day or two before a urine drug test, in this case the hair test will be negative and the urine will be positive. The detection of drugs in oral fluid above certain levels provides presumptive evidence of recent consumption of drugs but it is not possible to determine how recently they were consumed. The detection of drugs in urine provides an even weaker indicator of current drug intake.

A negative result, however, does not categorically mean that the person did not use drugs in the last 24 h or 3 days (urine and oral fluid) or the last 3 months (hair). It is possible for an individual to provide a negative hair result if they have used or been exposed to a drug infrequently or in low doses.

Consequently, it is erroneous to think that drug tests are always clear cut whichever of the matrices, urine, saliva, or hair, are used. It is important that the benefits and limitations of the tests are considered before the client assigns a contract for a drug screen. In reality a negative drug test only indicates that a drug was not found above the level of any cutoff level that may be applied.

A positive result from the analysis of drugs in any of the commonly used matrices such as urine, oral fluid, or hair, confirms that a person has used or been exposed to a drug.

Table 8.5 was adapted from Mieczkowski (2010) to explain the possible scenarios where a discrepancy might have occurred between urine testing

TABLE 8.5 Significance of the Parallel Urine and Hair Testing

	Preemployment Testing	Random Testing
Negative urine and negative hair	**More likely occurrence**	**More likely occurrence**
	Majority of job applicants	Most employees are not drug users
Negative urine and positive hair	**Hair detects 5 times more than urinalysis**	**Hair detects 3 times more than urinalysis**
	Expected outcome due to the longer retrospective capacity of hair to identify drug use	Expected outcome due to the longer retrospective capacity of hair to identify drug use
Positive urine and positive hair	**Possible occurrence**	**Less likely occurrence**
	More likely to be a chronic drug user	More likely to be a chronic drug user.
	A nondrug user would not show up for a job interview 2–3 days after using drugs	A nondrug user would not show up for a job interview 2–3 days after using drugs
Positive urine and negative hair	**Rare outcome**	**Rare outcome**
	Might occur when the hair is short and drug was used within a week before sampling, time taken for a drug to emerge out the scalp after ingestion	Most likely to be a cannabinoid slowly excreted drug and it can be detectable in urine up to weeks after last use and more difficult to be detected in hair
	Most likely to be a cannabinoid than a rapidly excreted drug such as cocaine	

Source: Adapted from Mieczkowski [42].

and hair testing results and the likelihood of the occurrence. The outcomes displayed in the table might also be applicable to oral fluid drug testing [42].

8.6.1 Can We Tell the Amount of Drug Used?

While drug levels detected in urine, oral fluid, or hair samples are not correlated to the amount of drug used, they are affected by it. The lack of interindividual correlation between drug concentration in hair and daily or cumulative dose is not surprising given the fact that correlation between dose and blood concentration between individuals is typically poor.

There are several factors that affect levels of drugs and metabolites in any biological sample, including hair. For example, the incorporation of drugs into hair varies from person to person, primarily due to differences in metabolism. Other factors affecting levels of drug and metabolites in hair are purity, frequency of drug use, hair color, and the use of cosmetic hair products [85].

One may assume that high levels detected in a hair sample may reflect higher use; however, this assumption can be misleading. The factors listed above can affect the numerical result of a hair test and the amount of drug used cannot be extrapolated. It is a dangerous oversimplification but often hair testing providers issue results with a firm statement about the levels of consumption, grading as a heavy user or a light user, comparing data from hair analysis among individuals or extrapolating the dose used.

Studies have shown a lack of correlation between dose and levels in the hair of both therapeutic drugs and illicit drugs, due to huge interindividual variation [99–104]. However, a low intraindividual correlation when analyzing multiple hair segments was observed [104].

Controlled studies assessing correlation between quantitative results of therapeutic drugs in hair found a wide variation between individuals using the same dose [99–104]. While intraindividual correlation was possible, enormous interindividual variations existed [103].

It is therefore concluded that it is not possible to know precisely from the results of hair analysis how much or how often an individual used drugs and classify them as a heavy user or a light user. While it is an indication of heavy use relative to light use, it is a highly subjective assessment.

8.6.2 Can We Compare Results Between Individuals?

While the incorporation of drugs in hair is dose related there is a large interindividual variation in drug users and levels in hair are within a wide range of variation [38,70,84,105,106]. This means that it is not possible to precisely compare the results of testing between individuals, although it is true that higher concentrations of drugs in hair may reflect higher doses used and/or more frequent use.

In general, ranges obtained in the analysis of hair samples for drugs are very wide, for example, the group of cocaine has a range that goes from 0.1 to 300 ng/mg [70,84,107].

8.6.3 It Is Possible to Grade the Amount of Drug Used Within the Same Individual

The levels of drugs detected in hair are currently best used as a guide to changes of use in the individual when sectional analysis is performed or two different periods are compared in the same individual. This attribute can be used to monitor drug use patterns, demonstrating increasing or decreasing doses being used by the same individual over longer time periods [70].

8.6.4 It Is Possible to Compare the Results with Data Accumulated by the Laboratory

It is, however, possible to interpret the results compared with the results of a population of previous positive results for guidance, the comparison of the results of hair analysis with statistical compilation of results of samples that the lab has measured [70,84].

8.6.5 Can We Tell the Frequency of Drug Use?

The hair test produces integrated results of drug use over weeks and months but it is not possible to tell how many times drugs were used within the period covered by the sample. To be more certain of the frequency of any use sectional analysis is more likely to reveal more intermittent use. Sequential sectional analysis can assume a minimum number of occasions when drugs were used over the integrated results over periods of weeks or months. However, a monthly sectional analysis is unable to give any indication as to whether the results indicate daily or weekly use as each result is an integrated value for the approximate 4-week period. This is because there is considerable variability in the area over which an incorporated drug can be distributed in the hair shaft and in the rate of axial distribution of the drug along the hair shaft [108]. Shuffling of the hair lock at the moment of collection and the telogen hair strands will contribute to the broadening of the band of positive hair.

When levels are detected in two sections, the results indicate drug use on at least one single occasion, not two. Nonetheless, it is possible to extrapolate from the results of serial sectional analysis of hair that an individual increased or decreased consumption or even ceased to take drugs. The results of multiple sections are evaluated against each other, whether increasing or decreasing, reflecting drug use patterns. However, the uncertainty of the measurement needs to be taken into account in the interpretation, particularly at lower levels.

Sectional analysis is required to demonstrate that an individual has decreased their use of drugs—head hair should be used. Monthly sectional analysis provides key information for specific situations. For example, hair analysis can help improve treatment efficacy by allowing a check of treatment compliance over the period covered by the sample or indicate whether the patient remained free from taking illicit drugs.

However, when levels are low they may reflect infrequent use of drugs or they may indicate use of drugs in an earlier period. The latter is an important factor when interpreting results because of the biology of the hair. A lock of hair can comprise of all three stages (anagen, catagen, and telogen) so it is possible that drugs taken in the earliest periods, furthest away from the scalp, may "leak" into the middle period, and then disappear in the most recent period.

8.6.6 Effect of Hair Cosmetics

In regularly shampooed hair, that is, not treated by aggressive cosmetic agents such as oxidative dyeing, bleaching, or permanent wave, drugs are usually well detected at least 1 year after intake. The extent of drug decline following cosmetic treatment is dependent on its initial concentration and the properties of the hair matrix. Hair products, such as bleaching and dyeing (or coloring), that affect the integrity of the hair shaft might affect the amount of drugs and analytes that are incorporated in the hair. Long-term effects of weather (sunshine, rain, wind) may cause damage to the hair shaft with subsequent impact on drug concentration. For example, cannabinoids in hair are particularly sensitive to sunlight [38].

8.6.7 Hair Color

The natural hair color also affects the incorporation of drugs. Dark hair incorporates relatively more drugs and metabolites than fair hair, mainly due to the binding of drugs to hair melanin [62,108]. Consequently, people with dark hair have a higher chance of having a positive result than people with light color hair using the same dose. This is particularly true of cationic drugs such as amphetamine, methamphetamine, and cocaine. The fact that melanin in hair does affect the incorporation of drugs in hair, particularly cationic drugs, has led to the assertion of a color bias in hair drug testing. However, with the analytical techniques currently available drug use can be detected in hair samples even where only a single dose has been taken. Detection of use is not biased but interpretation of the concentration of drug found does need very careful thought given to the ethnic origin of the donor, the color/melanin content of the hair, and any cosmetics that might have been used. Studies with a total of over 60,000 samples showed no significant relationship between the categories of hair color and the probability of a positive test [109]. We have personal knowledge of a case of a young man and woman who used drugs together and shared what they had. He had dark hair while she was a fine-haired blonde. Although their "habit" was identical, the test results were higher in him than in her. This was explicable by their hair color, not, as the court was tempted to conclude, by assuming a greater level of drug taking by the male. A report on the effect of drugs used in each case is important to supplement the raw drug test result.

8.6.8 Cutoffs: How Drug Test Results Are Interpreted to Clients

The cutoffs employed in hair analysis are usually analytical cutoffs established at the limit of quantification of the methods used. This is in contrast with the cutoffs employed in urinalysis, which are much higher than the limitation of the methods. In practice it means that results of urinalysis are reported as not detected when below the established cutoff even though the presence of drugs is clear. This is a very practical approach in the workplace setting, where it is important to know whether an individual is under the influence of drugs or not.

In the case of urine and saliva (oral fluid), the levels of cutoffs are used for the purpose of (1) minimizing the detection of drugs taken involuntarily (passively), as in the case of drugs that are smoked and (2) eliminating the detection of drugs used in the earlier periods of interest (e.g., eliminating the detection of drugs used over the weekend outside the workplace). However, the primary aim of cutoffs in hair analysis is to minimize detection of drugs used in previous periods and increase detection of current use. It is also used to minimize the effects of the eventual external contamination.

The total uncertainty of analytical methods for the measurements of drugs in human hair is derived mainly from the preanalytical and analytical variations that contribute to the total uncertainty of drug detection in hair [110]. The uncertainties of biological measurements should be considered when interpreting the resulting data. The uncertainties associated with the measurements and with the interpretations and any assumptions made during the process should be explicitly stated.

With hair analysis when results are nil or below cutoff, results are usually reported as "negative" or "not detected." However, a result below cutoff does not absolutely prove that an individual has not used drugs. Levels below cutoff are reported as "not detected," and low levels in hair are usually correlated to the small quantities of drugs used, although the minimal detectable dose that can be detected in hair is unknown [111].

While a positive result of a hair analysis indicates that a person has used or was exposed to a drug, a negative result does not refute use of or exposure to the drug because of the use of cutoffs by the testing laboratories. Clients may need to be offered the option of being informed when drugs are unequivocally detected at levels below cutoff in some critical industries.

8.7 HAIR ANALYSIS IN THE WORKPLACE

8.7.1 Hair Analysis Data for Preemployment in Brazil

The recruitment of military and civil staff for the police and fire brigade in Brazil follows a series of rigorous exams and tests, and hair analysis is now integrated in most of their recruitment. The first indispensable requirement for job acceptance is a mandatory negative hair test for drugs. This is to exclude the employment of drug users in activities which would be incompatible were they are to be employed.

The results of the analysis of 1,972 hair samples from candidates from eight federal states in Brazil revealed that some forces were inadequately advised when establishing the test requirements and that regulated procedural guidelines regarding substance abuse testing in hair could improve the service provided by drug testing laboratories [47]. The laboratory in São Paulo received hair samples from candidates for testing but did not take part in the set-up of the test requirements.

The different forces from the different states within the large country of Brazil publicly tender their requirements in terms of period covered by the hair samples, the window of detection, and the suite of drugs tested. On the whole, 90 or 180 days were commonly requested but often a window of detection of "365 days" was in the tenders, as body hair is mis-sold in Brazil as a "365-day test." Amphetamines, cannabis, opiates, and cocaine were common to all forces with a few forces requesting substances that are not analyzed in hair such as "hydrocarbons and solvents" or drugs that are unlikely to be used in Brazil such as PCP. In some cases it was required to state the level of drug consumption based on the drug levels detected in hair.

The testing over a period of a whole year to exclude drug users from the police forces might look an attractive proposition, however, it is ill advised. The length of the sample gives the period covered by hair samples with body hair following a similar biological cycle as head hair, but because of the increased proportion of resting (telogen) phase, body hair may represent a longer window of detection, on average of 5 months with a maximum of 9 months [53]. The average body hair length is approximately 2.2 cm, with some longer case exceptions up to 5–6 cm for axilla and pubic hair or 4–4.5 cm for leg and chest hair [112]. An assertion of a window of detection for body hair that is 365 days is misleading to the customers unless body hair is sufficiently long.

The results of this study showed a surprisingly high number of body hair samples (79%) with leg hair leading with 49% of the samples. Positive hair results for at least one group of drugs—cannabinoids, cocaine, or ecstasy—were found in 0.7% of the samples ($N = 13$). Cocaine was the prevalent drug with 13 cases, cannabinoids in three cases, and ecstasy (MDMA and methylenedioxyamphetamine (MDA)) in one case. During the analysis of the samples it was noted that approximately twice as many samples were shown to be above the analytical cutoff but below established test cutoff for cocaine, indicating that a greater number of candidates exposed to cocaine passed the test.

Testing for drugs using hair may have been a deterrence to some drug users as a number of candidates did not present themselves for the sample collection, but it is not possible to know whether that was because of the nature of the test or for other reasons. Furthermore, from the results of this study in view of the frequency of detection of cocaine below cutoff levels, there could be further benefits if forces were to monitor new employees randomly throughout their probationary period. Hair, urine, or oral fluid testing could be used for this random testing program.

8.7.2 Data from a Preemployment Setting in the United Kingdom, Brazil, and Australia

Hair samples from the workplace sector for preemployment and in-employment were received in our laboratory in the United Kingdom for analysis. All samples were extracted and analyzed by LC-MS/MS using methods accredited to ISO/IEC 17025 standards. The substances targeted included: amphetamines, benzodiazepines, cannabinoids, cocaine, opiates, opioids (UK, Brazil, and Australia samples), ketamine (UK samples), and PCP (Brazil samples). The length of head hair samples were analyzed according to clients' requests for period of detection and the whole length of body hair samples were analyzed.

Results are presented in Table 8.6. Hair samples from Brazil were largely body hair (87.6%), mostly leg hair (56%) and axilla hair (24%). The majority of samples collected in the United Kingdom and in Australia were predominantly head hair, 88.7% and 92.3%, respectively. Samples from Brazil were exclusively for preemployment while samples from the United Kingdom and Australia were a combination of pre- and in-employment cases. The window of detection requirements for samples from Brazil varied from 90 to 180 days and up to 365 days in the case of all body hair samples (see notes and commentaries in the previous section). A 3-month window of detection was typical for samples from the United Kingdom and Australia. The mean length of all body hair samples analyzed was 2.5 ± 0.9 (mean $\pm$ SD, $n = 2{,}813$) Table 8.7. Drug detection rates in the samples from the United Kingdom, Brazil, and Australia were 3.2%, 2.1%, and 9%, respectively, being codeine (1.7%) and cocaine (0.7%) more commonly detected in the United Kingdom, cocaine (1.4%) in Brazil, and cocaine (5.2%) and MDMA (4.5%) in Australia. Of the positive samples, the detection of single drug use was predominant, with 93% in the United Kingdom, 82% in Brazil, and 68% in Australia. In the latter country, two drug groups concomitantly, cocaine and amphetamines, were detected in 27% of the samples. Ketamine was only tested in the United Kingdom and detected in four samples (0.3%).

As different countries and services use hair testing as a tool to prevent recruitment of drug users, there would be benefits in standard criteria and a realistic drug panel, representative of the general population. Regulated procedural guidelines regarding substance abuse testing in hair could improve the service provided by drug testing laboratories.

From the survey using hair samples in the United Kingdom, Brazil, and Australia we make the following observations (Table 8.8).

TABLE 8.6 UK, Brazil, and Australia Results Samples from the Workplace: Preemployment and In-Employment

	United Kingdom		Brazil		Australia	
	N	(%)	*N*	(%)	*N*	(%)
Total number of samples	1,321	(100)	630	(100)	862	(100)
Negative for all drugs	1,254	(95)	610	(97)	718	(83)
Head hair	1,172	(89)	78	(12)	796	(92)
Amphetamine	Not detected		Not detected		1	(0.1)
MDA	Not detected		1	(0.2)	8	(0.9)
MDMA	2	(0.2)	2	(0.3)	39	(4.5)
Methamphetamine	Not detected		Not detected		3	(0.3)
Anhydroecgonine methyl ester	Not detected		Not detected		1	(0.1)
Cocaethylene	2	(0.2)	Not detected		21	(2.4)
Cocaine	8	(0.6)	9	(1.4)	43	(5.0)
Norcocaine	1	(0.1)	1	(0.2)	1	(0.1)
Benzoylecgonine	6	(0.5)	3	(0.5)	13	(1.5)
Chlordiazepoxide	1	(0.1)	Not detected		Not detected	
Clonazepam	1	(0.1)	Not detected		Not detected	
Diazepam	1	(0.1)	Not detected		Not detected	
Nitrazepam	2	(0.2)	Not detected		Not detected	
Nordiazepam	1	(0.1)	Not detected		Not detected	
Ketamine	4	(0.3)	Not tested		Not tested	
Tramadol	2	(0.2)	Not tested		Not tested	
Desmethyltramadol	2	(0.2)	Not tested		Not tested	
Codeine	22	(1.7)	Not detected		9	(1)
Dihydrocodeine	2	(0.2)	Not detected		Not detected	
THC	4	(0.3)	2	(0.3)	3	(0.3)
THC-COOH	2	(0.2)	Not detected		1	(0.1)
Cannabidiol	2	(0.2)	Not detected		Not detected	
Cannabinol	Not detected		2	(0.3)	1	(0.1)
PCP	Not detected		Not detected		Not tested	

TABLE 8.7 Mean ± Standard Deviation, in Centimeters, of Body Hair Length Showing the Number of Samples Analyzed (*n*) Per Type of Hair of Samples Received from Brazil, United Kingdom, and Australia

Hair Type	United Kingdom		Brazil		Australia	
	Mean	SD (*n*)	Mean	SD (*n*)	Mean	SD (*n*)
Arm	—	—	1.7	±0.6 (30)	—	—
Axilla	3.4	±1.1 (90)	2.5	±0.8 (152)	3.4	±0.9 (61)
Beard	2.3	±0.6 (4)	1.5	— (1)	3.8	±0.4 (2)
Body	2.5	±1.1 (10)	2.0	±0.7 (12)	—	—
Chest	3.2	±0.7 (40)	1.4	±0.9 (4)	2.2	±0.3 (3)
Head	2.8	±0.7 (1,172)	3.4	±3.2 (78)	2.2	±0.4 (796)
Leg	2.3	±0.5 (5)	1.8	±0.5 (352)	—	—
Pubic	—	—	1.2	— (1)	—	—
Grand mean			2.5 ± 0.8 (2,813)			

8.8 CONCLUSIONS

Drug testing in the workplace is a controversial subject. The drug testing industry promotes the advantages of testing as a means of reducing accidents, absenteeism, and improving on-job performance. The evidence for this is highly contested. However, the use of urine as a short-term assessment of drug use is generally accepted as a positive factor in safety critical occupations. Deterring drug and alcohol use, or at least their effect on safety and performance in the workplace, is the goal of testing with urine and oral fluid. The recent addition of hair testing into the armory of workplace drug testing has extended the surveillance from the workplace into the home and social life of the employee. Occupations such as the security services, police, and the military increasingly use the window of detection provided by hair testing to guard against unknowingly employing an individual with a drug habit. The very fact that these employers find the testing useful means that provision of the service is going to grow. The heavy manufacturing industries that have taken up urine testing as an in-employment deterrent will increasingly consider hair testing as a preemployment test. Some schools already use hair testing as a means of detecting drug misuse by their pupils.

The fact that hair analysis has arrived in the arena of workplace testing means that employers and the drug testing industry, as well as general

TABLE 8.8 Points of Consideration when Extrapolating Hair Results

Information from Hair Analysis Results	What *Can't* Be Extrapolated	What *Can* Be Extrapolated
Date when drugs were used	It is not possible to assign drug use detected by hair analysis to use on a particular day.	It is possible to assign drug use within approximate dates.
		A positive result confirms drug use made within an approximate period given by the length of the hair section analyzed.
Quantity of drug used	It is difficult to correlate the levels detected in hair with the amount of drug ingested because in-between individual variations.	The levels of drugs detected in hair are best used as a guide to changes of amount of drug used use in the same individual when sectional analysis is performed or two different periods are compared in the same individual.
		It is possible to interpret the results by comparing them with the results of a previous population of positive results by a laboratory.
		Different drugs incorporate in hair in different rates.
Frequency of drug use	It is not possible to know how many times with the period of hair analyzed from the analysis of one section of hair.	It is possible to establish regularity of use if serial segmental analysis is performed.
		Serial sectional analysis may demonstrate that an individual has decreased their use of drugs when head hair is used.
Window of detection	Body hair does not automatically cover 365 days.	Body hair covers longer period than head hair if not shaven.
		To cover 365 days a section of body hair has to be at least 6 cm long.
		The average body hair length is about 2 cm long.

(*Continued*)

TABLE 8.8 (Continued)

Information from Hair Analysis Results	What *Can't* Be Extrapolated	What *Can* Be Extrapolated
External contamination	It is not possible to know if the hair washing before analysis is 100% efficient and that it does not initiate the extraction of the drug in the hair sample.	Most external contamination is due to own use of smokable drugs.
		Most external contamination is removed by the wash.
		External contamination of hair can be assessed by examination of the wash residue.
Negative result	The donor has not used drugs	The donor might have used drugs but the levels were below the analytical cutoff.

society, are going to have to consider how to deal with the difficult issues that will arise.

A tenet commonly held by professionals engaged in workplace drug testing is that there is a need to protect the individual employee from the false accusation of working while under the influence of drugs. So detailed procedures have developed to control the interpretation of the data produced to try and protect the employee from adverse actions because of prescribed drug use, environmental contamination, and out-of-work use. In the workplace drug testing the primary element required to prove inappropriate drug use is the test result alone to ensure safety and productivity. It is time that this view should be reconsidered.

Since the original rules and regulations for workplace testing were formulated, principally in the United States of America, there have been many significant changes. Techniques have been developed that have added to the window of detection provided by urine testing. While oral fluid testing, sweat testing, and blood testing have similar windows of detection as urine, methodology has been developed that has made possible very significant extensions to their detection windows. And when using hair testing it is impossible to restrict the testing so as to detect only "on-the-job" presence. The link from presence of a drug to labeling an individual as under the influence of the drug is fraught with problems and perhaps it is time to reconsider the purpose of workplace drug testing. It is commonly asserted that most people addicted to or dependent on drugs are employed. Surely then it is appropriate to make use of the workplace to begin the process of weaning people off unnecessary and potentially harmful use; not just as a safety issue for coworkers but as a health issue for the individual.

Among the issues are:

- *"Human rights: to do what I like, especially in my own time."* This leads some employers to look for very short-term drug testing, such as saliva testing. The hope is that the test will not detect drug use outside of work and yet fulfill the requirement to detect anyone under the influence of drugs while in the workplace. But this argument ignores the risks to the individual or the "hangover" effect that may prejudice the safety of fellow workers and destroy productivity for the employer.
- *"Drug testing does not measure incapacity. It cannot determine whether an individual is under the influence, so why test?"*
- *"Hair testing cannot distinguish between use at work and use at home."* It is in essence a lifestyle test. Depending on the sensitivity of the tests used and the cutoff levels applied, a hair test can detect use of drugs several months prior to the test.
- *"The individual being tested may have forgotten that he/she had used."* On the other hand, such testing means that it is very much more difficult, near impossible, to defeat the test.

The case for testing for health and safety issues is agreed; nobody wants their airline pilot to be using drugs and there have been horrific accidents in the rail industry (London Underground) due to the use of drugs. The edict by Ronald Reagan that started workplace drug testing was precipitated by an accident on a US aircraft carrier that was found to have been caused by the use of cannabis. Where there is dangerous machinery that could cause harm to or death of the users or other workers, drug testing is noncontroversial.

There are employment situations where knowledge of prior drug use is relevant, especially when such use is not declared prior to the test. Some employers could regard the nondeclaration as an integrity issue that might invalidate the prospective candidate from employment.

In highly secure occupations prior drug use may leave the candidate open to blackmail when the use is not openly declared to the employer. And the employer will worry about what other information may have been suppressed. While some might argue that it is of no relevance to the employer there are occupations such as police, security services, and the prison service where drug use could put the individual at risk and cause them to act inappropriately.

The analysis of drugs in hair is a useful tool, enabling exclusion of drug users whose drug use would be incompatible with their activities were they are to be employed. Testing for drugs using hair may be a deterrent to some drug users but in view of the frequency of detection of drug below cutoff levels, new employees should be randomly tested using hair, urine, or oral fluid throughout their probationary period.

Guidelines for drug testing using hair samples do exist. They cover sample collection, advice as to suitable analytical procedures and even guidance as to best practice interpretation [45,46,58]. However, these guidelines are largely directed at laboratories. There is a need for laboratory best practice to be distributed into the wider world. The demands of marketing and sales departments tend to obscure the messages coming out of laboratories. Courts and tribunals look for absolute clarity in the measurements they use and are prone to take as irrefutable truth the assertions made in marketing literature from the laboratories. It must be the responsibility of the laboratory community to ensure that no user, whether an individual or a large organization, can misunderstand what the technique can deliver.

While the utilization of head hair is always recommended in workplace testing, the extraordinarily high proportion of body hair samples received from Brazil is likely to be the result of a misconception that any body hair can cover a period of detection of a year. As different countries and services use hair testing as a tool to prevent recruitment of drug users, there would be benefits in standard criteria and a realistic drug panel, representative of the general population. Regulated procedural guidelines regarding substance abuse testing in hair could improve the service provided by drug testing laboratories.

The workplace is the best environment to act against addiction to licit and illicit drugs, with benefits both to the employees and the company itself.

At present, the analysis of drugs in hair samples in preadmission examinations increases the likelihood of excluding individuals who may cause accidents or are not compatible with illegal activities. This is particularly important in excluding individuals linked to high-risk activities or where there is a mismatch of illicit drugs with the activities to be carried out. High-risk activities are also those in which both the general public and coworkers experience a high risk of accident due to the symptoms caused by recent or frequent use of drugs.

However, apart from its use in preemployment and within a company's existing workforce, the main objective of the analysis of drugs in hair samples should surely have in mind the public health good, the well-being of people in society as a whole.

REFERENCES

[1] Peat MA. Financial viability of screening for drugs of abuse. Clin Chem 1995;41:805–8.

[2] Carpenter CS. Workplace drug testing and worker drug use. Health Serv Res 2007;42:795–810.

[3] French MT, Roebuck MC, Alexandre PK. To test or not to test: do workplace drug testing programs discourage employee drug use? Soc Sci Res 2004;33:45–63.

[4] Hoffmann J, Larison C. Worker drug use and workplace drug-testing programs: results from the 1994 national household survey on drug abuse. Contemp Drug Probs 1999;26:331–54.

[5] SAMHSA—Substance Abuse and Mental Health Services Administration. Worker Drug Use and Workplace Policies and Programs: results from the 1994 and 1997 NHSDA. Office of Applied Studies Analytic Series Paper; 1999. A-11.

[6] Berge KH, Bush DM. The subversion of urine drug testing. Minn Med 2010;93:45–7.

[7] Wall PS. Drug testing in the workplace: an update. J Appl Bus Res 2011;8:127–32.

[8] van der Molen HF, Lehtola MM, Lappalainen J, Hoonakker PLT, Hsiao H, Haslam R, et al. Interventions to prevent injuries in construction workers. Cochrane Database Syst Rev 2012;12:CD006251.

[9] Ramirez M, Bedford R, Sullivan R, Anthony TR, Kraemer J, Faine B, et al. Toxicology testing in fatally injured workers: a review of five years of Iowa FACE cases. Int J Environ Res Public Health 2013;10:6154–68.

[10] Jungerman FS, Alves HNP, Carmona MJC, Conti NB, Malbergier A. Anesthetic drug abuse by anesthesiologists. Rev Bras Anestesiol 2012;62:380–6.

[11] Kintz P, Villain M, Dumestre V, Cirimele V. Evidence of addiction by anesthesiologists as documented by hair analysis. Forensic Sci Int 2005;153:81–4.

[12] Tetzlaff J, Collins GB, Brown DL, Leak BC, Pollock G, Popa D. A strategy to prevent substance abuse in an academic anesthesiology department. J Clin Anesth 2010;2:143–50.

[13] Agius R, Nadulski T, Kahl HG, Dufaux B. Ethyl glucuronide in hair—A highly effective test for the monitoring of alcohol consumption. Forensic Sci Int 2012;218:10–14.

[14] Frone MR. Does a permissive workplace substance use climate affect employees who do not use alcohol and drugs at work? A U.S. national study. Psychol Addict Behav 2009;23:386–90.

[15] Hickox SA. Drug testing of medical marijuana users in the workplace: an inaccurate test of impairment. Hofstra Lab Emp L J 2011;29:273–341.

[16] Moore J. Drug testing and corporate responsibility: the "ought implies can" argument. J Bus Ethics 1989;8:279–87.

[17] Quest Diagnostics Blog. <http://blog.employersolutions.com/wp-content/uploads/2013/11/dti25-press-release-11-18-2013.pdf>; 2013 [accessed 09.12.13].

[18] Olbina S, Hinze J, Arduengo C. Drug testing practices in the US construction industry in 2008. Constr Manage Econ 2011;29:1043–57.

[19] Fortner NA, Martin DM, Esen SE, Shelton L. Employee drug testing: study shows improved productivity and attendance and decreased workers' compensation and turnover. J Glob Drug Policy Pract 2011;5:1–22.

[20] Musshoff F, Driever F, Lachenmeier K, Lachenmeier DW, Banger M, Madea B. Results of hair analyses for drugs of abuse and comparison with self-reports and urine tests. Forensic Sci Int 2006;156:118–23.

[21] Tsanaclis LM, Wicks JF, Chasin AA. Workplace drug testing, different matrices different objectives. Drug Test Anal 2012;4:83–8.

[22] Pierce A. Regulatory aspects of workplace drug testing in Europe. Drug Test Anal 2012;4:62–5.

[23] Kintz P, Villain M, Cirimele V. Hair analysis for drug detection. Ther Drug Monit 2006;28:442–6.

[24] Kronstrand R, Scott K. Drug incorporation into hair. In: Kintz P, editor. Analytical and practical aspects of drug testing in hair. London: CRC Press; 2007. p. 1–23.

[25] Caplan YH, Goldberger BA. Alternative specimens for workplace drug testing. J Anal Toxicol 2001;25:396–9.
[26] Baumgartner WA, Hill VA, Blahd WH. Hair analysis for drugs of abuse. J Forensic Sci 1989;34:1433–53.
[27] Kintz P. Drug testing in addicts: a comparison between urine, sweat, and hair. Ther Drug Monit 1996;18:450–5.
[28] Sachs H, Kintz P. Testing for drugs in hair: critical review of chromatographic procedures since 1992. J Chromatogr B Biomed Sci Appl 1998;713:147–61.
[29] Barroso M, Gallardo E, Queiroz JA. Bioanalytical methods for the determination of cocaine and metabolites in human biological samples. Bioanalysis 2009;1:977–1000.
[30] Boumba VA, Ziavrou KS, Vougiouklakis T. Hair as a biological indicator of drug use, drug abuse or chronic exposure to environmental toxicants. Int J Toxicol 2006;25:143–63.
[31] Gallardo E, Barroso M, Queiroz JA. LC-MS: a powerful tool in workplace drug testing. Drug Test Anal 2008;1:109–15.
[32] Lund HME, Gjerde H, de Courtade SMB, Øiestad EL, Christophersen AS. A Norwegian study of the suitability of hair samples in epidemiological research of alcohol, nicotine and drug use. J Anal Toxicol 2013;37:362–8.
[33] Saito K, Saito R, Kikuchi Y, Iwasaki Y, Ito R, Nakazawa H. Analysis of drugs of abuse in biological specimens. J Health Sci 2011;7:472–87.
[34] Smolders R, Schramm KW, Nickmilder M, Schoeters G. Applicability of non-invasively collected matrices for human biomonitoring. Environ Health 2009;8:1–10.
[35] Bush DM. The US mandatory guidelines for federal workplace drug testing programs: current status and future considerations. Forensic Sci Int 2008;174:111–19.
[36] Barroso M, Gallardo E. Hair analysis for forensic applications: is the future bright? Bioanalysis 2014;6:1–3.
[37] Pragst F, Balikova MA. State of the art in hair analysis for detection of drug and alcohol abuse. Clin Chim Acta 2006;370:17–49.
[38] Cook RF, Bernstein AD, Andrews CM. Assessing drug use in the workplace: a comparison of self-report, urinalysis, and hair analysis. NIDA Res Monogr 1997;67:247–72.
[39] Hersch RK, McPherson TL, Cook RF. Substance use in the construction industry: a comparison of assessment methods. Subst Use Misuse 2002;37:1331–58.
[40] Ledgerwood DM, Goldberger BA, Risk NK, Lewis CE, Kato Price R. Comparison between self-report and hair analysis of illicit drug use in a community sample of middle-aged men. Addict Behav 2008;33:1131–9.
[41] Vignali C, Stramesi C, Vecchio M, Groppi A. Hair testing and self-report of cocaine use. Forensic Sci Int 2012;215:77–80.
[42] Mieczkowski T. Urinalysis and hair analysis for illicit drugs of driver applicants and drivers in the trucking industry. J Forensic Leg Med 2010;7:254–60
[43] Dufaux B, Agius R, Nadulski T, Kahl HG. Comparison of urine and hair testing for drugs of abuse in the control of abstinence in driver's license re-granting. Drug Test Anal 2012;4:415–19.
[44] Fulga N. Quality management and accreditation in a mixed research and clinical hair testing analytical laboratory setting-a review. Ther Drug Monit 2013;35:283–7.
[45] Agius R, Kintz P, European Workplace Drug Testing Society. Guidelines for European workplace drug and alcohol testing in hair. Drug Test Anal 2010;2:367–76.
[46] Cooper GA, Kronstrand R, Kintz P. Society of hair testing. Society of hair testing guidelines for drug testing in hair. Forensic Sci Int 2012;218:20–4.

[47] Andraus M, Tsanaclis L, Sodré C, Morales L, Pisaneschi C, Salvadori M, et al. Hair analysis in pre-employment of high-risk activities in Brazil. Paper presented at the 18th Annual Meeting of the Society of Hair Testing (SoHT); 2013 August 27–29. Geneva, Switzerland.

[48] Stramesi C, Polla M, Vignali C, Zucchella A, Groppi A. Segmental hair analysis in order to evaluate driving performance. Forensic Sci Int 2008;176:34–7.

[49] Harding H, Rogers G. Physiology and growth of human hair. Forensic Examination of Hair, T&F Foren; 1999. p. 1–77.

[50] LeBeau MA, Montgomery MA, Brewer JD. The role of variations in growth rate and sample collection on interpreting results of segmental analyses of hair. Forensic Sci Int 2011;210:110–16.

[51] Pianta A, Liniger B, Baumgartner MR. Ethyl glucuronide in scalp and non-head hair: an intra-individual comparison. Alcohol Alcohol 2013;1–8.

[52] Randall VA, Botchkareva NV. The biology of hair growth. Cosmetic applications of laser and light based systems. Dermatol Clin 2008;3–35.

[53] Seago SV, Ebling FJ. The hair cycle on the human thigh and upper arm. B J Dermatol 1985;113:9–16.

[54] Han E, Yang W, Lee J, Park Y, Kim E, Lim M, et al. Correlation of methamphetamine results and concentrations between head, axillary, and pubic hair. Forensic Sci Int 2005;47:21–4.

[55] Lee S, Han E, In S, Choi H, Chung H, Chung KH. Analysis of pubic hair as an alternative specimen to scalp hair: a contamination issue. Forensic Sci Int 2011;206:19–21.

[56] Felli M, Martello S, Marsili R, Chiarotti M. Disappearance of cocaine from human hair after abstinence. Forensic Sci Int 2005;154:96–8.

[57] Mieczkowski T. Distinguishing passive contamination from active cocaine consumption: assessing the occupational exposure of narcotics officers to cocaine. Forensic Sci Int 1997;84:87–111.

[58] SAMHSA (Substance Abuse and Mental Health Services Administration). Notice of proposed revisions to the mandatory guidelines for federal workplace drug testing programs. Fed Regist 2004;69:19673–732.

[59] Uhl M, Sachs H. Cannabinoids in hair: strategy to prove marijuana/hashish consumption. Forensic Sci Int 2004;145:143–7.

[60] Jurado C. Hair analysis of cocaine. In: Kintz P, editor. Analytical and practical aspects of drug testing in hair. London: CRC Press; 2007. p. 95–125.

[61] Aleksa K, Walasek P, Fulga N, Kapur B, Gareri J, Koren G. Simultaneous detection of seventeen drugs of abuse and metabolites in hair using solid phase micro extraction (SPME) with GC/MS. Forensic Sci Int 2012;218:31–6.

[62] Baumgartner MR, Guglielmello R, Fanger M, Kraemer T. Analysis of drugs of abuse in hair: evaluation of the immunochemical method VMA-T vs. LC-MS/MS or GC-MS. Forensic Sci Int 2012;215:56–9.

[63] Coulter C, Tuyay J, Taruc M, Moore C. Semi-quantitative analysis of drugs of abuse, including tetrahydrocannabinol in hair using aqueous extraction and immunoassay. Forensic Sci Int 2010;196:70–3.

[64] Hill V, Cairns T, Schaffer M. Hair analysis for cocaine: factors in laboratory contamination studies and their relevance to proficiency sample preparation and hair testing practices. Forensic Sci Int 2008;176:23–33.

[65] Huestis MA, Gustafson RA, Moolchan ET, Barnes A, Bourland JA, Sweeney SA, et al. Cannabinoid concentrations in hair from documented cannabis users. Forensic Sci Int 2007;169:129–36.

[66] Miyaguchi H, Takahashi H, Ohashi T, Mawatari K, Iwata YT, Inoue H, et al. Rapid analysis of methamphetamine in hair by micropulverized extraction and microchip-based competitive ELISA. Forensic Sci Int 2009;184:1–5.

[67] Musshoff F, Kirschbaum KM, Graumann K, Herzfeld C, Sachs H, Madea B. Evaluation of two immunoassay procedures for drug testing in hair samples. Forensic Sci Int 2012;215:60–3.

[68] Pujol ML, Cirimele V, Tritsch PJ, Villain M, Kintz P. Evaluation of the IDS One-Step ELISA kits for the detection of illicit drugs in hair. Forensic Sci Int 2007;170:189–92.

[69] de la Torre R, Civit E, Svaizer F, Lotti A, Gottardi M, Miozzo M. High throughput analysis of drugs of abuse in hair by combining purposely designed sample extraction compatible with immunometric methods used for drug testing in urine. Forensic Sci Int 2010;196:18–21.

[70] Tsanaclis L, Wicks JFC. Patterns in drug use in the United Kingdom as revealed through analysis of hair in a large population sample. Forensic Sci Int 2007;70:121–8.

[71] Angeli I, Minoli M, Ravelli A, Gigli F, Lodi F. Automated fast procedure for the simultaneous extraction of hair sample performed with an automated workstation. Forensic Sci Int 2012;218:15–19.

[72] Broecker S, Herre S, Pragst F. General unknown screening in hair by liquid chromatography–hybrid quadrupole time-of-flight mass spectrometry (LC-QTOF-MS). Forensic Sci Int 2012;218:68–81.

[73] Di Corcia D, D'urso F, Gerace E, Salomone A, Vincenti M. Simultaneous determination in hair of multiclass drugs of abuse (including THC) by ultra-high performance liquid chromatography–tandem mass spectrometry. J Chromatogr B Analyt Technol Biomed Life Sci 2012;899:154–9.

[74] Domínguez-Romero JC, García-Reyes JF, Molina-Díaz A. Screening and quantitation of multiclass drugs of abuse and pharmaceuticals in hair by fast liquid chromatography electrospray time-of-flight mass spectrometry. J Chromatogr B Analyt Technol Biomed Life Sci 2011;879:2034–42.

[75] Emídio ES, de Menezes Prata V, De Santana FJM, Dórea HS. Hollow fiber-based liquid phase microextraction with factorial design optimization and gas chromatography–tandem mass spectrometry for determination of cannabinoids in human hair. J Chromatogr B Analyt Technol Biomed Life Sci 2010;878:2175–83.

[76] Favretto D, Vogliardi S, Stocchero G, Nalesso A, Tucci M, Ferrara SD. High performance liquid chromatography–high resolution mass spectrometry and micropulverized extraction for the quantification of amphetamines, cocaine, opioids, benzodiazepines, antidepressants and hallucinogens in 2.5 mg hair samples. J Chromatogr A 2011;1218:6583–95.

[77] Imbert L, Dulaurent S, Mercerolle M, Morichon J, Lachâtre G, Gaulier JM. Development and validation of a single LC-MS/MS assay following SPE for simultaneous hair analysis of amphetamines, opiates, cocaine and metabolites. Forensic Sci Int 2014;234:132–8.

[78] Kim JY, Shin SH, Lee JI, In MK. Rapid and simple determination of psychotropic phenylalkylamine derivatives in human hair by gas chromatography–mass spectrometry using micro-pulverized extraction. Forensic Sci Int 2010;196:43–50.

[79] Lendoiro E, Quintela Ó, de Castro A, Cruz A, López-Rivadulla M, Concheiro M. Target screening and confirmation of 35 licit and illicit drugs and metabolites in hair by LC-MSMS. Forensic Sci Int 2012;217:207–15.

[80] Míguez-Framil M, Moreda-Piñeiro A, Bermejo-Barrera P, Álvarez-Freire I, Tabernero MJ, Bermejo AM. Matrix solid-phase dispersion on column clean-up/pre-concentration as

a novel approach for fast isolation of abuse drugs from human hair. J Chromatogr A 2010;1217:6342–9.

[81] Montesano C, Johansen SS, Nielsen MKK. Validation of a method for the targeted analysis of 96 drugs in hair by UPLC-MS/MS. J Pharm Biomed Anal 2014;88:295–306.

[82] Morini L, Vignali C, Polla M, Sponta A, Groppi A. Comparison of extraction procedures for benzodiazepines determination in hair by LC-MS/MS. Forensic Sci Int 2012;218:53–6.

[83] Pelander A, Ristimaa J, Rasanen I, Vuori E, Ojanperä I. Screening for basic drugs in hair of drug addicts by liquid chromatography/time-of-flight mass spectrometry. Ther Drug Monit 2008;30:717–24.

[84] Pragst F, Broecker S, Hastedt M, Herre S, Andresen-Streichert H, Sachs H, et al. Methadone and illegal drugs in hair from children with parents in maintenance treatment or suspected for drug abuse in a German community. Ther Drug Monit 2013;35:737–52.

[85] Rothe M, Pragst F, Thor S, Hungen J. Effect of pigmentation on the drug deposition in hair of grey-haired subjects. Forensic Sci Int 1997;84:53–60.

[86] Rust KY, Baumgartner MR, Dally AM, Kraemer T. Prevalence of new psychoactive substances: a retrospective study in hair. Drug Test Anal 2012;4:402–8.

[87] Stout PR, Ropero-Miller JD, Baylor MR, Mitchell JM. Morphological changes in human head hair subjected to various drug testing decontamination strategies. Forensic Sci Int 2007;172:164–70.

[88] Tsanaclis L, Wicks JF. Differentiation between drug use and environmental contamination when testing for drugs in hair. Forensic Sci Int 2008;176:19–22.

[89] Tsanaclis L, Nutt J, Bagley K, Bevan S, Wicks J. Differentiation between consumption and external contamination when testing for cocaine and cannabis in hair samples. Drug Test Anal 2014;6:37–41.

[90] Andraus M, Tsanaclis L, Ribeiro M, Laranjeira R, Pisaneschi C, Salvadori M, et al. O7: evaluation of hair analysis with socio-demographic profile of crack users in Brazil. Toxicol Analyt Clin 2014;26:S7–8.

[91] Nadulski T, Pragst F. Simple and sensitive determination of Δ9tetrahydrocannabinol, cannabidiol and cannabinol in hair by combined silylation, headspace solid phase microextraction and gas chromatography–mass spectrometry. J Chromatogr B Analyt Technol Biomed Life Sci 2007;846:78–85.

[92] Vearrier D, Curtis JA, Greenberg MI. Biological testing for drugs of abuse. In: Luch A, editor. Molecular, clinical and environmental toxicology. Basel: Birkhäuser; 2010. p. 489–517.

[93] Wada M, Ikeda R, Kuroda N, Nakashima K. Analytical methods for abused drugs in hair and their applications. Anal Bioanal Chem 2010;397:1039–67.

[94] Vincenti M, Salomone A, Gerace E, Pirro V. Role of LC-MS/MS in hair testing for the determination of common drugs of abuse and other psychoactive drugs. Bioanalysis 2013;5:1919–38.

[95] Eichhorst JC, Etter ML, Rousseaux N, Lehotay DC. Drugs of abuse testing by tandem mass spectrometry: a rapid, simple method to replace immunoassays. Clin Biochem 2009;42:1531–42.

[96] Jurado C, Sachs H. Proficiency test for the analysis of hair for drugs of abuse, organized by the Society of Hair Testing. Forensic Sci Int 2003;133:175–8.

[97] SoHT- Society of Hair Testing. Recommendations for hair testing in forensic, cases. Forensic Sci Int 2004;145:83–4.

[98] Musshoff F, Madea B. New trends in hair analysis and scientific demands on validation and technical notes. Forensic Sci Int 2007;165:204–15.

[99] Cirimele V, Kintz P, Gosselin O, Ludes B. Clozapine dose concentration relationships in plasma, hair and sweat specimens of schizophrenic patients. Forensic Sci Int 2000;107:289–300.

[100] Goullé JP, Noyon J, Layet A, Rapoport NF, Vaschalde Y, Pignier Y, et al. Phenobarbital in hair and drug monitoring. Forensic Sci Int 1995;70:191–202.

[101] Pragst F, Rothe M, Hunger J, Thor S. Structural and concentration effects on the deposition of tricyclic antidepressants in human hair. Forensic Sci Int 1997;84:225–36.

[102] Takiguchi Y, Ishihara R, Toni M, Kato R, Kamihara S, Uematsu T. Hair analysis of flecainide for assessing the individual drug-taking behavior. Eur J Clin Pharmacol 2002;58:99–101.

[103] Tracqui A, Kintz P, Mangin P. Hair analysis: a worthless tool for therapeutic compliance monitoring. Forensic Sci Int 1995;70:183–9.

[104] Williams J, Patsalos PN, Mei Z, Schapel G, Wilson JF, Richens A. Relation between dosage of carbamazepine and concentration in hair and plasma samples from a compliant inpatient epileptic population. Ther Drug Monit 2001;23:15–20.

[105] Cairns T, Hill V, Schaffer M, Thistle W. Amphetamines in washed hair of demonstrated users and workplace subjects. Forensic Sci Int 2004;145:137–42.

[106] Polettini A, Cone EJ, Gorelick DA, Huestis MA. Incorporation of methamphetamine and amphetamine in human hair following controlled oral methamphetamine administration. Anal Chim Acta 2012;726:35–43.

[107] Cairns T, Hill V, Schaffer M, Thistle W. Levels of cocaine and its metabolites in washed hair of demonstrated cocaine users and workplace subjects. Forensic Sci Int 2004;145:175–81.

[108] Kintz P. Issues about axial diffusion during segmental hair analysis. Ther Drug Monit 2013;35:408–10.

[109] Kelly RC, Mieczkowski T, Sweeney SA, Bourland JA. Hair analysis for drugs of abuse: hair color and race differentials or systematic differences in drug preferences? Forensic Sci Int 2000;107:63–86.

[110] Nielsen MKK, Johansen SS, Linnet K. Pre-analytical and analytical variation of drug determination in segmented hair using ultra-performance liquid chromatography–tandem mass spectrometry. Forensic Sci Int 2014;234:16–21.

[111] Kintz P. Value of the concept of minimal detectable dosage in human hair. Forensic Sci Int 2012;218:28–30.

[112] Nutt J, Tsanaclis L, Bevan S, Bagley K, Blackwell K, Wicks J. Hair analysis in the workplace: global harmonisation required. Paper presented at the 19th Annual Meeting of the Society of Hair Testing (SoHT); 2014 June 10–13. Bordeaux, France.

[113] Krause K, Foitzik K. Biology of the hair follicle: the basics. Semin Cutan Med Surg 2006;25:2–10.

Chapter 9

Forensic Applications of Hair Analysis

Carmen Jurado
National Institute of Toxicology and Forensic Sciences, Seville, Spain

9.1 INTRODUCTION

Toxicology is the study of adverse effects of drugs and chemicals on biological systems. Forensic toxicology involves the application of toxicology for the purposes of the law, or in a medicolegal context.

Hair analysis has been receiving increased attention in recent years and, currently, it has become the third most fundamental biological matrix used for drug testing in forensic toxicology, after blood and urine. Figure 9.1 shows the evolution in the number of cases analyzed in a forensic laboratory, from January 2004 to December 2013 [1]. A total of 6,000 cases were analyzed during the 10 years, and all of them were included in the study. An increase was noticeable from 2004, with 484 cases, to 2005, with 649, followed by a steady state during 2006 and 2007: 628 and 673 cases, respectively. In 2008 the number increased again to 792, which was maintained in 2009 with 780. Since them, the cases were going down during the last 3 years, with a short recovery in 2013.

Forensic toxicology involves a large setting of applications and in most of them hair analyses are useful and even, sometimes, they are essential. As a consequence a large variety of literature, related to forensic applications of hair testing, has been published to date.

The aim of this chapter is to review, summarize, and discuss different applications of hair analysis in forensic toxicology, based on the bibliography and on the experience of a forensic toxicological laboratory, which only performs analyses derived from Court procedures.

9.2 ADVANTAGES OF HAIR ANALYSIS

The main advantage of hair is the wide diagnostic window of detection allowed by this specimen. It is limited only by the length of the hair and ranges from weeks to months or even years.

Hair Analysis in Clinical and Forensic Toxicology.

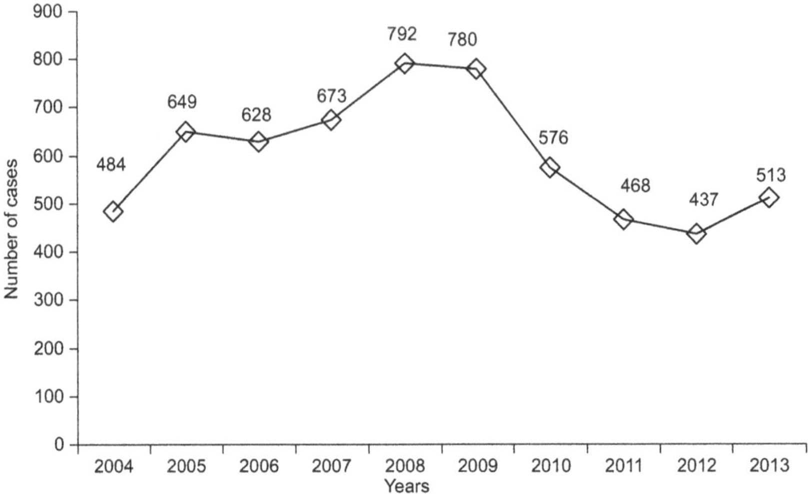

FIGURE 9.1 Evolution in hair analysis related cases during 10 years (2004–2013).

A particular characteristic of these analyses is the possibility of establishing the chronological profile. Assuming the ideal uniform hair growth of 1 cm/month, segmental hair analysis can provide the information about the time course of drug consumption.

Another advantage of hair analysis is that although quantitative correlation between drug measurements in hair and the quantity of drug consumed is not feasible, hair analysis provides information about the assiduity or severity of consumption, making possible to know if the person is a heavy, medium, or light consumer of drugs.

Several methods have been proposed to establish the ranges of concentrations corresponding to each one of the types of consumption. Pepin and Gaillard [2] compared the concentrations of cocaine and opiate compounds in hair with the amount of drug consumed (low, medium, or high) which was estimated on the basis of self-reported consumption (Table 9.1). The major problem of this protocol is that the doses consumed can be over- or underestimated and the establishment of drug purity, which is unknown in street drug samples.

A second method is based on retrospective population studies. Positive results are compared with the distribution of the concentrations obtained in the analysis of the different compounds [3–5]. The range with the highest incidence would correspond to medium consumption, lower and higher concentrations would be in the low and high range, respectively.

More recently, several authors have proposed statistical studies to establish the ranges of concentrations. The parameters studied were: mean and minimum concentrations, quartile 25, median concentration, quartile 75,

TABLE 9.1 Severity of Drug Consumption Based on Self-Reported Data [2] and Distribution of Concentrations [3,4,5]

Consumption	Drug Analyzed	Low	Medium	High	Reference
Heroin	6-MAM	0.5–5	5–10	>10	[3]
		0.5–2	2–10	>10	[2]
		0.5–5	5–15	>15	[4,5]
Cocaine	Cocaine	1–5	5–20	>20	[3]
		1–4	4–20	>20	[2]
		1–10	10–20	>20	[4,5]
Cannabis	THC	0.05–0.1	0.1–0.4	>0.4	[4,5]

6-MAM: 6-monoacetylmorphine; THC: Δ^9-tetrahydrocannabinol.

and maximum concentrations. Low consumption is considered when the concentration is higher than minimum and lower than quartile 25. In the practice it is recommended to apply the cutoff proposed by Society of Hair Testing (SoHT) [6] as the minimum concentration. Medium consumption is considered when the concentrations are in the range between quartile 25 and quartile 75. This range is quite large and could be divided into medium low (from quartile 25 to median concentrations) and medium high (from median to quartile 75). Finally, high consumption is considered when the concentrations are higher than quartile 75. Table 9.2 shows the data from three population studies. The ranges of concentrations are considerably different, probably due to differences in the population under study. They were performed in different countries—Switzerland [7], United Kingdom [8,9], and Spain [1]—and in different years, 2001 [7], 2004–2007 [8,9], and 2004–2013 [1], respectively.

For the correct interpretation of the results derived from these statistical studies, it is recommended that every laboratory should evaluate its own values based on its experience and population. In addition, it should be convenient to annually review the data because several papers have demonstrated changes in the evolution of drug concentrations. For example, Jurado and Soriano [1] determined the evolution in the incidence of cannabis-, heroin-, and cocaine-positive cases from January 2004 to December 2012. Figure 9.2 shows the evolution in the median concentrations of the three compounds per year. An increase in Δ^9-tetrahydrocannabinol (THC) concentrations was expected, since an increase in THC percentage in street drug samples has occurred during the later years. Nevertheless, the pattern of THC was similar during the 10-year period, median concentrations ranged

TABLE 9.2 Severity of Drug Consumption Based on Statistical Studies of Heroin, Cocaine, and Cannabis Consumers

	Jurado and Staub [7]	Cordero et al. [8] and Lee et al. [9]	Jurado and Soriano [1]
6-MAM concentrations (heroin consumption)			
Minimum	0.1	0.1	0.20
Percentile 25	1.3	0.9	1.8
Median	3.3	3.2	7.1
Percentile 75	6.3	12.5	11.1
Maximum	65.0	154.1	127.9
Cocaine concentrations (cocaine consumption)			
Minimum		10 ng total	0.5
Percentile 25		0.8	2.2
Median		4.0	9.5
Percentile 75		18.9	38.5
Maximum		384.7	815.8
THC concentrations (cannabis consumption)			
Minimum			0.05
Percentile 25			0.13
Median			0.30
Percentile 75			0.63
Maximum			24.7

6-MAM: 6-monoacetylmorphine; THC: Δ^9-tetrahydrocannabinol.
Source: Data from different authors [1,7–9].

from 0.28 ng/mg in 2004 to 0.30 ng/mg in 2012. A similar profile was found for 6-monoacetylmorphine (6-MAM), marker of heroin consumption. Median concentrations of 6-MAM suffered very low changes, ranging from 5.33 ng/mg in 2004 to 7.14 in 2012. Surprisingly, cocaine concentrations presented a noticeable decrease along with the time, from 26.89 ng/mg in 2005 to 4.91 ng/mg in 2010. A small recovery was found in 2011 and 2012, when the median concentration increased to 9.51 ng/mg.

In addition to these ranges of concentrations, which can be used as guidelines, all circumstances and aspects of the individual case has to be considered for the correct interpretation of hair results.

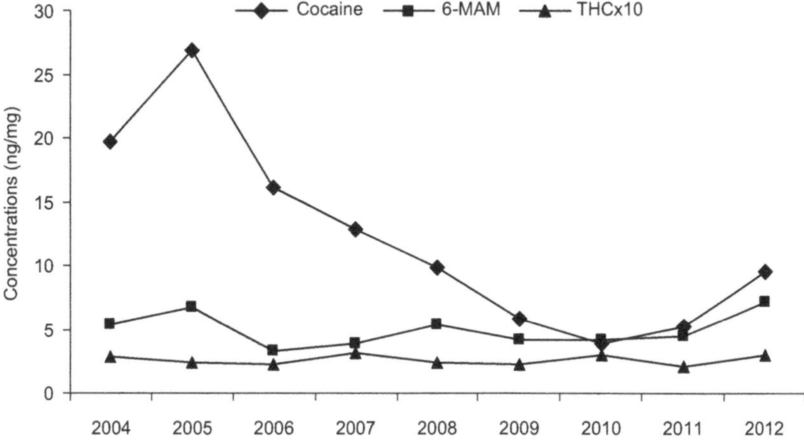

FIGURE 9.2 Evolution of cocaine, 6-monoacetylmorphine (6-MAM) and Δ^9-tetrahydrocannabinol (THC) median concentrations from 2004 to 2012 [1].

9.3 LIMITATIONS OF HAIR ANALYSIS

When interpreting hair analysis results, one of the most frequently asked questions is whether there is a predictable relationship between the amount of drug detected in the hair and the dose consumed. The papers published on this matter show major controversy between pros and cons.

Some authors do not find any relationship. Puschel et al. [10] compared the concentrations of drugs of abuse present in the hair of 13 drug abusers with self-reported consumption, but they were unable to establish any positive correlation. Similar results were obtained by Kintz et al. [11] in a study with 20 subjects taking part in a heroin-maintenance program. No correlation between the doses of administered heroin and the concentrations of total opiates in hair was observed ($r = 0.346$). However, when considering a single analyte, it was observed that the correlation coefficient seemed to be linked to its plasma half-life. The longer plasma half-life the better correlation coefficient, as $r = 0.12$, 0.25, and 0.64 for heroin, MAM, and morphine, respectively.

Opposite results were obtained by Musshoff et al. [12] with 47 subjects enrolled in another heroin-maintenance program. A correlation between the dose and the total opiate could be established for drug users who were administered prescribed doses of heroin ($r = 0.66$). Similarly to the previous study, when considering a single analyte, the coefficients of correlation increased with the respective plasma half-life ($r = 0.42$, $r = 0.58$, and $r = 0.69$ for heroin, MAM, and morphine).

Welp et al. [13] tried to assess the relationship between self-reported drug intake and the concentration of drugs and their metabolites in hair. The study was performed in Amsterdam with 95 subjects reporting consumption of cocaine, heroin, methadone, and/or amphetamines at least 3 days/week, the correlation coefficients ranged from 0.45 and 0.59, which increased to a range from 0.63 to 0.87 when parameters as hair color, sex, or race were considered.

Some individual circumstances may mediate as biasing factors in the dose-concentration relationship, including the hair growth cycle, which is not homogeneous, hair color, and cosmetic treatments. Differences in sweat and sebum secretions probably exert a certain influence, as well. Consequently, from the results obtained in hair analysis, we can deduce that, intraindividually, a higher concentration corresponds to a higher consumption, and vice versa. However, there is no interindividual correlation between frequency of drug use or drug dose and hair concentrations [14].

An important limitation of hair testing is that it cannot be used to prove impairment or to demonstrate that a person was under the influence in a specific moment.

In cases of sequential analysis, the choice of the length of the segments should depend on the requirements. Nevertheless, the segments should be at least 0.3–0.5 cm, which correspond to 0.3–0.5 months of drug consumption. Results from shorter segments are not reliable. Concentrations found in each of the segments represent the mean consumption during the period of time covered by the length of the segment. For example, quantitative results obtained in the 1 cm segment closest to the root would represent the mean concentrations of drugs consumed during the month before sampling. The most specific conclusion we can achieve is that during the month before sampling, the concentrations of the different compounds are in the range corresponding to high, medium, or low consumption, according to the population.

Finally the degree of drug addiction cannot be established by hair analysis. Although theses analyses provide information about the time and severity of consumption or can demonstrate the chronological or sequential profile of drug consumption, a physical or psychical addiction cannot be inferred from them.

9.4 APPLICATIONS OF HAIR ANALYSIS IN FORENSIC TOXICOLOGY

In this part of the chapter the following applications of hair analysis in forensic toxicology will be discussed:

- Postmortem toxicology
- Drug-facilitated crimes (DFCs)
- Divorce and child custody proceedings
- Follow-up of detoxification programs

9.4.1 Postmortem Toxicology

For the forensic toxicologist, the most important question to be addressed in postmortem cases is to know the substances consumed just before death and if they were the cause or could have contributed to the death. The first step consists in an adequate selection of the specimens to be analyzed. It may be based on case history and availability for a given case.

The autopsy allows the forensic pathologist a one-time opportunity to collect as many specimens as may be needed to complete the toxicological investigation. On the one hand, the ideal situation would be to perform toxicological analysis in all the specimens, since each one of them provides us with different information. But on the other hand, we have to be realistic and it is not possible to perform a complete study of all the available specimens in all the postmortem cases in the routine of a forensic toxicological lab.

Blood or urine are the matrices of choice in postmortem toxicology, since they provide short-term information of an individual's drug exposure and thus can give a more accurate picture of the situation at the time of death. Nevertheless, the large detection window achieved with hair analysis makes this matrix especially useful, or even essential, in some cases. For example:

- Putrefied or even mummified cadavers, when other biological samples are not available;
- Loss of tolerance after a period of abstinence;
- Long-term chronic consumption.

9.4.1.1 Putrefied and Mummified Cadavers

In cases of extreme putrefaction the physical state of the body will determine what specimens are available for collection [15]. Blood and urine are usually not available and body fluids in this type of cadavers are liquefied tissues, which provide very little information about the presence of drugs in the corpse. Consequently, muscular tissue, bones, and hair are the only useful specimens for toxicological analysis.

Although none of them allows the diagnosis of a lethal intoxication or acute poisoning, the analysis of hair root informs about consumption 1 or 2 days before death. Acute poisoning with methamphetamine (MA) was established by Nakahara et al. [16] by analyzing the hair roots of four men who died due to MA overdose. They detected MA at high concentrations, 61.0, 68.8, 97.6, and 134.6 ng/mg and its metabolite amphetamine was also detected at the concentrations of 1.2, 4.9, 4.1, and 9.0 ng/mg, respectively. They concluded that hair root is a good specimen to probe acute poisoning.

Jurado [5] presented the case of a putrefied cadaver which was found in the countryside. It was not possible to obtain any biological fluid or tissue for toxicological analysis, but a sample of scalp hair was collected at the autopsy and sent for analysis. Hair shafts were carefully removed from the scalp in order to maintain the roots. The lock of hair was sequentially analyzed in three segments: the first segment from the root to 0.5 cm; the second segment from 0.5 to 2 cm; and the last one from 2 to 4 cm. Cocaine, opiate, and cannabis compounds were detected in the first segment, demonstrating consumption of cocaine, heroin, and cannabis during 2 or 3 days before death. From the concentrations of the three families of drugs it was possible to establish heavy consumption of cocaine, heroin, and cannabis during the 4 months previous to the death.

Several studies have demonstrated the usefulness of hair to establish drug consumption in mummies. The first studies found benzoylecgonine (BE), major cocaine metabolite, in the hair of Chilean mummies dated from 2000 B.C. to 1500 A.D. [17]. Two years later, the same authors reported the presence of cocaine and its two metabolites BE and ecgonine methyl ester in hair of Peruvian coca leaf chewers dated around 1000 A.D. [18]. Wilson et al. [19] examined three frozen bodies, a 13-year-old girl and a girl and boy aged 4–5 years, separately entombed near the Andean summit of volcano Llullaillaco (Argentina). Hair analyses gave positive results for cocaine, BE, and cocaethylene, indicating that all three children had consumed cocaine and alcohol.

More recently, these analyses have been extended to alcohol markers. Musshoff et al. [20], within the "Sicily Mummy Project" analyzed hair samples from 38 mummies to determine the presence of ethyl glucuronide (EtG). All samples were analyzed in two segments. About 31 out of 76 segments were positive for EtG, with concentrations ranging from 2.5 to 531.3 pg/mg (mean 73.8, median 13.3 pg/mg). The authors concluded that EtG analyses can be performed on mummy hair samples even several hundred years after death to identify evidence for significant alcohol consumption during life.

9.4.1.2 Loss of Tolerance

Tolerance is a phenomenon in which a subject develops a neuroadaptation to a psychoactive substance requiring increasingly larger doses to achieve the same pharmacological effect. This occurs mainly for drugs of abuse, especially opiates, and rarely for prescribed drugs, with the exception of patients on long-term pain relief from morphine or those prescribed benzodiazepines for anxiety.

The development of tolerance to opiates is particularly important, since it affects the interpretation of blood concentrations. In heroin-related deaths blood morphine concentrations vary substantially, from nanograms to milligrams per liter. For example, in 179 postmortem blood samples of heroin users, morphine concentrations were in the range 0.01–2.29 mg/L (mean 0.29 mg/L and median 0.16 mg/L). Consequently, these data cannot be used in isolation to diagnose an overdose [21–23]. In cases where low blood morphine levels are found, it is often assumed that the victim either was a novice user, or had had a drug-free period prior to the final dose. In the absence of a reliable biomarker of opioid tolerance, a reliable alternative is to perform hair analysis for opiates.

Darke et al. [24] reported a positive correlation between hair and blood morphine levels in heroin fatalities and suggested that low blood morphine concentrations reflected a less frequent heroin use, and hence a reduced tolerance to opioids.

Opposite conclusions were achieved by Druid et al. [25] after examining the role of abstinence in drug-related deaths. They compared postmortem blood morphine concentrations with recent and past exposure to opioids, established by segmental hair analysis in 28 heroin fatalities. In 18 of them opioids were absent in the most recent hair segment, suggesting a recent opioid abstinence and, consequently, a reduced tolerance to opioids. However, the blood morphine concentrations were similar in both groups (median concentrations were 0.16 and 0.15 μg/g in abstinent and tolerant, respectively). These finding are in contradiction with the fact that abstinence, or loss of tolerance, is directly involved in heroin "overdose" death. Because, excluding interindividual variability, tolerant subjects died after the intake of similar doses to those that killed abstinent subjects.

A heroin-related death is shown in Figure 9.3 and Table 9.3. Toxicological analyses of blood and urine gave positive results for opiates (morphine and codeine), methadone and its metabolite 2-ethylidene-1,5-dimethyl-3,3-diphenylpyrrolidine (EDDP), and Δ^9-tetrahydrocannabinol carboxylic acid (THC-COOH) (Table 9.3). The ratio of morphine:codeine indicates recent heroin consumption. Nevertheless, the death could not be attributed to heroin overdose since opiates concentrations were lower than the lethal range. Segmental analysis was performed in a lock of hair 12 cm in length.

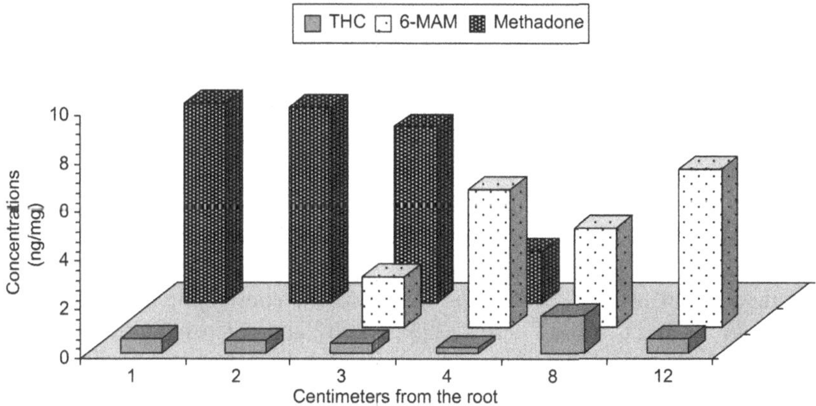

FIGURE 9.3 Chronological profile of Δ^9-tetrahydrocannabinol (THC), 6-monoacetylmorphine (MAM), and methadone concentrations, in a death due to loss of tolerance to heroin.

TABLE 9.3 Blood and Urine Analysis Results in a Heroin-Related Death

	Concentrations (mg/L)	
Compound	**Blood**	**Urine**
Ethanol	ND	ND
Morphine	0.16	0.48
Codeine	0.07	0.10
Methadone	0.10	1.88
EDDP	0.02	3.11
THC-COOH	ND	0.03

ND: Not detected; EDDP: 2-ethylidene-1,5-dimethyl-3,3-diphenylpyrrolidine; THC-COOH: Δ^9-tetrahydrocannabinol carboxylic acid.

Results corroborated the history of heroin, methadone, and cannabis consumption. The chronological profile showed fairly uniform MAM concentrations in the segments from 3 to 12 cm further to the root, while the two proximal segments were negative (Figure 9.3), demonstrating heroin abstinence during the last 2–3 months. Heroin addicts do not know the loss of tolerance and usually adjust the dose of heroin to that used before the abstinence.

9.4.1.3 Multiorgan Damage Caused by Long-Term Consumption

Long-term abuse of alcohol, drugs of abuse or, even, medical drugs can lead to pathogenic alterations, which favor and exacerbate previous diseases, leading sometimes, even, to the death.

From a pathological point of view, chronic abuse of stimulants, for example amphetamines and cocaine, has been associated with coronary and myocardium abnormalities [26,27], such as myocardial infarction, myocarditis, cardiac hypertrophy, and dilated cardiomyopathy [28]. In addition, repeated cocaine administration may accelerate the development of atherosclerosis in humans and animals. In a study performed in Southern Spain, 3.1% of the sudden unexpected deaths were related to cocaine consumption [29].

Hair analysis was useful to demonstrate the associated cause of death in a 30-year-old man, who was found dead in his bed. He was healthy and did not have any known disease. Apparently, the case was a sudden unexpected death. Nevertheless, his friends suspected cocaine consumption. Autopsy and histology findings showed cardiac hypertrophy and atherosclerotic coronary disease. Both abnormalities have been associated to chronic cocaine abuse, but could also be related to other cardiovascular risk factors. Toxicological analyses were performed in blood, urine, and hair. Blood and urine analyses were negative for alcohol, drugs of abuse (including cocaine compounds), and medical drugs. The lock of hair, 6 cm in length, was sequentially cut in three segments of 2 cm. Cocaine and its major metabolite, BE, were detected in the three segments. Cocaine concentrations (48.32, 49.50, and 42.10 ng/mg) were in the high range, demonstrating chronic and heavy consumption of this drug during, approximately, 6 months before death. From the toxicological findings in hair it was possible to conclude that the pathological alterations in the coronary system were related to cocaine consumption.

In other situations toxicological results in biological fluids are positive, but inconclusive and do not allow us to establish the cause of death. This was the case of a 44-year-old man with a history of alcoholism and drug abuse. He was found dead at home. Pathological findings were cirrhosis and gastric hemorrhage. Toxicological analyses were performed in blood, urine, and hair. Urinalysis revealed codeine consumption: codeine (0.12 mg/L) and morphine (0.03 mg/L). Blood gave positive results for codeine (traces) and ethyl alcohol (0.12 g/L). From the results in biological fluids it was not possible to establish the cause of death. Nevertheless, hair analysis was positive for EtG (0.73 ng/mg). This concentration is in the range of chronic excessive alcohol consumption, which could have been the cause of the liver disease.

Although the majority of these studies have been performed with drugs of abuse, hair testing has proved to be useful to document chronic consumption of other types of compounds. For example, a death due to chronic anabolic consumption was assessed by Kintz [30] in a 36-year-old body builder. Autopsy and histology findings were consistent with anabolic abuse. Toxicological analyses were negative in blood and the results in urine were indicative of exogenous testosterone consumption. Hair analysis gave positive results for nandrolone, methenolone, stanozolol, and testosterone acetate and enantholate, thus confirming chronic exposure to anabolics.

9.4.1.4 Pitfalls in Hair Analysis in Postmortem Toxicology: External Contamination with Biological Fluids

External contamination of hair and the possibility of reporting false-positive results have been a major concern and discussion since the beginning of hair analysis.

In living people, external contamination can come mainly from the environment, but in postmortem cases we have to add the possibility of drug incorporation into the hair from the individual body fluids, such as blood, sweat, or putrefactive fluid. With this premise it is reasonably easy to understand that while in living people the major source of external contamination is cocaine or other drugs such as cannabis easy to be found in the environment, postmortem hair can be contaminated with any drug present in the biological fluids.

External contamination of hair is deeply discussed in Chapter 3 of this book. For this reason, here we will focus on the contamination with body fluids from the cadaver, which is the specific problem of postmortem hair samples.

Two different "*in vitro*" experiments have been performed to verify if there is any possibility that long-term contact between hair and organic material containing a drug can produce a false-positive result. In the first one [31], brown drug-free hair was soaked in blood, containing from 0.05 to 3.0 μg/mL cocaine and BE, for 5 min. It was then removed, washed, and left at room temperature. Analyses were performed 6 h, 1, 2, 4, and 7 days after contamination. Both cocaine and BE were absorbed into hair in significant concentrations when the concentration in the blood was 0.5 mg/mL or greater; cocaine was more readily absorbed than BE. The concentrations of both compounds in hair did not increase over time, probably due to the fact that the time of soaking was 5 min in all the experiments.

Other experiment was performed with benzodiazepines [32]. Before incubation, the cut end of the hair was sealed by wax. Then, hair was totally incubated in 10 mL whole blood, spiked with 7-aminoflunitrazepam (100 ng/mL), and maintained at different temperatures for 7, 14, and 28 days. All the samples tested positive for 7-aminoflunitrazepam, irrespective of the period and the temperature of incubation. Concentrations ranged from 233 to 742 pg/mg. With these findings the authors could explain the results obtained in the hair of a skeleton which tested positive for this benzodiazepine. Some hair fibers (4 cm, dark), attached to the skull were collected and cut in two segments. The 7-aminoflunitrazepam concentrations were as follows: 15 pg/mg (0–2 cm) and 19 pg/mg (2–4 cm). Investigations from the police revealed that there was no medical prescription for Rohypnol. Consequently, the presence of 7-aminoflunitrazepam could be due to postmortem contamination.

Similar findings have been also described from other real cases involving different drugs of abuse and medical drugs. For example, cyamemazine (3,660 ng/mL) was found in femoral blood of a subject who was never prescribed for this drug. Upon request of the judge, the body was exhumed 18 months after burial. Putrefaction was massive and the hair was in contact with putrefactive liquid. Segmental hair analysis gave positive results for cyamemazine: 3.1 ng/mg (0–2 cm), 2.9 ng/mg (2–4 cm), and 3.1 ng/mg (4–6 cm). The authors [32] conclude that the cyamemazine in hair could be the result from contact with the putrefactive organic material in a case of acute overdose for suicidal purposes.

When no changes in the concentrations are observed after segmental analysis, the possibility of potential contamination from an individual's body fluids or tissues should be considered.

Results of hair analysis from postmortem cases should be interpreted with caution. It must be emphasized that a single segment of hair should not be used to discriminate long-term exposure to a drug. It is recommended to perform multisectional analysis of hair, because homogenous results can be indicative of contamination.

Nevertheless, in some corpses it is not possible to perform sequential hair analysis because the corpse was badly preserved and only a tangle of hair is available. The strategy to differentiate if the drug present originates from consumption or was incorporated from external sources could be based on the profile of the compounds detected and the ratio of drug:metabolite. One of the characteristic of hair is that the parent drug is the majority compound detected in this matrix. In fact, the SoHT, among other criteria for obtaining a positive hair test result, recommends the application of the following metabolite-to-parent drug ratios: BE:cocaine higher than 0.05 in the case of cocaine and 6-MAM:morphine higher than 1.3 for heroin [33]. In 286 hair samples positive for cocaine in 2013, the ratio of BE:cocaine ranged from 0.08 to 2.08 (mean 0.54, median 0.49).

The profile of compounds detected in blood samples after drug consumption is different to hair. BE is the majority compound, while cocaine concentrations are considerably lower, even, in some cases, the latter is not found. Consequently, in hair samples contaminated with biological fluids, BE concentrations should be higher than cocaine concentrations. This was the case of a 40-year-old polydrug abuser who was found at home several months after death. The corpse was putrefied and only a lock of hair 3 cm in length was collected during the autopsy. The analysis revealed the presence of cocaine (0.65 ng/mg), BE (9.80 ng/mg), and methadone (13.70 ng/mg). From the ratio of BE:cocaine (15.07) we can conclude that a postmortem contamination with biological fluids has occurred.

Up to now, we do not have enough data to be statistically representative, but we can propose that when the ratio of BE:cocaine is higher than 5, an external contamination with body fluids could be assumed.

9.4.2 Drug-Facilitated Crimes

DFCs are criminal acts carried out by means of administering a substance to a person with the intention of impairing behavior, perceptions, or decision-making capacity. It also extends to taking advantage of an impaired person, without their consent, after their voluntary intake of an incapacitating substance [34].

Most substances used in DFCs are potent fast-acting central nervous system depressants with effects that mimic severe alcohol intoxication or general anesthesia.

An adequate selection of the biological specimens is of paramount importance for the correct interpretation of the results. It will depend on the delay in reporting the crime. In spite of each case having special features, urine is the most useful specimen in the majority of DFC investigations [35]. Sample collection should be performed as soon as possible in order to be able to detect the drugs that are quickly eliminated from the body, since urine samples are not useful if collected later than 4 or 5 days after the alleged crime [34,35]. While the majority of the drugs involved in these crimes are eliminated in less time, few of them can remain at very low concentrations, making possible the detection if sensitive instrumentation is used. Blood specimens are suitable when the crime has occurred within 24 h of collection [36,37]. United Nations Office on Drugs and Crime (UNODC) increases the available time until 48 h postincident [34]. If possible, the combination of blood and urine specimens is recommended, since it may provide a clearer picture as to the window of exposure to the drug [35,37].

In some cases, and because of several reasons, DFCs are lately reported, at one time when natural metabolic processes have eliminated the drug from blood and urine.

In these cases hair is the most helpful sample to identify the compound used to commit the crime. In fact this matrix is gaining recognition in DFC.

For practical purposes the three matrices complement each other. Blood and urine are the matrices of choice if they can be obtained within a suitable timeframe, since they provide short-term information of an individual's drug exposure, thus giving more accurate information of the compounds present in the body at the time of the alleged crime. Hair is especially useful in cases of late reporting of the alleged assault or when it is necessary to discriminate between a single exposure and chronic consumption.

Although the number of DFC complaints is considerably increasing during the last years, these crimes continue being underreported. A study was performed, within a national telephone household probability sample, with 441 women who had experienced but did not report a rape to police, to extract a dominant set of reasons for not reporting, and to reduce the set of dependent variables. Results of all the statistical studies indicated three unique factors: not wanting others to know, no acknowledgment of rape, and criminal justice concerns [38].

9.4.2.1 Substances Used

Drugs used in DFC must produce the following effects: sedation and the induction of sleep, alteration of the victim's behavior, anterograde amnesia, and the creation of a helpless state that the criminal can deliberately exploit [39]. In addition, they must be odorless and tasteless and must dissolve readily in alcoholic or other beverages.

Drugs usually used in DFC are fast-acting (within 30 min), have a short plasma half-life, and generally require a low dose to be effective, with few exceptions (ethanol, gamma hydroxybutyric acid (GHB), and related compounds) [34].

UNODC, in the guidelines for the forensic analysis of the drugs facilitating sexual assault and other criminal acts [34], provides a list of substances that should be targeted for in the analysis of urine. In spite of the list being comprehensive, nevertheless, each laboratory needs to select the substances which are most commonly used in their region and/or country. The availability of the drug to the perpetrator is a very important criterion to select a drug to commit DFCs.

Chèze and Gaulier [40] in the chapter devoted to hair analysis in the last book about DFCs [41] provided a large table with the substances which should be first targeted in DFC analysis. More than 200 case reports, concerning 320 victims and 50 substances were reported and reviewed by Dumestre-Toulet and Eysseric-Guérin [42] in the same book.

Approximately 100 substances have been described to be potentially active in DFC [43]. The substances and family of drugs most commonly used in DFC are:

- Ethanol. It is the most common substance involved in DFCs either alone or in combination with other drugs.
- Benzodiazepines. They are used in therapy as tranquilizers, anticonvulsants, hypnotics, and sedatives. In DFC these are used because they can cause drowsiness, memory impairment, amnesia, and transient euphoria, above all when taken in combination with alcohol.
- Z-drugs (hypnotics). Zopiclone, zolpidem, and zaleplone are members of the latest generation of nonbenzodiazepine hypnotic agents. They have anxiolytic, sleep-promoting, and muscle-relaxing effects. Their use in DFC derives from their capacity to rapidly impair the individual (within 10–30 min) and because they have a short half-life and can only be detected for a short time. Moreover, due to its amnesic properties, the victims are less able to accurately recall the circumstances under which the offence occurred.
- Antihistamines. First-generation H1 antihistamines readily enter the central nervous system and are used as antiemetic and as sleep-promoting medications. They are used in DFC because they reduce alertness and induce somnolence. In addition, the ethanolamines (carbinoxamine, clemastine, diphenhydramine, dimenhydrinate, and doxylamine) induce sedation [44].
- Drugs of abuse. Drugs of abuse—cannabis, cocaine, amphetamines, opiates, etc.—are also used in DFC. Cannabis is the most frequent drug of abuse incriminated in DFCs in France, usually in combination with other substances [45]. Amphetamines, above all MDMA (3,4-methylenedioxymethamphetamine), due to its effects as stimulant and hallucinogenic, are used in DFC. The National Institute on Drug Abuse (NIDA) found that young women under the influence of MDMA were more likely to have sex with men they had not intended to [46]. Opiates are able to produce sedation and muscle relaxation. Some of them produce a pleasant drowsy state, which along with decreased sensation of pain can decrease resistance from an intended victim of DFCs. Cocaine is used clinically as an anesthetic agent, but it is used illicitly for its psychotropic effects and has been found in some DFCs. Other substances considered drugs of abuse have been also detected in DFCs, such as scopolamine, ketamine, phencyclidine, etc.

- GHB. It is a naturally occurring, endogenous compound found in most mammalian tissues, including the brain, which has been used clinically as an anesthetic and hypnotic agent since the early 1960s [45]. In the late 1990s, GHB started to be used as "party drug." Over the past decade, GHB and related products have been used to commit DFCs [47]. The half-life of GHB ranges from 20–60 min and it is totally cleared from the body within 4–6 h after ingestion, regardless of the dose [48]. At low doses GHB causes induction of a pleasant state of relaxation and tranquility, placidity, sensuality, emotional warmth, reduced inhibitions, and pleasant drowsiness. When taken in higher doses, like those probably involved in DFCs (up to 2.5 g), it leads at first, like alcohol, to a heightening of mood and drive, sometimes also of sexual desire. At still higher doses, it is heavily sleep-inducing. Overdoses can cause a sudden, deep sleep from which the affected person can hardly be aroused [39].

9.4.2.2 Analytical Strategies

The special characteristics of these crimes with respect to the other types of cases, usually found in forensic toxicology and discussed in this chapter, make necessary the analytical strategies be discussed in a separate section.

Collection of hair samples: An adequate sample collection is of paramount importance in DFC. The protocol for the investigation of these crimes should start with interview of the victim, subsequent examination by a health care professional, and then collection of biological specimens for analysis [34].

Biological samples must be collected ideally before any medication is administered to the victim, but if this is not possible, all the medications must be registered.

It is very important that hair is sampled in a strict manner by properly trained personnel. Hair must be collected from the posterior vertex as close as possible to the scalp. Since segmental analysis is mandatory in DFC, the lock of hair has to be perfectly aligned and secured and the root tip identified.

Although United Nations gives the possibility of using body hair when head hair is not available, it is not advisable to segment body hair. In fact, there is no report in the literature involving segmental analysis of body hair, irrespective of its anatomical origin (armpit, pubis, chest, or leg).

Hair collection in DFCs has some special features:

Quantity of hair: One of the differences lies on the quantity of hair required for analysis. United Nations, in the previously cited document, recommend the collection of at least two hair samples, with the thickness of a pencil. Other authors, including the French Society of Analytical Toxicology, increase the number of samples collected to four in order to be able to perform all the required analyses in DFC [40,49].

Time of collection: When a single exposure has to be established, the time of hair collection is top priority. The first question to be addressed is the time elapsed between the crime and the complaint, in order to decide the appropriate date of sample collection. Scalp hair growth is not homogeneous and different authors propose different ranges: from 0.6 to 3.36 cm/month [40], or from 0.7 to 1.4 cm/month. Nevertheless, for forensic cases it is admitted at an average growth rate of 1 cm/month [50]. It takes between 7 and 10 days for the growing hair to reach the surface of the scalp. Therefore, the sample taken up to 5–7 days after the DFC will not contain any drug ingested at the time of the crime. For this reason, samples should be collected 1 month [51] or, more specifically, 4–6 weeks after the alleged crime [6,34]. In addition, the SoHT [6] recommends that when hair analyses are positive, a second hair sample must be collected to corroborate the results.

Segmentation of hair samples: Segmental hair analysis provides information on the chronological profile of drug consumption. As previously stated, it is generally performed in DFCs, because it provides the only possibility to differentiate between single exposure and chronic consumption or to corroborate a single exposure over the time of the incident.

There are some pitfalls which have to be considered for the correct interpretation of segmental hair analysis. All of them will be discussed later in this section. Nevertheless, assuming the ideal situation of absence of migration along the hair shaft and steady and uniform growth, a single exposure should be confirmed by the presence of the drug in the segment corresponding to the period of the crime while not detected in the other segments. A useful approach for segmentation, mainly accepted by the scientific community, is to cut the lock of hair into three segments of 2 cm, with the exception of GHB analysis, which requires shorter segments.

9.4.2.3 Drug-Facilitated Sexual Assaults

Drug-facilitated sexual assaults (DFSAs) have been defined as offences in which victims are subjected to nonconsensual sexual acts, while they are incapacitated or unconscious due to the effects of alcohol and/or drugs and are therefore prevented from resisting or are unable to consent [52]. More recently, it has been suggested that DFSA should be divided into proactive DFSA, when victim is administered of an incapacitating or disinhibiting substance for the purpose of sexual assault, and opportunistic DFSA, when victim is profoundly intoxicated by his or her own actions to the point of near or actual unconsciousness [51].

DFSAs are the most prevalent DFCs [47]. In a study performed in France over a 9-year period, 473 cases met the criteria to be considered as DFC. A wide range of crimes were found, but they were dominated by sexual assaults, which accounted more than one half (51%), both in isolation (41%) or, more rarely, combined with theft (4%) [43].

The utility of hair testing in DFSA lies in the long detection time achieved with this specimen and the possibility of performing sequential analysis.

With both premises hair analysis is increasingly used in these crimes, as demonstrated by the large bibliography on the matter. Although specific and isolated applications can occur, the majority of the applications are related to: (a) cases of late reporting of the alleged assault; and (b) to differentiate single exposure and chronic consumption in cases of false denounce.

DFSA could be demonstrated in two women, 5 and 17 months after the crime [53]. Both women reported to the police that they had been sexually abused after accepting a drink from the same man, at different times. Thereafter, they allegedly lost memory and consciousness for several hours. After these two reports, the man was prosecuted. For the purposes of the investigation, the two women were asked to give hair samples 5 and 17 months, respectively, after the alleged abuse. Hair samples were also collected from the accused, as well as his wife in order to verify if they were habitual consumers of any psychoactive drug. The results revealed the presence of low concentrations of zolpidem in 3 out of 11 segments of hair specimen obtained from the first of the alleged victims, offering plain evidence of single or sporadic exposure, whereas the agent was detected in high concentrations in the hair collected from the wife, coherently with therapeutic administration.

Another case refers to a 24-year-old girl who was sexually assaulted after administration of GHB and morphine [54]. She had been living in an international college for foreign students for about 1 year and often complained of a general unhealthy feeling in the morning. At the end of the college period she returned home and received some video clips where she was having sex with a boy she met when she was studying abroad. A lock of hair, 20-cm long, was collected for toxicological analysis. It was sequentially analyzed in 2–3 cm segments. Morphine and GHB were detected in hair segments related to the period of time she was abroad. A higher value of GHB was found in the period associated with the possible criminal activity and was also associated with the presence of morphine in the same period.

In the context of sex crimes, drugs are sometimes also given with the intention of increasing sexual desire and lowering inhibitions (amphetamines, cocaine). Among these drugs, a case involving sildenafil has been reported [55]. A female 15–17 years old alleged having been subjected to sexual assaults by her stepfather over a 2-year period. There was some suspicion that drugs may have been administered to facilitate the attacks. Hair was sequentially analyzed. The proximal segment tested positive for sildenafil at 38 pg/mg, and all others were negative, which was in agreement with the victim's claim.

False denounce of DFSA and deliberate contamination of hair with scopolamine was demonstrated in a woman trying to explain her marital infidelity [56]. A middle-aged woman, upon the discovery by her husband that she had engaged in an affair over the past 2 years, argued in her defense that her lover had chronically intoxicated her during their affair, probably with scopolamine. The secret affair finished 2 months before the analysis request, and therefore, hair was the only biological sample collected. The lock of hair was divided in three segments of 4.5–5 cm, scopolamine was detected in all three segments, indicating an apparent exposure to scopolamine. However, all the concentrations were below the limit of quantification established at 10 pg/mg. Decontamination of the hair was carried out with two dichloromethane washes. Scopolamine was detected in both washes in much higher concentrations than in hair. The scopolamine wash:hair ratios ranging from 74 to 709 for the first wash, and from 4.6 to 42 for the second wash. From these findings together with the orientation of the toxicological analysis to look for scopolamine, it was possible to conclude that there was a deliberate contamination of the hair by the victim herself.

Similar situation was found with a 19-year-old woman who went to the police to declare a rape after having a drink contaminated with MDMA [57]. Analysis of urine, which was collected at the medicolegal unit of the hospital, gave positive results for MDMA and its metabolite 3,4-methylenedioxyamphetamine (MDA) at 1,852 and 241 ng/mL. She claimed that she never took ecstasy and directly gave the name of the rapist, who was rapidly arrested and sent to jail. As the circumstances were unclear, the Judge requested the analysis of hair, which revealed the presence of MDMA (21.3 ng/mg), 3,4-methylenedioxy-*N*-ethylamphetamine (MDEA) (31.6 ng/mg), and MDA (6.7). These concentrations indicate chronic consumption and not single exposure. Later she admitted that it was a false notification, that no rape had occurred, and that it was a revenge on the alleged rapist.

9.4.2.4 DFC in the Elder People

Elder abuse is internationally recognized as a growing problem. It can occur in different settings, but mainly at the nursing homes and within the family context. A prospective observational study in the psychogeriatric unit of an acute psychiatric hospital demonstrated that 30% of the patients were physically restrained, confirming that physical abuse remains a common practice in this type of centers. The highest incidence (48%) was found in elderly patients with severe cognitive impairments, dementia, and delirium [58]. Unfortunately, a high percentage of elder adults experiences abuse at the hands of family, either immediate or extended. In a study performed in Portugal, 12.3% of elder adults experienced abuse by the family: psychological, 6.3%; financial, 6.3%; physical, 2.3%; neglect, 0.4%; and sexual, 0.2% [59].

Although abuse occurs across all socioeconomic, racial, and religious lines [60], several studies have demonstrated that elders from groups traditionally considered being economically, medically, and sociodemographically vulnerable are more exposed [61]. It has been demonstrated that education level, age, and functional status are significantly associated with elder abuse [59].

In addition to the previously mentioned abuses, DFCs committed in elder people have to be considered, even if the literature is scarce on this matter. Stankova et al. [62] present a series of eight cases of acute combined poisonings with benzodiazepines and opiates in people over 70 years old. All the crimes were similar in the circumstances of the exposure, clinical course of the poisonings, and the identified toxic substances. In fact, it could be demonstrated that they were committed by a group of criminals with the aim of robbery.

As in other forensic applications, hair analysis is a valuable tool for the identification of DFC in elder persons, above all for the possibility of discriminating between single and chronic exposure. Kintz et al. have demonstrated the utility of hair to evidence DFCs in the elderly [63]. They could demonstrate two cases of elder abuse in nursing homes. The first one was an 87-year-old man living in a retirement house. The family notices a strange behavior and incoherent speech. Segmental analysis of a 6-cm lock of hair revealed the presence of promazine at 9, 2, and 6 pg/mg. Consequently, exposure to promazine could be established over the previous 6 months. In the other case, the daughter of an 81-year-old woman in a nursing home noticed marked somnolence. Diphenhydramine and doxylamine, two antihistamine drugs with sedative properties, were detected in the hair 3 cm in length. The nurse admitted giving the drugs to sedate the woman and reduce her workload.

Within the family setting, the body of a 64-year-old man was found dead by his wife at home [63]. Toxicological investigations detected ethanol at 2.31 g/L and the presence of 7-aminoflunitrazepam in blood samples. Since the subject had not been treated with flunitrazepam, a lock of 3 cm of hair was analyzed to discriminate between single and multiple exposures and 7-aminoflunitrazepam was detected at 78 pg/mg. The wife admitted repetitive administration in the evening soup to decrease the libido of her husband.

9.4.2.5 DFC in Children

Although Chapter 6 is devoted to hair analysis in children, the specific problem of DFC in the childhood will be discussed here.

A retrospective study performed in United States from 2000 to 2008 with the aim of studying all pharmaceutical exposures involving children younger than 7 years old [64] demonstrated that a total of 1,439 cases met the criteria for which the reason for exposure was coded as "malicious." The mean number of cases per year was 160 (range, 124–189) and showed an increase over time. The median age was 2 years. About 9.7% of cases involved more than one exposed substance. In 51% of cases there was an exposure to at least one sedating agent. There were 18 (1.2%) deaths. Of these, 17 (94%) were exposed to sedating agents, including antihistamines (8 cases) and opioids (8 cases).

DFCs in children have two major objectives: to have the children quiet or to sexually assault them [65]. As in elder people, they occur either at the nursery school or within the family context.

Sedation could be demonstrated in two children (7 and 13 years old) from the same family who were living with the stepmother [66]. Drowsiness, ataxia, sedation, muscular weakness, and marked somnolence were noted for both kids at school and during the weekend. These symptoms were present for at least for 3 months. Hair analysis of the two subjects revealed the presence of trimeprazine at concentrations in the range 23–339 pg/mg. The stepmother admitted the use of trimeprazine as a sedative drug.

Child sexual abuse is sexual contact with a child which occurs as a result of force or in a relationship where it is exploitative (age differences of caretaking responsibility) [67]. Although no child may be excluded from the possibility of being sexually abused, some characteristics are associated with greater risk: girls more than boys, preadolescents and early adolescents, having a stepfather, living without a natural parent, having an impaired mother, poor parenting, or witnessing family conflict. Nevertheless, class and ethnicity appear not to be associated with risk [68].

Several papers have confirmed the usefulness of hair testing to assess the use of incapacitating drugs in DFC in children, above all when there is a delay in reporting the crime. In fact, results obtained in hair analysis helped to demonstrate the surreptitious administration of diphenhydramine to eight victims of sexual abuse [69]. Among these victims, a 9-year-old girl was assaulted and the incident was filmed by the two alleged assailants. Examination of mobile phones found references to a pharmaceutical preparation containing diphenhydramine. The child was never under medical prescription of the drug. A single strand of hair, 32 cm in length, from the victim was obtained approximately 7 weeks after the alleged incident. Segmental analysis of the first 5 cm of hair revealed the presence of diphenhydramine with the following concentrations: 37 pg/mg (0–1 cm), 39 pg/mg (1–3 cm), and 33 pg/mg (3–5 cm). These results confirmed an exposure to diphenhydramine over a period of at least 5 months prior to the hair sampling. This evidence was not challenged during the court trial.

It is very important to consider that hair analysis not always provides conclusive results, above all in children because there are many differences between their hair and those from adults. For example, it is thinner and more porous, the ratio of anagen:catagen phases is not maintained, and the growth rate is different than in adults. In addition the literature about hair analysis in children is very scarce, making the interpretation of the results more difficult.

9.4.2.6 *Pitfalls in Hair Analysis in DFCs: Sequential Hair Analysis*

The major pitfall for the utility of hair analysis in DFC derives from the segmental analysis that is required for the correct interpretation of the results.

Segmental analysis of hair involves cutting the samples into predetermined lengths in order to extrapolate the length of the segment with the corresponding period of time. Although for the interpretation of forensic cases, it can be assumed that hair growths 1 cm/month and the variability in growth rates may influence the interpretation of the analytical findings.

It is well known that hair grows in cycles of various phases: anagen, growth phase; catagen, regressing phase; and telogen, resting phase. In addition, there are a several factors and circumstances that may affect the growth rate of hair, including genetic factors such as age [70], pregnancy [71] or hormones [72], and external variables as season of the year [73]. Consequently, there is a wide range of published growth rates observed in the literature. Lebeau et al. [74] have reviewed 8 papers, published between 1951 and 2007, dealing with growth rates of human hair. The wider rate was reported to be from 0.65 to 2.2 cm/month [75]. After combining all the data, the mean values ranged from 0.86 cm/month [76] to 1.12 cm/month [77]. The average was calculated as 1.06 cm/month with a relative standard deviation of 6% [74]. This calculated value is very close to the growth rate of 1 cm/month proposed by the SoHT [50], thus corroborating the acceptability of this value.

When interpreting segmental hair analysis it is assumed that the cut end of the hair is obtained directly next to the scalp, which is an oversimplification. In the same previously cited study, Lebeau et al. [74] evaluated the ability of 14 collectors, 9 novices and 5 experts, to cut hair next to the scalp. The length of the hair left on the scalp ranged from 0.4 cm to 1.4 cm with an average of 0.8 ± 0.1 cm, which, translated into time, indicates that an average of, approximately, 3 weeks of hair growth are left on the scalp after sampling. There were no significant differences in the length of the remaining hair after sampling by the expert than by the novice collectors. This factor has to be considered when extrapolating length of the hair and time of consumption. After these studies, the authors recommend a minimum delay of 8 weeks between suspected drug exposure and collection of hair.

Finally, but not least important, the possible diffusion along the hair shaft has to be considered. When performing segmental analysis to determine single exposure, the ideal situation occurs when the drug is detected only in the segment corresponding with the alleged time of the offense [57]. But sometimes the drug is detected in two or even three consecutive segments [78–80]. This finding does not necessarily mean multiple exposures.

According to Kintz [81] there are several possible explanations for having more than one positive hair segment after a single drug exposure. They include: (a) chronic consumption instead of single exposure, (b) differences in hair growth rate and ability of sample collection, (c) external contamination, (d) incorporation of the drug and its metabolites through sweat, (e) quality of the hair, that is, damaged, cosmetic treatment [82,83], porosity, etc.

All the authors reporting positive results in several segments after single exposure agree in the fact that the concentration in the segment of interest is considerably higher than in the contiguous segments. In order to establish some mathematical correlation, Kintz [81] calculated the ratio between the concentration in the segment corresponding with the crime and the concentration in the adjacent segments in several cases from the literature. The ratios ranged from 3 in a DFC related to clonazepam [78] to 8 in another crime related to zolpidem, where the concentrations in two consecutive 2-cm segments were 0.1 and 0.8 pg/mg [80].

After these studies and to qualify for a single exposure in hair, the author proposes to consider that the highest drug concentration must be detected in the segment corresponding to the period of the alleged event (calculated with a hair growth rate at 1 cm/month) and that the measured concentration be at least three times higher than those measured in the previous or the following segments. This must only be done using scalp hair after cutting the hair directly close to the scalp.

In addition, some other circumstances have to be considered: (a) the segments should be 2–3 cm in length, while segmentation centimeter by centimeter is not recommended; and (b) single exposure should not be documented over periods of time longer than 6 months or even 3 months in case of GHB to minimize displacement by radial migration of the hair [81].

9.4.3 Divorce and Child Custody Proceedings

Toxicological analyses are required in civil cases, such as divorce and child custody proceedings, when one of the parts adduced drug consumption as a cause of divorce or in the context of child custody cases when one parent accuses the other of using drugs and exposing the child passively or actively.

As previously stated, toxicological analysis can be performed in both biological fluids and hair. At the beginning only blood and urine were analyzed in civil cases, but the experience has demonstrated that positive results can be avoided in these fluids when sampling is done per appointment. For this reason, hair analysis is currently requested in these cases and it has demonstrated a higher efficacy, in identifying drug consumption, than biological fluids. It was corroborated in a divorce proceeding where both parts adduced drug consumption of the couple as a cause of divorce, and both denied any type of consumption. Toxicological analyses were performed in blood, urine, and hair. Biological fluids (blood and urine) gave negative results in both, wife and husband. Husband's hair analysis (3 cm in length) continued giving negative results. The lock of hair of the wife (12 cm) was cut in 6 × 2 cm segments and the 5 segments closer to the root were positive for cocaine and its metabolite BE (Table 9.4).

The results were in agreement with the statement of the husband, no drug consumption during the previous 3 months. Hair analysis of wife hair

TABLE 9.4 Hair Analysis Results in a Wife Involved in a Divorce Proceeding

	Cocaine[a]	Benzoylecgonine[a]
Segment 0–2 cm	3.1	0.9
Segment 2–4 cm	4.4	1.1
Segment 4–6 cm	4.2	1.3
Segment 6–8 cm	1.8	0.4
Segment 8–10 cm	1.5	0.3
Segment 10–12 cm	ND	ND

ND: Not detected.
[a]*Concentrations in ng/mg.*

revealed consumption of cocaine over the previous 10 months and the concentrations were in the range of light consumers of cocaine. If the subject does not consume drugs during the 3 or 5 days before sampling positive results can be avoided in blood and urine, but not in hair. During the trial she admitted sporadic cocaine consumption at the weekends.

Another case is related to a woman involved in a divorce and child custody proceeding who denies drug and alcohol consumption [5]. She was enrolled in a detoxification program with methadone. Samples of blood, urine, and hair were collected for analysis. Blood analysis gave positive results only for methadone and traces of EDDP. Urine was positive for cocaine compounds BE (48 ng/mL) and ecgonine methyl ester (50 ng/mL); THC-COOH (30 ng/mL) and methadone (1,880 ng/mL) and its metabolite EDDP (3,110 ng/mL).

The lock of hair, 12 cm in length, was cut into 3×4 cm segments. From the results (Table 9.5), it was possible to conclude that she consumed cocaine, heroin, methadone, cannabis, and alcohol, the latter was established due to the presence of cocaethylene and EtG. The chronological profile showed an increase in cocaine concentration in segment 2, opiate (6-MAM, morphine, and codeine) concentrations decreased over time, and methadone concentrations were homogeneous. EtG concentrations indicated chronic excessive alcohol consumption during the whole period under study.

In child custody proceedings not only the parents have to be analyzed, but also the children to demonstrate that they are exposed to inappropriate substances. Toxicological analyses can be requested by one of the parents or by other entities such as Judges, General Administration, hospitals, etc.

The following case illustrates this situation. It is related to a father seeking custody of the 4-year-old child [84]. Parents were divorced and the child

TABLE 9.5 Sequential Hair Analysis of a Woman Involved in a Divorce and Child Custody Proceeding

	Segment 0–4 cm	Segment 4–8 cm	Segment 8–12 cm
Cocaine[a]	33.80	50.90	30.40
Benzoylecgonine	16.07	21.12	11.29
Cocaethylene	4.61	4.40	2.13
6-Monoacetylmorphine	2.54	2.79	3.32
Morphine	1.23	1.14	1.31
Codeine	0.73	0.64	0.62
THC	0.11	ND	ND
CBD	0.20	ND	ND
CBN	0.25	ND	ND
Methadone	1.35	1.36	1.40
EtG	0.05	0.05	0.06

ND: Not detected; THC: Δ^9-tetrahydrocannabinol; CBD: cannabidiol; CBN: cannabinol; EtG: ethyl glucuronide.
[a]*Concentrations in ng/mg.*

resided with the mother. Hair samples from the father and son were submitted for cocaine testing. At birth, the baby's urine tested positive for cocaine, and the father acknowledged that 4 years ago he was snorting cocaine on a regular basis together with the child's mother. However, according to his report, 2 years ago he stopped cocaine consumption. Sequential analysis of father's hair (20 cm in length) tested negative, reflecting an estimated 20 months cocaine-free period and corroborating his story. The son's hair was positive for cocaine. Cocaine analysis was repeated 1 month later, and the 1 cm proximal segment was again positive. The data indicate that the child was exposed to cocaine.

Two different cases derived from complaints of the hospital [85]. Two children 16 (Child 1) and 18 (Child 2) months old were taken to Pediatric Emergencies because they presented red eyes and drowsiness. Preliminary toxicological analyses, performed at the hospital, gave positive results for cannabis in both cases. After a complaint of the hospital, the Judge asked for toxicological analyses in hair samples and biological fluids (blood and/or urine). Child ♯ 1 tested positive for THC-COOH in urine at a concentration of 36.6 ng/mL, while blood analysis gave negative results. Hair analysis (4 cm in length) gave positive results for cannabis: THC (0.11 ng/mg), cannabidiol (CBD; 0.18 ng/mg), and cannabinol (CBN; 0.23 ng/mg). Child ♯ 2 tested positive for THC-COOH in blood at a concentration of 20.5 ng/mg. The entire hair sample (2.5 cm in length) was also positive for cannabis: THC (0.49 ng/mg), CBD (0.74 ng/mg), and CBN (0.42 ng/mg). Although there is no statistical reference to interpret cannabis concentrations in hair of children, if we consider the data from adults, THC concentrations are in the medium and high ranges in case ♯ 1 and 2, respectively. The results demonstrated that both children were chronically exposed to cannabis.

9.4.3.1 *Hair versus Urine to Estimate Drug Abuse*

Several studies have confirmed the higher efficacy of hair versus urine to estimate drug consumption. Jurado et al. [86] compared hair and urine samples in order to ascertain which matrix better estimates drug consumption, and they concluded that: (a) both matrices can be used as biomarkers of drug consumption; nevertheless, hair analysis appeared to present higher sensitivity in identifying drug use in cases of negative urine results; (b) discrepancies between both samples mainly occur in light, or sporadic, consumers of drugs. The study was performed with 165 adult arrestees with a history of drug consumption. Head hair and urine were simultaneously collected in a blind setting. Hair revealed that 82.4% of the arrestees had consumed drugs; in contrast, urine identified only 70.3%. There was agreement in 69.1% positive and 16.4% negative cases in both samples. Hair analysis revealed drug consumption in 13.3% cases which urine failed to detect and, contrariwise, urine identified only 1.2% cases undetected in hair. To determine the influence of the severity of drug consumption, the study population was classified in heavy, moderate, and light consumers of drugs, based on concentrations in the hair. The major discrepancies were found in light consumers, since only 39%, 25%, and 53% of cocaine, cannabis, and opiate were confirmed by urinalysis in light consumers; while from 76% to 100% of the moderate consumers were confirmed, and 100% of the heavy consumers gave positive results in both samples.

Similar results were obtained by Dufaux et al. [87] with subjects trying to regain their revoked driver's license after a drug- or alcohol-related traffic offence. About 14,000 urine and 3,900 hair samples were analyzed. Positivity rates for the majority of the compounds, including EtG, ranged from 1.7 (for MA) to 9.3-fold (for MAM) higher in hair than in urine. In contrast, the detection rate for benzodiazepines was higher in urine than in hair (oxazepam, 0.21% vs 0%, nordiazepam 0.10% vs 0.03%).

Identification of cocaine use based on a urine test may miss many cases because of the short elimination half-life of the drug. In seven subjects attending the Local Medical Commission for driving license regranting, with negative urinalysis over several months, hair analyses gave positive results for cocaine (0.51–2.23 ng/mg) and BE (0.08–1.70 ng/mg) in all the cases [88].

The different studies have demonstrated that both matrices, hair and urine, are useful biomarkers of drug consumption. Nevertheless, hair is able to assess chronic drug consumption, and therefore gives a more real picture of drug use than that provided by urine, not only by confirming positive results, but also by avoiding negative analytical results due to temporary abstention.

9.4.4 Follow-Up of Detoxification Programs

Some Court proceedings require the evaluation of drug abstinence after a long period of drug consumption or the effectiveness of detoxification programs.

For example, in Spain, the article 87 of the Penal Code related to the "Cancellation of imprisonment sentences," more or less, states: "In case of crimes committed due to the dependency to drugs of abuse, the Judge or Court can decide the suspension of the imprisonment sentence when the convicted has stopped drug consumption or has joined a detoxification program." The Judge requests a report to the forensic coroner to certify that the person has stopped drug consumption, and a toxicological analysis for drugs of abuse is necessary to support the psychomedical examination.

Drug tests are usually performed by either urine analyses (regularly over several months) or by hair analysis after several months of abstinence.

Noncompliance of a detoxification program was demonstrated in one male reporting drug consumption since he was 18 years old. Three months before sampling he enrolled in a detoxification program with methadone [5].

Toxicological analyses were performed in urine and hair. Urinalysis gave positive results for opiate (6-MAM, morphine, and codeine) and cocaine (cocaine, BE, and ecgonine methyl ester) compounds and methadone and its metabolite EDDP.

The lock of hair, 6 cm in length was cut in 1 cm segments and the analysis demonstrated that this person was a heavy consumer of cocaine and heroin before enrolling in the detoxification program and that point and during the following 3 months, he started the consumption of the prescribed methadone, but continued consuming cocaine and heroin, although a decrease in the concentrations was noticeable.

Opposite conclusion was achieved in a 34-year-old man, who was in jail accused of drug trafficking. He reported chronic cocaine and cannabis consumption over the last 6–7 years. After remaining abstinent since imprisonment, 6 months ago, he asked for the application of the article 87 of the Penal Code. A lock of hair 13 cm was sequentially analyzed in two segments 0–5 cm and 6–13 cm, respectively. Results (Table 9.6) were positive for cocaine and cannabis compounds in the segment further to the root, while the proximal segment tested negative, reflecting an estimated 5 months cocaine- and cannabis-free period and corroborating the story of drug consumption.

TABLE 9.6 Hair Results from a Woman Enrolled in a Detoxification Program, Trying to Demonstrate Abstinence for 6 Months

	Segment 0–5 cm	Segment 6–13 cm
Cocaine[a]	ND	2.50
Benzoylecgonine	ND	2.15
Cocaethylene	ND	1.15
THC	ND	0.30
CBN	ND	0.18
CBD	ND	0.36

ND: Not detected; THC: Δ^9-tetrahydrocannabinol; CBD: cannabidiol; CBN: cannabinol.
[a]*Concentrations in ng/mg.*

9.4.4.1 Window of Detection in Hair Samples

The first question to be addressed, when interpreting compliance with abstinence of former drug users, is to know how long the different compounds can be detected in the hair after consumption has ended.

Some years ago, several experimental studies were performed with animals to establish these data. Thus Ferko et al. [89] established that the disappearance followed first-order kinetics. They administered rats with different intraperitoneal doses of cocaine during a 28-day period. About 25–30 days were required to eliminate cocaine and BE from hair after ending the highest-dose administration. Similarly, Jurado et al. in experiments performed with rabbits [90] and rats [4] concluded that the detection time increased with the dose administered. Rabbits received a single intraperitoneal dose of 5 mg/kg of cocaine, while rats were divided into two lots, and each one was administered a single dose of 40 mg/kg and 60 mg/kg of cocaine. Cocaine was detected for 9 days in the rabbits, and for 11 and 14 days in the rats administered with low and high dose of cocaine, respectively.

More recently, several papers have been published dealing with the disappearance of opiates and cocaine from human hair. Shen et al. [91] were the first to investigate the disappearance of opiates, 6-MAM, morphine, and codeine from human scalp hair after the discontinuation of drug use. The study was performed with 32 healthy women (ages 21–51 years) with a known history of heroin abuse, who went to a rehabilitation center and ceased heroin consumption (for 4–5 months). Assuming a rate of hair growth of 1 cm/month, the mean hair elimination half-lives of 6-MAM, morphine, and codeine were 0.88 months, 0.73 months, and 0.61 months, respectively. They suggest that to evaluate the discontinuation of heroin abuse after a 6-month period of abstinence, the results from a 3-cm proximal hair segment should be free of 6-MAM at the proposed 0.2 ng/mg cutoff level. Garcia-Bournissen et al. [92] evaluated the kinetics of disappearance of cocaine and its metabolite, BE, from hair after discontinuation of drug use. About 137 subjects were included in the study because the concentrations of both compounds had gradually decreased in sequential hair samples. Elimination of half-life of cocaine and BE in hair was calculated using standard kinetics calculations. The median half-life of cocaine and BE strongly correlated. They were 1.5 months for both compounds in males and females. It implies that, assuming first-order elimination, approximately 3–4 months have to pass for hair testing to become negative in the segment proximal to the scalp. Similarly, Felli et al. [93] studied the disappearance of cocaine in a woman after 1 year of abstinence. Cocaine decreases to 32% of the original concentration after 3 months of abstinence and it takes another few months to completely disappear from the segment closest to the root.

In conclusion, all of the studies agreed that the disappearance is dose-dependent, since the larger the dose, the longer the time period of detection in the hair.

REFERENCES

[1] Jurado C, Soriano T. Changes in patterns of drug-abuse consumption evidenced by hair analysis: study over the last decade. Presented at 18th Meeting of the Society of Hair Testing. Geneva, Switzerland; 2013.

[2] Pepin G, Gaillard Y. Concordance between self-reported drug use and findings in hair about cocaine and heroin. Forensic Sci Int 1997;84:37–41.

[3] Kintz P, Mangin P. What constitutes a positive result in hair analysis: proposal for the establishment of cut-off values. Forensic Sci Int 1995;70:3–11.

[4] Jurado C. El pelo como matriz biológica en el diagnóstico toxicológico. Doctoral Thesis, University of Sevilla, Spain; 1999.

[5] Jurado-Montoro C. Análisis de drogas de abuso en muestras de pelo. Diagnóstico del consumo crónico. Trastornos adictivos 2007;9(3):172–83.

[6] Cooper GAA, Kronstrand R, Kintz P. Society of Hair Testing guidelines for drug testing in hair. Forensic Sci Int 2012;218:20–4.

[7] Jurado C, Staub C. Interpretation of hair analysis: opiates. Presented at the Society of Hair Testing Workshop "Interpretation of Hair Analysis." Bordeaux, France; 2001.

[8] Cordero R, Lee S, Paterson S. Distribution of concentrations of cocaine and its metabolites in hair collected postmortem from cases with diverse causes/circumstances of death. J Anal Toxicol 2010;34:543–8.

[9] Lee S, Cordero R, Paterson S. Distribution of 6-acetylmorphine and morphine in head and pubic hair from heroin-related deaths. Forensic Sci Int 2009;183:74–7.

[10] Puschel K, Thomasch P, Arnold W. Opiate levels in hair. Forensic Sci Int 1983;21:181.

[11] Kintz P, Bundeli P, Brenneisen R, Ludes B. Dose-concentration relationships in hair from subjects in a controlled heroin-maintenance program. J Anal Toxicol 1998;22(3):231–6.

[12] Musshoff F, Lachenmeier K, Wollersen H, Lichtermann D, Madea B. Opiate concentrations in hair from subjects in a controlled heroin-maintenance program and from opiate-associated fatalities. J Anal Toxicol 2005;29:345–52.

[13] Welp EA, Bosman I, Langendam MW, Totté M, Maes RA, van Ameijden EJ. Amount of self-reported illicit drug use compared to quantitative hair test results in community-recruited young drug users in Amsterdam. Addiction 2003;98(7):987–94.

[14] Pragst F, Balikova MA. State of the art in hair analysis for detection of drug and alcohol abuse. Clin Chim Acta 2006;370:17–49.

[15] Drummer OH. Postmortem toxicology of drugs of abuse. Forensic Sci Int 2004;142 (2–3):101–13.

[16] Nakahara Y, Kikura R, Yasuhara M, Mukai T. Hair analysis for drug abuse XIV. identification of substances causing acute poisoning using hair root. I. Methamphetamine. Forensic Sci Int 1997;84:157–64.

[17] Cartmell LW, Aufderheide A, Weems C. Cocaine metabolites in pre-Columbian mummy hair. J Okla State Med Assoc 1991;84:11–12.

[18] Springfield AC, Cartmell LW, Aufderheide AC, Buikstra J, Ho J. Cocaine and metabolites in the hair of ancient Peruvian coca leaf chewers. Forensic Sci Int 1993;63(1–3):269–75.

[19] Wilson AS, Brown EL, Villa C, Lynnerup N, Healey A, Ceruti MC, et al. Archaeological, radiological, and biological evidence offer insight into Inca child sacrifice. PNAS 2013;110(33):13322–7.

[20] Musshoff F, Brockmann C, Madea B, Rosendahl W, Piombino-Mascali D. Ethyl glucuronide findings in hair samples from the mummies of the Capuchin Catacombs of Palermo. Forensic Sci Int 2013;232:213–17.

[21] Steentoft L, Worm K, Pedersen CB, Sprehn M, Mogensen T, Sørensen MB, et al. Drugs in blood samples from unconscious drug addicts after the intake of an overdose. Int J Legal Med 1996;108:248–51.

[22] Meissner M, Recker S, Reiter A, Friedrich HJ, Oehmichen M. Fatal versus non-fatal heroin “overdose”: blood morphine concentrations with fatal outcome in comparison to those of intoxicated drivers. Forensic Sci Int 2002;130:49–54.

[23] Drummer OH. Recent trends in narcotic deaths. Ther Drug Monit 2005;27:738–40.

[24] Darke S, Hall W, Kaye S, Ross J, Duflou J. Hair morphine concentrations of fatal heroin overdose cases and living heroin users. Addiction 2002;97:977–84.

[25] Druid H, Strandbergm JJ, Alkass K, Nyström I, Kugelberg FC, Kronstrand R. Evaluation of the role of abstinence in heroin overdose deaths using segmental hair analysis. Forensic Sci Int 2007;168:223–6.

[26] Egred M, Davis GK. Cocaine and the heart. Postgrad Med J 2005;81:568–71.

[27] Yu Q, Larson DF, Watson RR. Heart disease, methamphetamine and AIDS. Life Sci 2003;73:129–40.

[28] Karch SB. Cocaine. Karch's pathology of drug abuse. 4th ed. Boca Raton, FL: CRC Press; 2009. p. 1–207.

[29] Lucena J, Blanco M, Jurado C, Rico A, Salguero M, Vazquez R, et al. Cocaine-related sudden death: a prospective investigation in south-west Spain. Eur Heart J 2010;31 (3):318–29.

[30] Kintz P. Value of hair analysis in postmortem toxicology. Forensic Sci Int 2004;142:127–34.

[31] Paterson S, Lee S, Cordero R. Analysis of hair after contamination with blood containing cocaine and blood containing benzoylecgonine. Forensic Sci Int 2010;194:94–6.

[32] Kintz P. Segmental hair analysis can demonstrate external contamination in postmortem cases. Forensic Sci Int 2012;215:73–6.

[33] Society of Hair Testing. Statement of the Society of Hair Testing concerning the examination of drugs in human hair. Forensic Sci Int 1997;84:3–6.

[34] UNODC (United Nations Office on Drug and Crime). Guidelines for the forensic analysis of the drugs facilitating sexual assault and other criminal acts; 2013.
[35] LeBeau M, Andollo W, Hearn WL, Baselt R, Cone E, Finkle B, et al. Recommendations for toxicological investigations of drug facilitated sexual assaults. J Forensic Sci 1999;44:227–30.
[36] LeBeau M. Toxicological investigations of drug-facilitated sexual assaults. Forensic Sci Commun 1999;1:12.
[37] LeBeau MA. Guidance for improved detection of drugs used to facilitate crimes. Ther Drug Monit 2008;30:229–33.
[38] Cohn AM, Zinzow HM, Resnick HS, Kilpatrick DG. Correlates of reasons for not reporting rape to police: results from a national telephone household probability sample of women with forcible or drug-or-alcohol facilitated/incapacitated rape. J Interpers Violence 2013;28(3):455–73.
[39] Madea B, Mußhoff F. Knock-out drugs: their prevalence, modes of action, and means of detection. Dtsch Arztebl Int 2009;106(20):341–7.
[40] Chèze M, Gaulier JM. Drugs involved in drug-facilitated crimes (DFC). Analytical aspects: 2. Hair. In: Kintz P, editor. Toxicological aspects of drug-facilitated crimes. Academic Press–Elsevier; 2014. p. 181–222.
[41] Kintz P. Toxicological aspects of drug-facilitated crimes. Academic Press–Elsevier; 2014.
[42] Dumestre-Toulet V, Eysseric-Guérin H. Case reports in drug-facilitated crimes. From A for alprazolam to Z for zopiclone. In: Kintz P, editor. Toxicological aspects of drug-facilitated crimes. Academic Press–Elsevier; 2014. p. 223–54.
[43] Djezzar S, Richard N, Deveaux M. Epidemiology of drug-facilitated crimes and drug facilitated sexual assaults. In: Kintz P, editor. Toxicological aspects of drug-facilitated crimes. Academic Press–Elsevier; 2014. p. 11–46.
[44] Lemaire-Hurtel AS, Alvarez JC. Drugs involved in drug-facilitated crime—Pharmacological aspects. In: Kintz P, editor. Toxicological aspects of drug-facilitated crimes. Academic Press–Elsevier; 2014. p. 47–92.
[45] Shbair MKS, Eljabour S, Lhermitte M. Drugs involved in drug-facilitated crimes: part I: alcohol, sedative-hypnotic drugs, gamma-hydroxybutyrate and ketamine. A review. Ann Pharm Fr 2010;68:275–85.
[46] Freese TE, Miotto K, Reback CJ. The effects and consequences of selected club drugs. J Subst Abuse Treat 2002;23:151–6.
[47] LeBeau MA, Mozayani A. Drug-facilitated sexual assault. A forensic handbook. London: Academic Press; 2001.
[48] Gahlinger PM. Club drugs: MDMA, gamma-hydroxybutyrate (GHB). Rohypnol and Ketamine. Am Fam Physician 2004;69:2619–26.
[49] Kintz P. Bioanalytical procedures for detection of chemical agents in hair in the case of drug-facilitated crimes. Anal Bioanal Chem 2007;388:1467–74.
[50] Society of Hair Testing. Recommendations for hair testing in forensic cases. Forensic Sci Int 2004;145:83–4.
[51] Shbair MKS, Lhermitte M. Drug-facilitated crimes: definitions, prevalence, difficulties and recommendations. A review. Ann Pharm Fr 2010;68:136–47.
[52] Payne-James J, Rogers D. Drug-facilitated sexual assault, "ladettes" and alcohol. JRSM 2002;95:326–7.
[53] Salomone A, Gerace E, Di Corcia D, Martra G, Petrarulo M, Vincenti M. Hair analysis of drugs involved in drug-facilitated sexual assault and detection of zolpidem in a suspected case. Int J Legal Med 2012;126:451–9.

[54] Rossi R, Lancia M, Gambelunghe C, Oliva A, Fucci N. Identification of GHB and morphine in hair in a case of drug-facilitated sexual assault. Forensic Sci Int 2009;186:9–11.

[55] Kintz P, Evans J, Villain M, Chatterton C, Cirimele V. Hair analysis to demonstrate administration of sildenafil to a woman in a case of drug-facilitated sexual assault. J Anal Toxicol 2009;33:553–6.

[56] de Castro A, Lendoiro E, Quintela O, Concheiro M, López-Rivadulla M, Cruz A. Hair analysis interpretation of an unusual case of alleged scopolamine-facilitated sexual assault. Forensic Toxicol 2012;30(2):193–8.

[57] Villain M. Applications of hair in drug-facilitated crime evidence. In: Kintz P, editor. Analytical and practical aspects of drug testing in hair. Boca Raton, FL: CRC Taylor & Francis; 2007. p. 255–72.

[58] Bredthauer D, Becker C, Eichner B, Koczy P, Nikolaus T. Factors relating to the use of physical restraints in psychogeriatric care: a paradigm for elder abuse. Z Gerontol Geriatr 2005;38(1):10–18.

[59] Gil AP, Kislaya I, Santos AJ, Nunes B, Nicolau R, Fernandes AA. Elder abuse in Portugal: findings from the first national prevalence study. J Elder Abuse Negl 2014;14:1–22.

[60] Paris BE, Meier DE, Goldstein T, Weiss M, Fein ED. Elder abuse and neglect: how to recognize warning signs and intervene. Geriatrics 1995;50:47–53.

[61] Peterson JC, Burnes DP, Caccamise PL, Mason A, Henderson Jr CR, Wells MT, et al. Financial exploitation of older adults: a population-based prevalence study. J Gen Intern Med 2014;29(12):1615–23.

[62] Stankova E, Gesheva M, Hubenova A. Age and criminal poisonings. Przegl Lek 2005;62 (6):471–4.

[63] Kintz P, Villain M, Cirimele V. Chemical abuse in the elderly: evidence from hair analysis. Ther Drug Monit 2008;30:207–11.

[64] Yin S. Malicious use of pharmaceuticals in children. J Pediatr 2010;157(5):832–6.

[65] Kintz P. Interpretation of hair findings in children: about a case involving carbamazepine. Drug Test Anal 2014;6:2–4.

[66] Kintz P, Marion M, Cirimele V. Determination of trimeprazine-facilitated sedation in children by hair analysis. J Anal Toxicol 2006;30:400–2.

[67] Finkelhor D. Child sexual abuse. In: Rosenberg ML, Fenley MA, editors. Violence in America—a public health approach. Oxford University Press; 1991. p. 79–94.

[68] Finkelhor D. Epidemiological factors in the clinical identification of child sexual abuse. Child Abuse Negl 1993;17(1):67–70.

[69] Kintz P, Evans J, Villain M, Salquebre G, Cirimele V. Hair analysis for diphenhydramine after surreptitious administration to a child. Forensic Sci Int 2007;173:171–4.

[70] Myers RJ, Hamilton JB. Regeneration and rate of growth of hairs in man. Ann N Y Acad Sci 1951;53:562–8.

[71] Lynfield YL. Effect of pregnancy on the human hair cycle. J Invest Dermatol 1960;35:323–7.

[72] Randall VA. Androgens hair growth. Dermatol Ther 2008;21:314–28.

[73] Randall VA, Ebling FJ. Seasonal changes in human hair growth. Br J Dermatol 1991;124:146–51.

[74] LeBeau MA, Montgomery MA, Brewer JD. The role of variations in growth rate and sample collection on interpreting results of segmental analyses of hair. Forensic Sci Int 2011;210:110–16.

[75] Potsch L. A discourse on human hair fibers and reflections on the conservation of drug molecules. Int J Legal Med 1996;108:285–93.
[76] Pecoraro V, Astore I, Barman J, Ignacioaraujo C. The normal trichogram in the child before the age of puberty. J Invest Dermatol 1964;42:427–30.
[77] Miyazawa N, Uematsu T. Analysis of ofloxacin in hair as a measure of hair growth and as a time marker for hair analysis. Ther Drug Monit 1992;14:525–8.
[78] Xiang P, Sun Q, Shen B, Liu W, Shen M. Segmental hair analysis using liquid chromatography-tandem mass spectrometry after a single dose of benzodiazepines. Forensic Sci Int 2011;204:19–26.
[79] Villain M, Chèze M, Dumestre V, Ludes B, Kintz P. Hair to document drug-facilitated crimes: four cases involving bromazepam. J Anal Toxicol 2004;28:516–19.
[80] Kintz P, Villain M, Dumestre-Toulet V, Ludes B. Drug-facilitated sexual assault and analytical toxicology: the role of LC-MS/MS. A case involving zolpidem. J Clin Forensic Med 2005;12:36–41.
[81] Kintz P. Issues about axial diffusion during segmental hair analysis. Ther Drug Monit 2013;35(3):408–10.
[82] Yegles M. Pitfalls in hair analysis: cosmetic treatment. Ann Toxicol Anal 2005;17:275–8.
[83] Jurado C, Kintz P, Menéndez M, Repetto M. Influence of cosmetic treatment of hair on drug testing. Int J Legal Med 1997;110:159–63.
[84] Klein J, Karaskov T, Koren G. Clinical applications of hair testing for drugs of abuse—the Canadian experience. Forensic Sci Int 2000;107:281–8.
[85] Soriano T, Moreno E, Huertas T, Rodríguez IM, Jurado C. The utility of hair testing to determine chronic cannabis exposure in young children. TIAFT; 2014.
[86] Jurado C, García S, Rodríguez IM, Soriano T. Hair versus urine analyses to estimate drugs of abuse consumption in an arrestee population. TIAFT; 2012.
[87] Dufaux B, Agius R, Nadulski T, Kahl HG. Comparison of urine and hair testing for drugs of abuse in the control of abstinence in driver's license re-granting. Drug Test Anal 2012;4(6):415–19.
[88] Polla M, Stramesi C, Pichini S, Palmi I, Vignali C, Dall'Olio G. Hair testing is superior to urine to disclose cocaine consumption in driver's license regranting. Forensic Sci Int 2009;189(1–3):41–3.
[89] Ferko AP, Barbieri EJ, DiGregorio GJ, Ruch EK. The accumulation and disappearance of cocaine and benzoylecgonine in rat hair following prolonged administration of cocaine. Life Sci 1992;51:1823.
[90] Jurado C, Rodriguez-Vicente C, Menéndez M, Repetto M. Time course of cocaine in rabbit hair. Forensic Sci Int 1997;84:61.
[91] Shen M, Xiang P, Sun Y, Shen B. Disappearance of 6-acetylmorphine, morphine and codeine from human scalp hair after discontinuation of opiate abuse. Forensic Sci Int 2013;227(1–3):64–8.
[92] Garcia-Bournissen F, Moller M, Nesterenko M, Karaskov T, Koren G. Pharmacokinetics of disappearance of cocaine from hair after discontinuation of drug use. Forensic Sci Int 2009;189:24–7.
[93] Felli M, Martello S, Marsili R, Chiarotti M. Disappearance of cocaine from human hair after abstinence. Forensic Sci Int 2005;154:96–8.

Chapter 10

Doping, Applications of Hair Analysis

Detlef Thieme and Patricia Anielski
Institute of Doping Analysis, Kreischa, Germany

10.1 INTRODUCTION

The definition of "doping" is mainly correlated to potential performance enhancement in sports. Its interpretation in terms of hair analysis is affected by a number of administrative definitions. The most recent versions of the Prohibited List updated annually by the World Anti-Doping Agency (WADA [1]) discriminates between classes of substances prohibited at all times, prohibited in competition, or prohibited in particular sports.

The resulting differentiation is basically due to the fact that certain compounds need to be bioavailable at therapeutic levels to provide acute performance enhancement (e.g., stimulants) whereas the indirect biological effects of other drugs (anabolic agents) persists after their elimination from the body. The resulting strategically analytical consequences are similar to the forensic discrimination between quantitative estimation of biologic effects (e.g., impairment) and qualitative identification of drug administration (e.g., abstinence control).

Most of the traditional abused drugs, for example, stimulants, narcotics, and cannabinoids, are prohibited in competition only. The in-competition period is typically defined to begin at 12 h prior to the start of the event but different definitions may hold, in particular for multiday events. In none of the cases a clear discrimination between prohibited administration in-competition and permitted administration out-of competition may be achieved based on hair testing. Therefore, respective substance classes which are exclusively prohibited in-competition, namely, S6: stimulants, S7: narcotics, S8: cannabinoids, and S9: corticosteroids, are not suitable for routine hair testing in doping. However, complementary hair analyses to collect additional evidences in cases of positive urine tests were already applied to differentiate between long-term abuse versus illegal single administration of stimulants. Hair collection and the application of hair testing is legally acknowledged by the International Standard of Laboratories (ISL [2]) but respective results from alternative specimens "shall not be used to counter Adverse Analytical Findings or Atypical Findings from urine." This means that that a negative hair test is not suitable to overrule a positive testing in urine, what seems to be justified due to the higher sensitivity and sharper focus on competition times. On the other hand, hair testing could significantly contribute to exculpation of athletes by demonstrating alternative administration pathways or contamination risks (e.g., unintended clenbuterol incorporation).

The main target compounds of hair testing in doping control are therefore all substance group prohibited at-all-times, that is, S0: nonapproved substances, S1: anabolic agents, S2: peptide hormones, growth factors, and related substances, S3: beta-2 agonists, S4: hormone and metabolic modulators, and S5: diuretics and masking agents. Resulting analytical challenges in hair analyses are due to their neutral (steroids) or acidic (diuretics) analytical properties or the high molecular mass of peptide hormones, resulting in low incorporation rates into hairs.

An intensified application of hair tests in doping control was recently suggested by WADA [3]. However, it is not likely that mandatory routine hair collections in athletes testing can be enforced, due to legal as well as aesthetic reasons. Moreover, the traditional urine testing procedure benefits from availability of large sample volume, relatively high concentrations of xenobiotics and the improved knowledge of urinary ultra-long-term metabolites. It appears therefore likely that hair testing will gain increasing relevance for specific testing of anabolic agents, nonapproved substances, beta-2 agonists, and metabolic modulators (e.g., anti-estrogens). Moreover, there is a large potential in animal doping control, where limitations of drug administration appear to be much stricter.

10.2 HAIR TESTING FOR POSITIVE IDENTIFICATION OF PROHIBITED SUBSTANCES

10.2.1 S0: Nonapproved Substances

This group was established to prohibit the administration of all pharmaceutical compounds which are not approved by relevant authorities, for example, drugs under preclinical or clinical development, drugs with discontinued approval, designer drugs, or substances approved only for veterinary use. The benefit of hair testing for nonapproved substances is clearly due to the preferential incorporation of parent compounds into the hair matrix. In contrast, urinary excretion is often limited to polar biotransformation products, which are unknown or not available when a new research drug enters the black market.

This was demonstrated in the past, when compounds like the agonist of the peroxisome proliferator-activated δ-receptor GW501516 [4] or several selective androgen receptor modulators (SARMs, e.g., andarine and ostarine; Figure 10.1 [5]) entered the black market. [These compounds were chosen to illustrate the principle and potential of hair testing although they were meanwhile reevaluated by WADA [1] as "other anabolic agents" (ostarine) or "metabolic modulators" (GW501516).] Parent compounds could be synthesized or became rapidly available but their individual metabolism had to be investigated before the compounds could be implemented into routine urine screening procedures. The necessity to examine biotransformation patterns and synthesize relevant metabolites or carry out excretion studies lead to significant delays in designing sensitive assays for urine testing. This phenomenon is similar to the problems when dealing with synthetic cannabinoids [6,7]. The multitude of structural modifications, rapid changes of availability on the black market, poor international harmonization of legal classification, and a delayed availability of urinary metabolites does not permit a prompt inclusion of these substances into urine screening.

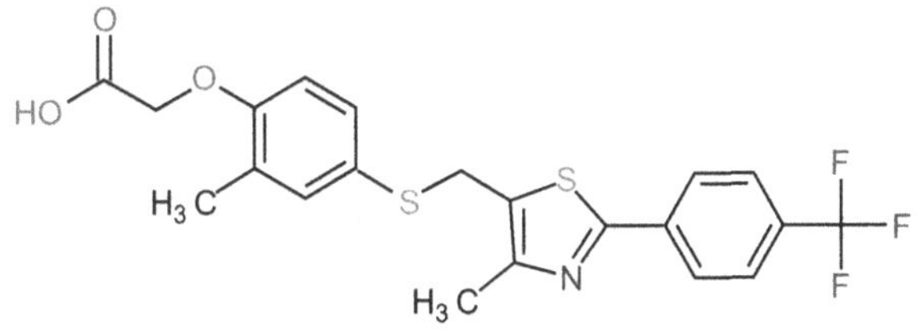

GW501516

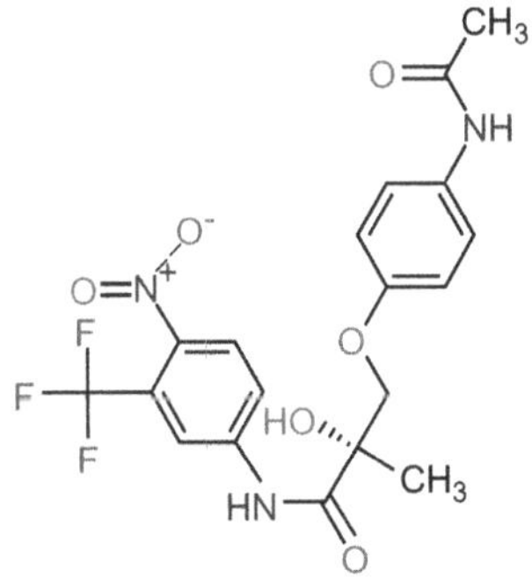

Andarine

FIGURE 10.1 Nonapproved substances (e.g., the PPARδ-receptor agonist, GW501516, or the selective androgen receptor modulator, andarine) are prohibited in sports. Their detection in urine requires an initial elucidation of their biotransformation whereas parent compounds are more likely to be incorporated into hair.

10.2.2 S1.1.a: Synthetic Anabolic Androgenic Steroids

Anabolic androgenic steroids (AASs) appeared to be the most potent and prevalent doping agents during the past decades of the twentieth century. Their moderate dosages and complex biotransformation patterns posed a significant analytical challenge in doping control. Development of efficient derivatization procedures prior to GC-MS detection, access to LC-MS as complementary detection technique for unsaturated steroids (e.g., trenbolone and its analogues), and the detection of urinary long-term metabolites of many popular steroids (e.g., metandienone [8], oxandrolone [9], or dehydrochloromethyltestosterone [10]) were milestones to improve its sensitive detection in urine. Even single therapeutic dosages became typically detectable for several weeks. This has significantly reduced the potential to misuse synthetic anabolic steroids in sports at least as far as out-of-competition doping controls were conducted. As a consequence there was an apparent shift from abusing these compounds to the misuse of endogenous hormones (e.g., testosterone and growth hormone).

Synthetic anabolic steroids have attracted significant scientific attention as target compounds in hair analysis of doping and forensic cases. Testosterone and nandrolone esters [11,12]; testosterone undecanoate [13]; stanozolol [12,14–16]; methenolone and mesterolone [12]; metandienone [17], methyltestosterone, nandrolone, and dehydromethyltestosterone [18] were successfully identified in either human or in equine testing [11,19]. The clear benefits of hair testing are due to the long-term storage of parent compounds. In particular the incorporation of steroid esters may provide a clear differentiation of an external administration of intact testosterone esters from endogenous testosterone. However, typical detection limits of 0.1–5 pg/mg correspond to multiple or even long-term administration of steroids in high dosages which are typically associated with permanent and excessive positive results in conventional urine testing. The majority of cases deal with steroid administration in bodybuilding or powerlifting, resulting hair concentrations are rather high (e.g., stanozolol ranging from 5 to 86.3 pg/mg [14,16], metandienone from 7 to 108 pg/mg [17]) and do not reflect crucial detection challenges in doping analysis. However, continuous improvement of detection limits may well permit the detection of doping relevant dosages in near future.

Owing to the lack of systematic studies of hair testing in sports, the following data result from forensic cases. Respective hair donors had a presumptive background in bodybuilding and were typically suspected of possession or trafficking of doping agents. In most cases, additional evidences were collected from corresponding urine samples and confiscated steroids which could verify the conclusiveness of most hair tests.

The analytical strategy is governed by different requirements of hair sample preprocessing and analytical detection. Most of the anabolic steroids are neutral compounds and are thought to be (weakly) incorporated into hair via sweat and extracted best by methanol extraction. A washout of steroids was recurrently observed and hair segmentation is certainly not expedient when dealing with neutral anabolic steroids. However, the basic compounds clenbuterol and stanozolol require an additional sodium hydroxide digestion of the remaining hair fibers to improve their recovery.

The analytical strategy (Figure 10.2) is mainly characterized by the two subgroups of highly lipophilic anabolic steroids (in particular steroid esters) which are not suitable for LC-MS detection and therefore covered by a GC-MS/MS detection. The slightly more polar steroids (metandienone, stanozolol, trenbolone, and its derivatives), clenbuterol, and all relevant anti-estrogens are detected best by LC-MS/MS. Boldenone and metandienone require a multi-stage LC-MS3 experiment to be identified at sufficient detection limits.

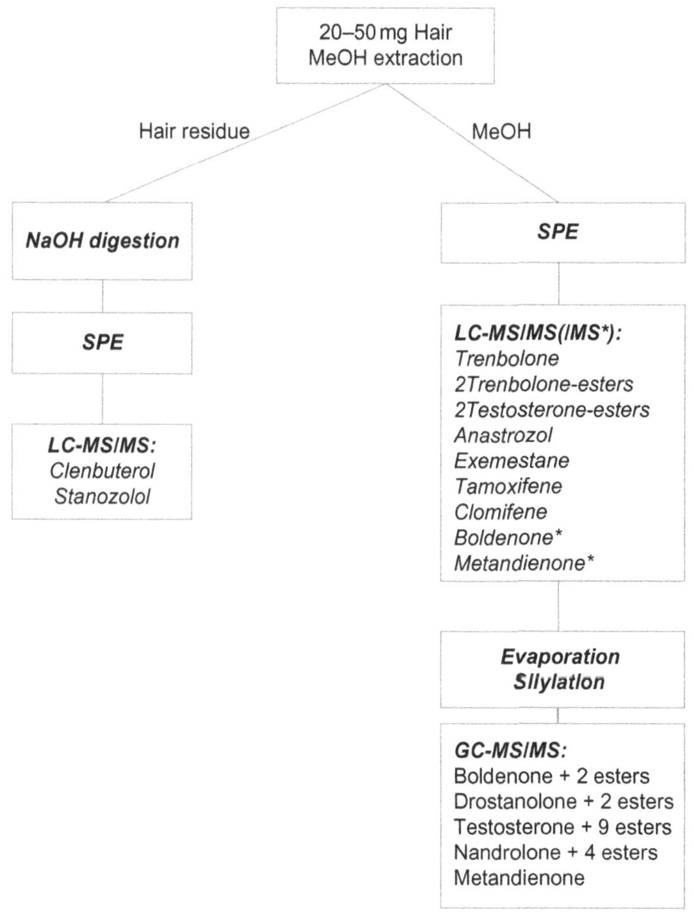

FIGURE 10.2 Analytical strategy of doping agents in hair.

Earlier attempts to identify biotransformation products (e.g., epimetandienone or hydroxyl-stanozolol [20]) as incorporation marker were not maintained due to their low hair concentration. Moreover, unintentional contamination or passive consumption of doping agents, which are typically administered orally or intramuscularly, seems to be no priority issue.

The data summarized in Table 10.1 demonstrate impressively that a large number of doping agents were detected in hair samples of presumptive AAS users. The majority of these findings were in good accordance with corresponding findings in urine samples (Table 10.2). A significant number of additional steroids findings (i.e., not detected in urine) could be attributed to hair tests. This included steroids typically used by oral administration (e.g., metandienone and stanozolol) as well as steroid esters (trenbolone acetate, trenbolone enantate, testosterone propionate, and testosterone enantate) which seem to reflect the frequency of respective steroids on the black market rather than biochemical or analytical issues. Typical hair concentrations of the most abundant anabolic steroids testosterone propionate, testosterone enantate, and metandienone examined in abuse cases reached maxima of

TABLE 10.1 Number of Findings of AAS in Hairs Collected from Presumptive Users

Anabolic Steroid	No. of Cases
Testosterone enantate	29
Metandienone	24
Testosterone propionate	24
Clenbuterol	8
Trenbolone acetate	8
Stanozolol	7
Trenbolone enantate	7
Drostanolone propionate	4
Boldenone undecylenate	3
Nandrolone decanoate	3
Testosterone decanoate	3
Testosterone phenylpropionate	3
Drostanolone enantate	2
Nandrolone phenylpropionate	1
Testosterone cypionate	1

TABLE 10.2 Findings of Anabolic Agents and Hormone Modulators in Hairs of 96 Presumptive Steroid Users

Case No.	Hair Characteristics (Gender = Male)	Results	Concentration (pg/mg)
1	24 years	Nandrolone	15
	Black		
	0–3.5 cm	Testosterone	5
2	17 years	Testosterone	30
	Black		
	0–6 cm		
4	31 years	Metandienone*	50
	Dark brown	Tamoxifen*	1,300
	0–3 cm	Testosterone enantate*	10
		Testosterone*	5
6	45 years	Clenbuterol	15
	Bleached	Testosterone cypionate	12
	0–6 cm	Testosterone propionate	15
		Testosterone	3
7	37 years	Testosterone propionate*	290
	Black		
	0–4 cm	Testosterone*	10
8	49 years	Testosterone enantate	25
	Black	Testosterone propionate	25
	0–3 cm	Testosterone	10
9	33 years	Stanozolol	12
	Brown	Testosterone	5
	0–3 cm		
10	30 years	Testosterone propionate	175
	Light brown		
	0–4 cm	Testosterone	14
11	42 years	Testosterone propionate	22
	Black	Boldenone	10
	0–1.5 cm	Testosterone	20

(*Continued*)

TABLE 10.2 (Continued)

Case No.	Hair Characteristics (Gender = Male)	Results	Concentration (pg/mg)
15	39 years	Metandienone	20
	Light brown		
	0–2 cm	Testosterone	2
17	28 years	Testosterone enantate	20
	Black		
	0–1.5 cm	Testosterone	1
18	32 years	Metandienone	30
	Bleached	Testosterone	n.d.
	0–3 cm	Stanozolol	1
19	32 years	Stanozolol	19
	Light brown	Testosterone	n.d.
	0–3.5 cm		
21	22 years	Nandrolone decanoate*	50
	Dark brown	Testosterone enantate*	40
	0–3 cm	Testosterone*	15
		Nandrolone*	10
22	27 years	Testosterone decanoate	30
	Black	Testosterone phenylpropionate	15
	0–3 cm	Testosterone	5
23	41 years	Metandienone*	80
	Dark brown	Stanozolol*	10
	0–1.5 cm	Nandrolone decanoate*	150
		Testosterone decanoate*	50
		Testosterone enantate*	450
		Testosterone phenylpropionate*	15
		Nandrolone*	10
		Testosterone*	20

(*Continued*)

TABLE 10.2 (Continued)

Case No.	Hair Characteristics (Gender = Male)	Results	Concentration (pg/mg)
24	29 years	Testosterone decanoate*	20
	Blond	Testosterone enantate*	500
	0–1 cm	Trenbolone acetate#	8
		Testosterone*	10
25	43 years	Testosterone*	10
	Bleached		
	0–1 cm		
27	32 years	Nandrolone*	10
	Blond	Trenbolone enantate#	15
	0–1 cm	Testosterone enantate#	30
		Testosterone#	5
29	30 years	Testosterone enantate	2
	Black	Clenbuterol	90
	0–5 cm	Stanozolol	20
		Testosterone	5
30	26 years	Testosterone enantate	14
	Black		
	0–2 cm	Testosterone	2
32	46 years	Testosterone*	10
	Light brown	Nandrolone*	20
	0–3 cm	Metandienone*	4
		Testosterone propionate*	20
34	40 years	Testosterone*	20
	Brown	Metandienone*	40
	0–2 cm	Testosterone enantate*	180
36	Age (n.s.)	Testosterone enantate*	300
	Brown	Stanozolol#	60
	0–2 cm	Metandienone*	250
		Testosterone*	100

(*Continued*)

TABLE 10.2 (Continued)

Case No.	Hair Characteristics (Gender = Male)	Results	Concentration (pg/mg)
41	30 years	Testosterone*	1
	Light brown	Stanozolol*	10
	0–3 cm		
42	25 years	Testosterone*	5
	Brown		
	0–2.5 cm		
43	35 years	Testosterone*	20
	Brown		
	0–3 cm		
44		Testosterone*	10
	27 years		
	Brown	Metandienone*	50
	0–3 cm	Clenbuterol*	0.7
45	32 years	Testosterone*	10
	Brown		
	0–0.7 cm		
46	26 years	Testosterone*	50
	Light brown	Testosterone propionate*	10
	0–2.5 cm	Boldenone*	30
		Stanozolol#	1
47	23 years	Testosterone#	10
	Brown	Testosterone propionate#	15
	0–4 cm	Boldenone#	50
49	42 years	Testosterone#	10
	Dark brown		
	0–3 cm	Testosterone propionate#	35
50	22 years	Testosterone*	30
	Brown	Testosterone enantate*	40
	0–1 cm	Boldenone*	30
		Trenbolone*	190
		Clenbuterol*	10

(*Continued*)

TABLE 10.2 (Continued)

Case No.	Hair Characteristics (Gender = Male)	Results	Concentration (pg/mg)
51	25 years	Testosterone*	5
	Brown	Testosterone enantate*	15
	0–4.5 cm	Boldenone*	5
		Trenbolone*	45
		Metandienone#	10
52	20 years	Testosterone	1
	Brown	Boldenone*	20
	0–2 cm		
		Trenbolone*	30
53	22 years	Testosterone	1
	Brown		
	0–6 cm	Metandienone#	7
55	34 years	Testosterone*	50
	Brown	Testosterone enantate*	10
	0–3 cm	Testosterone propionate*	15
		Boldenone*	60
		Trenbolone*	290
		Trenbolone enantate*	235
		Metandienone*	50
56	33 years	Testosterone	1
	Blond		
	0–3 cm	Boldenone*	8
57	30 years	Testosterone	10
	Brown	Clenbuterol	20
	0–1 cm		
58	41 years	Testosterone*	35
	Brown		
	0–4 cm	Testosterone propionate*	500
61	35 years	Testosterone*	10
	Brown		
	0–3 cm		

(*Continued*)

TABLE 10.2 (Continued)

Case No.	Hair Characteristics (Gender = Male)	Results	Concentration (pg/mg)
62	22 years	Testosterone	5
	Dark brown		
	0–4 cm		
65	23 years	Testosterone	1
	Light brown	Boldenone	6
	0–1.5 cm		
66	29 years	Clenbuterol#	2
	Bleached	Boldenone*	3
	0–3 cm	Boldenone undecylenate*	400
		Drostanolone*	4
		Drostanolone enantate*	30
		Metandienone*	220
		Nandrolone phenylpropionate#	145
		Stanozolol*	43
		Testosterone enantate*	500
		Trenbolone*	15
		Trenbolone acetate*	10
		Trenbolone enantate*	270
		Tamoxifen#	150
		Anastrozole*	85
		Testosterone*	500
67	27 years	Metandienone	50
	Black	Trenbolone	145
	0–1.5 cm	Trenbolone acetate	30
		Trenbolone enantate	57
		Anastrozole	1
		Testosterone propionate	98
		Testosterone enantate	120
		Testosterone	25

(*Continued*)

TABLE 10.2 (Continued)

Case No.	Hair Characteristics (Gender = Male)	Results	Concentration (pg/mg)
68	Age (n.s.)	Boldenone	35
	Black	Metandienone	100
	0–1.5 cm	Nandrolone	15
		Trenbolone	2
		Trenbolone enantate	97
		Testosterone propionate	105
		Testosterone	16
69	34 years	Drostanolone propionate#	4
	Dark brown		
	0–4 cm	Testosterone	2
70	44 years	Clenbuterol	1
	Dark brown		
	0–6 cm	Testosterone	5
71	40 years	Trenbolone acetate	300
	Dark brown	Drostanolone propionate	30
	0–2 cm	Testosterone propionate	9
		Testosterone enantate	65
		Testosterone	19
72	62 years	Metandienone*	107
	Brown	Boldenone*	5
	0–6 cm	Boldenone undecylenate*	trace
		Testosterone propionate#	9
		Testosterone enantate#	140
		Testosterone#	8
73	25 years	Testosterone propionate*	6
	Brown	Testosterone enantate*	500
	0–3 cm	Tamoxifen*	100
		Testosterone*	24

(*Continued*)

TABLE 10.2 (Continued)

Case No.	Hair Characteristics (Gender = Male)	Results	Concentration (pg/mg)
74	44 years	Drostanolone	2
	Gray/brown	Drostanolone enantate	2
	0–2 cm	Testosterone propionate	25
		Testosterone enantate	48
		Testosterone	14
75	30 years	Metandienone[#]	60
	Brown	Trenbolone acetate*	4
	0–2 cm	Nandrolone decanoate*	2
		Testosterone propionate*	230
		Testosterone enantate*	10
		Testosterone*	15
78	33 years	Stanozolol[#]	7
	Black	Metandienone*	64
	0–4 cm	Trenbolone acetate[#]	4
		Testosterone enantate*	16
		Testosterone*	7
79	28 years	Metandienone	45
	Black	Trenbolone	24
	0–3 cm	Boldenone	10
		Testosterone	6
80	Age (n.s.)	Testosterone propionate	5
	Black	Testosterone enantate	20
	0–2 cm	Testosterone	10
81	44 years	Testosterone propionate[#]	14
	Dark brown		
	0–4 cm	Testosterone	1
82	33 years	Testosterone*	2
	Brown		
	0–2 cm	Metandienone*	8

(Continued)

TABLE 10.2 (Continued)

Case No.	Hair Characteristics (Gender = Male)	Results	Concentration (pg/mg)
83	Age (n.s.)	Testosterone enantate*	50
	Black		
	0–1 cm	Testosterone*	10
84	23 years	Testosterone enantate*	180
	Dark brown		
	0–1.5 cm	Testosterone*	20
85	26 years	Metandienone#	18
	Black		
	0–3 cm	Testosterone	15
86		Metandienone#	70
	Age (n.s.)	Trenbolone#	50
	Dark brown	Trenbolone acetate#	80
	0–4 cm	Trenbolone enantate#	290
		Anastrozole*	10
		Drostanolone*	17
		Drostanolone propionate*	120
		Testosterone propionate*	450
		Testosterone enantate*	390
		Testosterone*	35
87	22 years	Testosterone propionate	50
	Light brown	Testosterone	5
	0–3 cm		
88	Age (n.s.)	Metandienone#	7
	Dark brown	Testosterone enantate*	20
	0–1.5 cm	Testosterone*	20
89	Age (n.s.)	Trenbolone*	100
	Dark brown	Clenbuterol*	50
	0–2 cm	Testosterone propionate*	18
		Testosterone enantate*	130
		Testosterone	10

(Continued)

TABLE 10.2 (Continued)

Case No.	Hair Characteristics (Gender = Male)	Results	Concentration (pg/mg)
93	20 years	Testosterone	2
	Brown	Trenbolone	290
	0–1 cm	Metandienone	7
	Vellus hair (leg)	Stanozolol	2
96	52 years	Stanozolol#	19
	Gray/brown	Metandienone*	120
	0–4 cm	Trenbolone*	330
	Vellus hair (chest)	Trenbolone acetate*	65
		Trenbolone enantate*	145
		Boldenone*	20
		Boldenone undecylenate*	8
		Drostanolone*	2
		Drostanolone propionate*	10
		Testosterone propionate*	150
		Testosterone phenylpropionate*	7
		Testosterone enantate*	360
		Testosterone*	80

A total of 29 cases (data not shown) were found to be negative in urine and/or hair testing. Results labeled by asterisks were analytically confirmed or supported by corresponding urine analyses. Findings in excess to corresponding urine tests were marked by hash tags.
n.d., not detected; n.s., not specified

500 pg/mg and there is no apparent correlation to administration route or half-life of the drugs (Figure 10.3).

10.2.3 S1.1.b: Endogenous AASs

This substance group comprises a positive list of dedicated compounds which are considered as prohibited. There was a number of so-called prohormones (androstenedione, androstenediols) legally available in the 1990th and became very popular as nutrition supplements. These drugs suffered from poor bioavailability and became widely irrelevant after its annexation into

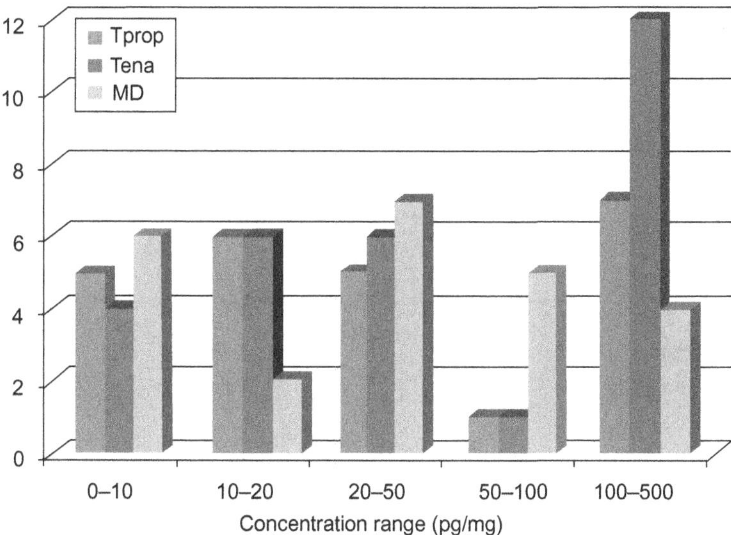

FIGURE 10.3 Hair concentration distribution of positive findings of metandienone (MD), testosterone enantate (Tena), and testosterone propionate (Tprop) as detected in forensic abuse cases.

the list of scheduled III drugs (US Controlled Substance Act [21]). The endogenous anabolic steroid with the most significant abuse potential appears to be testosterone, which is available as injectable esters (acetate, propionate, cypionate, isocaproate, isobutyrate, phenylpropionate, enantate, phenylpropionate, decanoate, undecanoate), as injectable testosterone suspension, testosterone gel and patches, and oral capsules of testosterone undecanoate. Besides, there is only a minor importance of dehydrotestosterone and dehydroepiandiosterone (DHEA), which is still legally available in some countries, including the United States.

Like in all cases dealing with endogenous compounds, there is a challenge to differentiate basal levels from abuse. An evaluation of endogenous steroids based on threshold limits is assumed to be less sensitive and potential subject of court disputes than direct identifications of exogenous compounds.

Typical endogenous testosterone concentrations have been published in several papers [17,22] and were proposed to range from 0.8 to 24.2 pg/mg for testosterone. Endogenous threshold should always be adjusted to the topical sample preprocessing (extraction and clean-up) of the hair samples, and the discrimination of elevated levels from normal values is potentially affected by degradation (e.g., due to UV radiation [23]) and washout due to cosmetic treatment [24]. According to the proposed analytical strategy (Figure 10.2) a threshold of 5 pg/mg was applied to identify suspicious cases. There seem to be no significant influence of hair color on endogenous hair

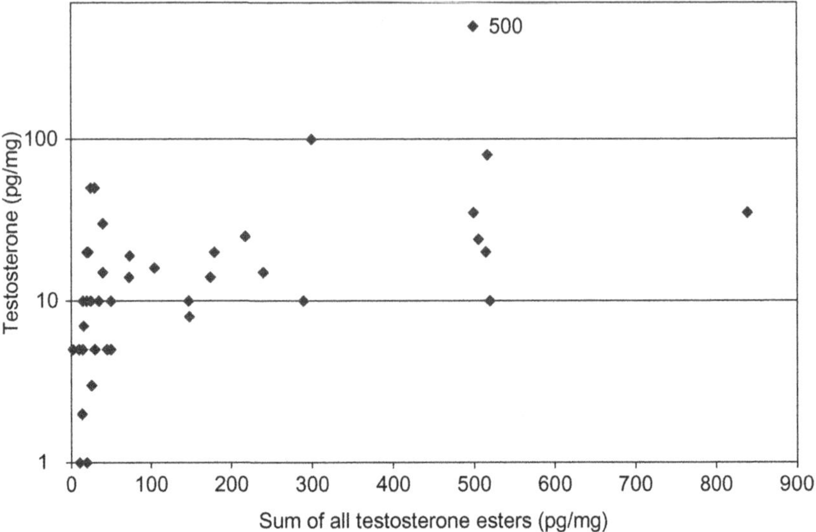

FIGURE 10.4 Correlation of hair concentrations of testosterone esters and free testosterone in hair. The intramuscular injection of depot preparations results in moderate increase of free testosterone in hair. The outlier exhibiting 500 pg/mg free testosterone was due to the (very atypical) injection of free testosterone.

concentrations. In addition to the hair levels of free testosterone, any amount of testosterone esters can be accepted as clear evidence for steroid abuse. The identification of intact esters of testosterone appears to be a better diagnostic criterion than testosterone threshold concentrations. There is only a weak correlation between total concentrations of all testosterone esters to free testosterone in hair (Figure 10.4). Half of the 42 cases positive for testosterone esters showed hair concentration of free testosterone below 10 pg/mg. The highest case exhibiting 500 pg/mg of free testosterone in hair was due to the fact that the person had injected testosterone suspensions extra to testosterone enantate. This represents an extremely unusual and ineffective administration pathway of testosterone because of the very short half-life of injected free testosterone.

10.2.4 S1.2: Other Anabolic Agents

The most common nonsteroidal anabolic agent is certainly clenbuterol and its detectability in hair is well documented [25–27]. It has a long record of being abused in human sports, horseracing, and is—whether legally or illegally—still widely (ab)used in meat-producing industry, leading to potential contaminations following the incorporation of affected meat. There are numerous confirmed cases of unintentional administration, and discrimination between a late abuse stage and an accidental intake cannot be achieved based on urinary concentrations. This became a serious issue when detection limits of conventional doping controls had improved. Respective MRPL (minimum required performance levels) were 2 ng/mL as long as analytical assays were based on GC-MS but has decreased significantly by introduction of HPLC-(HR)MS/MS. Nowadays, urinary threshold concentrations of 10 pg/mL can be easily detected under routine conditions, leading to an overflow of respective cases and warnings issued by numerous sports federations and authorities [28].

A differentiation between illegal administration and unintended intake may be easily achieved by quantitative hair analysis. Typical concentrations corresponding to clenbuterol abuse were reported to range from 15 to 122 pg/mg [25]. Positive clenbuterol findings within our routine analysis of human hair (were submitted by police, prosecution and customs, unpublished data) are in accordance. Clenbuterol was detected in five hair samples of bodybuilders exhibiting concentrations between 10 and 90 pg/mg within the last 2 years. In contrast, an examination of athletes from a high contamination risk area (Mexican football players) revealed positive findings in 89% of the tested athletes. Hair concentrations ranged from 0.02 and 1.88 pg/mg (median: 0.16 pg/mg) clenbuterol in hair [29] and were hence significantly lower than typical abuse levels. This quantitative discrimination is restricted to pigmented hair, due to the high melanin binding of clenbuterol.

There are many other and anecdotal reports of using brombuterol, zeranol, or zilpaterol which may easily be detected in hair but did not gain sufficient relevance so far.

10.2.5 S3: Beta-2 Agonists

All beta-2 agonists are prohibited in sports at all times, except for the therapeutic inhalation of salbutamol, formoterol, and salmeterol. Therapy must comply with the manufacturers' recommended dosages. The discrimination of this therapeutic use from overdose or systemic administration is hardly possible by means of hair analysis. However, a proof of other (prohibited) compounds by hair testing (e.g., terbutaline, pirbuterol, metaproterenol, or fenoterol) could be established [30,31]. This substance group seems to be predestined as target for screening tests due to their considerable dosages, unique chemical properties, excellent detection by HPLC-MS/MS, and very good incorporation into hair matrix.

10.2.6 S4: Hormone and Metabolic Modulators

This substance class is somewhat heterogeneous and contains peptide hormones (e.g., insulins) which are certainly not suitable for hair testing. The most interesting compounds in this group are anti-estrogens, which are very frequently used in conjunction with anabolic steroids to treat side effects of anabolic steroids (e.g., gynecomastia), formation of edema, and suppression of endogenous testosterone synthesis. In spite of their indirect mechanism and moderate dosages (even in bodybuilding only therapeutic dosages of anastrozole, exemestane, tamoxifen, and clomifene were suggested) there is a high potential of detecting these compounds in hair. The incorporation rate into hair and the stability seem to be higher compared to most anabolic steroids and detection is more sensitive due to a high structural uniqueness. These compounds represent therefore another group of very sensitive markers for detection of doping in hair (Table 10.2).

10.3 HAIR TESTING AS ADDITIONAL EVIDENCE IN PRESUMPTIVE DOPING CASES

10.3.1 Case 1: Clenbuterol

Owing to the relevance of the substance and its outstanding incorporation rate into hair, clenbuterol represents the most interesting target analyte for hair testing in doping control. Numerous positive cases of clenbuterol were found in hair of human athletes or equine. Hair testing is equally efficient for doping control, breeding surveillance (stallion licensing), monitoring of meat production industries, or abstinence control. Detection limits of 0.02 pg/mg permitted the proof of exposure of athletes to clenbuterol. Figure 10.5 shows a case of a cyclist who was tested positive with trace amounts of clenbuterol in urine after returning from a training camp in Mexico. Segmentation of a hair strand demonstrated that positive finding was limited to the proximal segment (corresponding to the duration of stay in Mexico), the absence of clenbuterol in the adjacent distal segment and very low hair concentrations were compatible with exposure to contaminations rather than medication or abuse. The case was dismissed by the relevant sports federations. In contrast, temporal clenbuterol profiles of permanent residents (Mexican football players) showed constant levels of clenbuterol concentrations along the hair strand (Figure 10.6) and revealed a permanent exposure to contaminated food.

10.3.2 Case 2: MDMA

Follow-up investigations of controversial analytical findings may require access to retrospective analytical data, which cannot be provided by retesting of urine or plasma samples, because the typical time span between initial testing and result management includes several weeks. Therefore, retesting of hair represents the only option to collect additional analytical evidence to improve case interpretations. A negative hair sample is principally not eligible to overrule positive urine tests, because the latter is typically more sensitive and focused to a short time period. Negative hair tests may well be in good

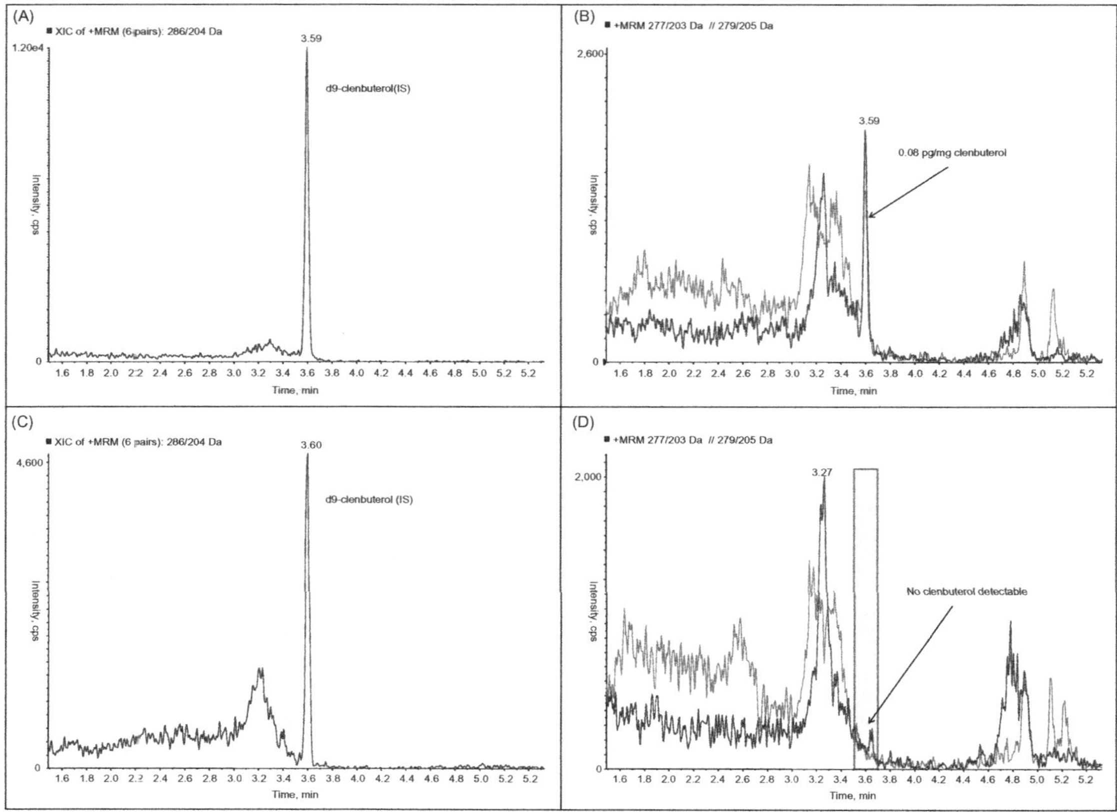

FIGURE 10.5 LC-MS/MS-chromatogram of the positive clenbuterol finding in the proximal hair segment (B: 0–2 cm) of a male cyclist. The distal (D: 2–4 cm) segment proved to be free of clenbuterol residues. The chromatograms of corresponding deuterated internal standards are shown in parts A and C.

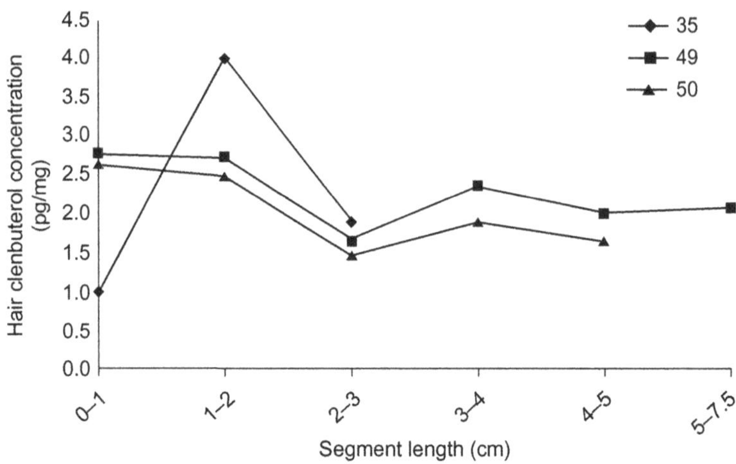

FIGURE 10.6 Clenbuterol-concentration profile of three Mexican football players representing long-term exposure to clenbuterol contamination.

accordance with positive urine findings and vice versa. However, a potential long-term abuse may well be discriminated from a single administration by means of hair testing. This approach was successfully applied when players of two national hockey teams were complaining about various health issues (tachycardia, fatigue, and tremor) after a welcome dinner which was held prior to an Olympic qualification tournament. Two of the players were later got drawn for doping controls and were tested positive for ecstasy (methylenedioxymethamphetamine, MDMA). MDMA is a mild stimulant with predominant entactogen psychoactive effects, belongs to the category S6b, "specified stimulants" which may potentially be used due to medical or social motivations and is considered less potent as doping agent. However, its positive identification in-competition at urinary concentrations higher than 50 ng/mL constitutes a doping offence. Owing to the unusual and suspicious circumstances, a follow-up investigation was intended. By retrospective retesting of hair, the intoxication of the whole delegation (athletes and representatives) with MDMA could be demonstrated [32]. The concentration profiles indicated a single administration rather than a typical repetitive use. Therefore, group intoxication was found to be the presumptive origin for two apparent MDMA doping cases and the case was dismissed by the international federation.

10.4 HAIR TESTING FOR DOPING CONTROL IN ANIMALS

The aforementioned principles of hair testing can generally be extended to doping controls in animals [33]. However, legal and biochemical particularities need to be taken into consideration. The list of prohibited compounds relevant for greyhounds or equines is much longer to prevent medical treatment during competition. The Association of Racing Commissioners [34] distinguished five classes of prohibited compounds depending on its classification as controlled substances, performance enhancing potential, and their acceptance for medical treatment in the racing horse. The most relevant class 1 includes stimulants and depressant drugs that have the highest potential to affect performance and scheduled drugs (opiates, synthetic opioids, psychoactive drugs, amphetamines, and amphetamine-like drugs). Most of these compounds are well known target analytes in hair testing and could easily be detected in hairs of equine or other animals (e.g., the detection of morphine) [35]. However, medical treatment may be granted for some of these prohibited substances, if applied out-of competition and in compliance with appropriate waiting periods. These legal regulations may result in crucial temporal discrimination requirements similar to human sports. Moreover, biochemical particularities (e.g., the endogenous biosynthesis of nandrolone), which may depend on age, gender, and fertility, need to be taken into consideration.

The most potent applications of hair in racing anti-doping analysis is therefore devoted to the identification of anabolic agents, in particular clenbuterol, anabolic steroids, and corresponding esters [11,36,37].

REFERENCES

1. 2014 Prohibited List. <https://www.wada-ama.org/en/resources/science-medicine/prohibited-list#.VEJ9rRZ8sct>; 2013 [accessed 01.10.14].
2. International Standard for Laboratories (ISL). <http://www.wada-ama.org/en/Science-Medicine/Anti-Doping-Laboratories/Technical-Documents/>; 2013 [accessed 01.10.14].
3. New WADA chief Craig Reedie wants to test hair to identify drug cheats. <http://www.theguardian.com/sport/2013/dec/31/wada-craig-reedie-drug-cheats-doping-hair>; 2013 [accessed 01.10.14].
4. Thevis M, Beuck S, Thomas A, Kortner B, Kohler M, Rodchenkov G, et al. Doping control analysis of emerging drugs in human plasma—identification of GW501516, S-107, JTV-519, and S-40503. Rapid Commun Mass Spectrom 2009;23:1139.
5. Thevis M, Thomas A, Moller I, Geyer H, Dalton JT, Schanzer W. Mass spectrometric characterization of urinary metabolites of the selective androgen receptor modulator S-22 to identify potential targets for routine doping controls. Rapid Commun Mass Spectrom 2011;25:2187.
6. Hutter M, Kneisel S, Auwarter V, Neukamm MA. Determination of 22 synthetic cannabinoids in human hair by liquid chromatography-tandem mass spectrometry. J Chromatogr B Analyt Technol Biomed. Life Sci 2012;903:95.
7. Salomone A, Luciano C, Di Corcia D, Gerace E, Vincenti M. Hair analysis as a tool to evaluate the prevalence of synthetic cannabinoids in different populations of drug consumers. Drug Test Anal 2014;6:126.
8. Gomez C, Pozo OJ, Garrostas L, Segura J, Ventura R. A new sulphate metabolite as a long-term marker of metandienone misuse. Steroids 2013;78:1245.
9. Guddat S, Fussholler G, Beuck S, Thomas A, Geyer H, Rydevik A, et al. Synthesis, characterization, and detection of new oxandrolone metabolites as long-term markers in sports drug testing. Anal Bioanal Chem 2013;405:8285.

10. Sobolevsky T, Rodchenkov G. Detection and mass spectrometric characterization of novel long-term dehydrochloromethyltestosterone metabolites in human urine. J Steroid Biochem Mol Biol 2012;128:121.
11. Anielski P. Hair analysis of anabolic steroids in connection with doping control—results from horse samples. J Mass Spectrom 2008;43:1001.
12. Thieme D, Grosse J, Sachs H, Mueller RK. Analytical strategy for detecting doping agents in hair. Forensic Sci Int 2000;107:335.
13. Strano-Rossi S, Castrignano E, Anzillotti L, Odoardi S, De-Giorgio F, Bermejo A, et al. Screening for exogenous androgen anabolic steroids in human hair by liquid chromatography/orbitrap-high resolution mass spectrometry. Anal Chim Acta 2013;793:61.
14. Deshmukh N, Hussain I, Barker J, Petroczi A, Naughton DP. Analysis of anabolic steroids in human hair using LC-MS/MS. Steroids 2010;75:710.
15. Kintz P, Cirimele V, Ludes B. Discrimination of the nature of doping with 19-norsteroids through hair analysis. Clin Chem 2000;46:2020.
16. Cirimele V, Kintz P, Ludes B. Testing of the anabolic stanozolol in human hair by gas chromatography-negative ion chemical ionization mass spectrometry. J Chromatogr B Biomed Sci Appl 2000;740:265.
17. Bresson M, Cirimele V, Villain M, Kintz P. Doping control for metandienone using hair analyzed by gas chromatography-tandem mass spectrometry. J Chromatogr B Analyt Technol Biomed Life Sci 2006;836:124.
18. Deng XS, Kurosu A, Pounder DJ. Detection of anabolic steroids in head hair. J Forensic Sci 1999;44:343.
19. McKinney AR. Modern techniques for the determination of anabolic-androgenic steroid doping in the horse. Bioanalysis 2009;1:785.
20. Deshmukh NI, Zachar G, Petroczi A, Szekely AD, Barker J, Naughton DP. Determination of stanozolol and 3′-hydroxystanozolol in rat hair, urine and serum using liquid chromatography tandem mass spectrometry. Chem Cent J 2012;6:162.
21. Schedules of controlled substances. <http://www.fda.gov/RegulatoryInformation/Legislation/ucm148726.htm>; 2014 [accessed 01.10.14].
22. Shen M, Xiang P, Shen B, Bu J, Wang M. Physiological concentrations of anabolic steroids in human hair. Forensic Sci Int 2009;184:32.
23. Li J, Xie Q, Gao W, Xu Y, Wang S, Deng H, et al. Time course of cortisol loss in hair segments under immersion in hot water. Clin Chim Acta 2012;413:434.
24. Dettenborn L, Tietze A, Bruckner F, Kirschbaum C. Higher cortisol content in hair among long-term unemployed individuals compared to controls. Psychoneuroendocrinology 2010;35:1404.
25. Dumestre-Toulet V, Cirimele V, Ludes B, Gromb S, Kintz P. Hair analysis of seven bodybuilders for anabolic steroids, ephedrine, and clenbuterol. J Forensic Sci 2002;47:211.
26. Gleixner A, Sauerwein H, Meyer HH. Detection of the anabolic beta 2-adrenoceptor agonist clenbuterol in human scalp hair by HPLC/EIA. Clin Chem 1996;42:1869.
27. Schlupp A, Anielski P, Thieme D, Muller RK, Meyer H, Ellendorff F. The beta-agonist clenbuterol in mane and tail hair of horses. Equine Vet J 2004;36:118.
28. China issues clenbuterol warning to national athletes. <http://www.ukad.org.uk/news/article/china-issue-clenbuterol-warning-to-national-athletes>; 2012 [accessed 01.10.14].
29. Thieme D, Krumbholz A, Anielski P, Gfrerer L, Graw M, Geyer H, et al. Statistical significance of hair analysis of clenbuterol to discriminate therapeutic use from contamination. Drug Test Anal 2014;6:1108–16.

30. Kintz P, Dumestre-Toulet V, Jamey C, Cirimele V, Ludes B. Doping control for beta-adrenergic compounds through hair analysis. J Forensic Sci 2000;45:170.
31. Nielen MW, Lasaroms JJ, Essers ML, Oosterink JE, Meijer T, Sanders MB, et al. Multiresidue analysis of beta-agonists in bovine and porcine urine, feed and hair using liquid chromatography electrospray ionisation tandem mass spectrometry. Anal Bioanal Chem 2008;391:199.
32. Segura J. Is anti-doping analysis so far from clinical, legal or forensic targets?: the added value of close relationships between related disciplines. Drug Test Anal 2009;1:479.
33. Dunnett M, Lees P. Equine hair analysis: current status and future prospects. Equine Vet J 2004;36:102.
34. Uniform Classification Guidelines for Foreign Substances. <http://www.dfa.arkansas.gov/offices/racingCommission/Documents/drug_list.pdf>; 2009 [accessed 01.10.14].
35. Whittem T, Davis C, Beresford GD, Gourdie T. Detection of morphine in mane hair of horses. Aust Vet J 1998;76:426.
36. Anielski P, Thieme D, Schlupp A, Grosse J, Ellendorff F, Mueller RK. Detection of testosterone, nandrolone and precursors in horse hair. Anal Bioanal Chem 2005;383:903.
37. Gray BP, Viljanto M, Bright J, Pearce C, Maynard S. Investigations into the feasibility of routine ultra high performance liquid chromatography-tandem mass spectrometry analysis of equine hair samples for detecting the misuse of anabolic steroids, anabolic steroid esters and related compounds. Anal Chim Acta 2013;787:163.

Chapter 11

Detection of New Psychoactive Substances

Alberto Salomone
Centro Regionale Antidoping e di Tossicologia "A. Bertinaria", Orbassano (TO), Italy

11.1 INTRODUCTION

For many decades there have been few additions to the spectrum of drugs abused. In recent years, there has been a huge upsurge in new psychoactive substances (NPS), also known as "legal highs," "designer drugs," "herbal highs," or "research chemicals," finding a wide and efficient distribution through the "e-commerce" or specialized shops [1–3]. The term "designer drugs" had traditionally been used to identify synthetic substances but recently has been broadened to include other psychoactive substances that mimic the effects of illicit drugs and are produced by introducing slight modifications to the chemical structure of controlled substances to circumvent drug controls [4]. In 2005, the European Union defined the term "new psychoactive substances" as a new narcotic or psychotropic drug, in pure form or in a preparation, that is not scheduled under the Single Convention on Narcotic Drugs of 1961 or the Convention on Psychotropic Substances of 1971, but which may pose a public health threat comparable to that posed by substances listed in those conventions (Council of the European Union decision 2005/387/JHA) [5]. In common usage, the word "new" often stands either for newly synthesized and newly misused; indeed, nearly all of the substances encountered were first synthesized many years ago [6]. These NPS are usually sold "not for human consumption" to evade legislation, and are marketed as plant foods, bath salts, herbal incenses, etc., often delivered in colorful and attractive packages with imaginative names, while they are in fact intended for human consumption. Therefore, once one compound becomes regulated, new derivatives appear on the market in order to meet consumers need and at the same time to avoid legislations. In this "cat-and-mouse game" [7], most NPS tend to be rapidly replaced, making intervention measures particularly challenging in this area.

Hair Analysis in Clinical and Forensic Toxicology.

The information and analysis presented throughout this chapter is based on the identification in hair samples of three main groups of substances present in the market and mostly abused, among the NPS, that is, ketamine, synthetic cathinones, and synthetic cannabinoids (SCs). For consistency, the terms synthetic cathinones and SCs are only used. Reported drugs in different categories were regrouped in an attempt to obtain similar groupings.

11.2 THE CHALLENGE OF NPS DETECTION

The misuse of NPS often leads governments to prohibit them, but once these drugs have been banned, their chemical structure is slightly altered to create legal drugs with similar properties [1]. Although many of the latest substances continue to be stimulants of the central nervous system, their chemical structures present different forms. This fast growth and structure variability creates further problems at both the analytical and legislative levels. The absence of reference standards for principal drugs and their metabolites has been a long-standing problem, and has led to increasing challenges to forensic and clinical laboratories in the identification and quantification of NPS. The traditional route and timescale of reference material production of 6–24 months is no longer appropriate in this rapidly changing environment, and alternative approaches are required to produce materials of a suitable quality to satisfy the requirements of a robust quality system [2,8]. Another reason which makes the identification of NPS difficult is the lack of immunoassays. As a matter of fact, the selectivity of the immunoassays relies on the affinity of antibodies for the immunogen (in general, the target analyte) used for their production. In the case of multiple compounds, such as the NPS, the target analyte is hard to identify [3]. Finally, serious challenges to detect their presence in biological matrices, especially urine, are posed by the extensive, yet not exhaustively investigated, metabolic transformation that these substances undergo once introduced into the body, and the consequent, limited availability of pure standards of metabolites of NPS. These considerations apply to both SCs and cathinones, although the latter exhibit a less-extensive metabolism than the former [3].

As long as these new classes of substances are not routinely screened in roadside control and workplace testing, an increasing risk exists that further classes of drug consumers (including police, transportation workers, or military personnel) will be induced to substitute the traditional cannabis products or "old" stimulants, namely cocaine and amphetamines, with these new synthetic substances [9–13]. The replacement of "old" drugs with "new" drugs seems to be attractive in other conditions involving regular urine drug screening, for instance in driver's license recovery or in forensic psychiatry settings [14].

To circumvent the issues of NPS identification in urine, the detection of the parent drugs in hair samples has been proposed. Unlike in urine, the parent drug usually represents the target analyte in the keratin matrix that incorporates it from the sweat and/or the bloodstream, the sebum, and from external environment. The corresponding analytical strategy is facilitated by the wide availability of reference standards, which in turn allows a rapid upgrading of the analytical methods to detect them. Finally, the few existing studies describing the detection of NPS in hair have given some preliminary information about the current diffusion of NPS among the population and on the characteristics of the users of these synthetic drugs.

11.3 KETAMINE

Ketamine is a powerful anesthetic drug used in both human and veterinary surgery since the early 1960s, and can thus be considered one of the oldest NPS. Its abuse was recognized in the United States since the beginning of the 1980s and started to be noticed in Europe in the 1990s [4]. In pharmaceutical preparations, ketamine is normally found as an injectable solution, but it can also be produced as a powder or tablet. In this form, it can be administered orally, inhaled, snorted, or smoked. Since the misuse of this drug has been reported in many countries worldwide, its determination in hair samples is offered as a specialist test by hundreds of laboratories, using different analytical techniques [15–33]. Norketamine, generated by N-demethylation of ketamine, reaches concentrations in serum similar to those of ketamine itself. However, the concentration of ketamine in hair is expected to be higher than that of norketamine, because the parent drug is less polar than the metabolite, and the extent of hair deposition notoriously correlates with the lipophilicity of the drug. In the following summary of the inherent published literature, only articles presenting real hair samples are included. The analytical techniques most frequently used to detect ketamine and norketamine in hair were GC-MS [15,16,18,27,34,35] and LC-MS/MS (or LC-HRMS) [17,20,32]. In one case, a specific method using HPLC-Chip-MS/MS [36] was developed. A summary of the reviewed methods is presented in Table 11.1. All results are summarized in Table 11.2.

11.3.1 Analysis of Real Samples

In most cases, the analytical detection of ketamine in hair is rather straightforward. However, the interpretation of concentration levels can be challenging, due to the lack of internationally accepted cutoffs and the limited number of existing data. In the following paragraphs a miscellaneous of case reports and analytical results, either obtained in our laboratory or from peer-reviewed literature, are presented.

11.3.1.1 Case Study 1

A 20-year-old male, black haired, declared intake of ketamine for recreational purpose. He estimated to have consumed ketamine about 10 times in the 6–8 months before the sampling. His hair length was 4.5 cm. Samples

TABLE 11.1 Analytical Procedures for Ketamine and Norketamine Determination in Hair Samples

Sample Preparation				Instrumentation		Detection Limits				Reference
Hair Amount	Pretreatment	Extraction	Derivatization	Technique	Column	LOD (ng/mg)		LOQ (ng/mg)		
						K	NK	K	NK	
50 mg	MeOH 15 h at 55°C	–	–	LC-ESI-MS/MS	Acquity UPLC BEH C18	0.004	0.003	0.010	0.013	Personal data
25 mg	HCl 0.5 M overnight at 45°C	SPE at pH 7 C18–Bond Elut	–	GC-EI-MS	J&W Scientific HP-5	0.4	0.4	0.6	0.8	[18]
10 mg	HCl 0.1 M 4 h at 45°C	LLE at pH > 10 CH_2Cl_2: *n*-hexane 9:1	–	LC-Chip-ESI-MS/MS	Zorbax 80 SB-C18	0.5	0.5	1	1	[24]
25 mg	MeOH overnight at 25°C	SPE at pH 6 Bond Elut	HFBA	GC-EI-MS	J&W Scientific DB-5	0.05	0.05	0.08	0.08	[35]
25 mg	MeOH:TFA 8.5:1.5 overnight at 25°C	SPE at pH 6 Bond Elut	HFBA	GC-EI-MS	J&W Scientific HP-5MS	0.05	0.05	0.08	0.08	[15,16]
				GC-NCI-MS		0.25[a]	0.025[a]	1[a]	0.08[a]	
30 mg	Methanolic HCl 0.25 M ultrasonication 2 h at 45°C	–	TFAA MBTFA	GC-EI-MS	J&W Scientific DB-5MS	0.03	0.01	0.11	0.05	[27]

20 mg	Formic acid 0.01% ultrasonication 4 h	–	–	LC-ESI-MS/MS	Synergi Polar HPLC	0.1	0.1	0.5	0.5	[20]
10 mg	Phosphate buffer 0.1 M (pH 5) 18 h at 45°C	MISPE at pH 5	–	LC-ESI-MS/MS	Synergi Hydro RP	0.10	0.14	0.18	0.23	[17]
50 mg	HCl 0.1 M 4 h at 45°C	LLE at pH > 10 Diethyl ether	–	GC-EI-MS	J&W Scientific HP-5	0.02	0.02	0.05	0.05	[34]
10 mg	0.5 mL methanol–TFA (8.5:1.5, v/v)	Microwave oven, 700 W, 3 min.	–	LC-ESI-MS/MS	Kinetex C-18	0.0005	0.0005	0.002	0.002	[32]

[a] *In pg/mg.*

TABLE 11.2 Comparison of Ketamine and Norketamine Concentrations in Hair Obtained from Published Literature

Case Study	Subjects (*N*)	K (ng/mg)	K_{Mean} (ng/mg)	K_{Median} (ng/mg)	NK (ng/mg)	NK_{Mean} (ng/mg)	NK_{Median} (ng/mg)	NK/K	NK/K_{Mean}	Comments	Reference
1	1	1.87	–	–		–	–			Self-reported K consumer (*Italy*)	[37]
2	6	0.11–11.4	2.09	0.26	0.02–0.71	0.15	0.04	0.06–0.29	0.14	Driving relicensing (*Italy*)	[37]
3	8	0.32–7.22	2.75	1.96	0.06–0.65	0.31	0.25	0.09–0.26	0.19	Driving relicensing (*Italy*)	[37]
4	51	0.6–489	49	n/a	0.8–196.3	12.1	n/a	0.05–0.84	n/a	Self-reported/ suspected K abusers (*Singapore*)	[18]
5	10	11.07–1548	341	51.9	1.67–130.74	59.8	22.6	0.08–1.13	0.29	Drug abusers (*China*)	[24]
6	1	4.5	–	–	0.35	–	–	0.08	–	Suspected K abuser (*South Korea*)	[27]
7	4	0.2–5.7	3.12	3.3	0.1–1.2	0.52	0.18	0.04–0.5	0.26	Hair from drug misuse prevention center (*Malaysia*)	[17]

8	6	0.63–371.8	83.6	47.7	0.56–6.58	2.86	2.51	0.01–1.46	0.31	Multidrug abusers (*Spain*)	[20]
9	3	0.67–2.40	1.61	1.76	0.10–0.66	0.34	0.26	0.11–0.38	0.21	Hair from regional prevention centers for drug abuse (*Taiwan*)	[15]
10	4	0.46–28.15	12.3	10.2	0.02–5.28	2.12	1.58	0.04–0.27	0.16	Hair from regional prevention centers for drug abuse (*Taiwan*)	[16]
11	14	0.8–92.3	2.94	1.6	0.8–7.7	2.94	1.6	0.03–0.88	0.32	K abusers (*China*)	[34]
12	1	0.141	–	–	0.063	–	–	0.45	–	Infant exposed to K during gestation (*Taiwan*)	[35]
13	4	0.022–23.6	12.69	13.57	ND-2.90	1.64	1.34	0–0.20	0.09	Drug abusers (*Taiwan*)	[32]
14	5	<LOD-0.13	0.12	0.12	<LOD-0.06	0.05	0.05	0.32–0.80	0.56	Medical cases	[37]

n/a: not applicable;
ND: not detected;
LOD: Limit of Detection

were taken and analyzed in our laboratory, according to the procedure previously described. Calculated concentration of ketamine (K) was 1.87 ng/mg, norketamine (NK) was 0.11 ng/mg, and the NK/K ratio was 0.06 [37].

11.3.1.2 Case Study 2

During the year 2013, six subjects (five males, one female) were found positive for ketamine (range 0.11–11.4 ng/mg, mean 2.09 ng/mg, median 0.26 ng/mg). Norketamine was detected in all samples (range 0.02–0.71 ng/mg, mean 0.15 ng/mg, median 0.04 ng/mg). The NK/K ratios were in the range 0.06–0.29. Age range was 20–29 years (mean 24.2) [37].

11.3.1.3 Case Study 3

During the year 2014, eight subjects (seven males, one female) were found positive for ketamine (range 0.32–7.22 ng/mg, mean 2.75 ng/mg, median 1.96 ng/mg). Norketamine was detected in all samples (range 0.06–0.65 ng/mg, mean 0.31 ng/mg, median 0.25 ng/mg). The NK/K ratios were in the range 0.03–0.53. Age range was 17–32 years (mean 24.1) [37].

11.3.1.4 Case Study 4

In this study [18], hair samples were obtained from either self-confessed or suspected ketamine abusers. The authors fixed an arbitrary cutoff level at 1 ng/mg. Fifty-one hair samples resulted positive for ketamine, with concentrations varying in the range 0.6–489.0 ng/mg (mean 49.0 ng/mg), whereas the concentration range of norketamine was 0.8–196.3 ng/mg (mean 12.1 ng/mg). The NK/K ratio varied from 0.05 to 0.84 (mean 0.33). According to voluntary confessions from the abusers who snorted the drug about once a week, the resulting hair concentration of ketamine was in the range 1.1–42.7 ng/mg (mean 9.9 ng/mg).

11.3.1.5 Case Study 5

Ten hair specimens from drug abusers were analyzed [24]. The resulting concentrations were relatively high, if compared to other studies. Ketamine ranged from 11.07 to 1,548.12 ng/mg, while norketamine ranged from 1.67 to 130.74 ng/mg. The NK/K ratios were in the range 0.08–1.13.

11.3.1.6 Case Study 6

This method was applied to the analysis of a hair sample from a suspected K abuser [27]. The concentrations measured in this sample were 4.50 ng/mg for K and 0.35 ng/mg for NK, their ratio being 0.08.

11.3.1.7 Case Study 7

Four hair samples were supplied by a drug misuse prevention center in Malaysia [17]. Ketamine ranged from 0.2 to 5.7 ng/mg, while norketamine ranged from 0.1 to 1.2 ng/mg. The NK/K ratios were in the range 0.04–0.5.

11.3.1.8 Case Study 8

Real hair samples from 25 multidrug abusers were analyzed [20]. Ketamine was found in 13 samples, only 6 of which also contained the metabolite norketamine. In the latter subjects, ketamine ranged from 0.63 to 371.8 ng/mg, while norketamine ranged from 0.56 to 6.58 ng/mg. The NK/K ratios were in the range 0.01–1.46.

11.3.1.9 Case Study 9

Authentic hair samples were collected from regional prevention centers for drug abuse [15] and analyzed with GC-MS using two ionization methods. The average from the two results is presented hereafter. Ketamine was found in four subjects, but norketamine reached measurable levels only in three specimens. The parent drug ranged from 0.67 to 2.40 ng/mg, while the metabolite was in the range 0.10–0.66 ng/mg. The NK/K ratios for these subjects were 0.11, 0.15, and 0.38. The authors also presented a real case in which passive exposure to ketamine could be claimed. In the corresponding hair sample, ketamine concentration was 0.18 ng/mg, norketamine was 0.007 ng/mg, and, notably, their NK/K ratio was 0.04.

11.3.1.10 Case Study 10

Authentic hair samples were collected from regional prevention centers for drug abuse [16]. Ketamine and norketamine were found in four samples, in the respective range of 0.46–28.15 ng/mg and 0.02–5.28 ng/mg. Their NK/K ratio was consequently in the interval 0.04–0.27.

11.3.1.11 Case Study 11

Hair samples were collected from 15 ketamine abusers in various entertainment places [34]. In one case, norketamine was not detected. In the remaining 14 specimens, ketamine levels were in the range 0.8–92.3 ng/mg (mean 21.7 ng/mg, median 8.1 ng/mg). Norketamine levels were in the range 0.8–7.7 ng/mg (mean 2.94 ng/mg, median 1.6 ng/mg). The NK/K ratios varied in the range 0.03–0.88 (mean 0.32).

11.3.1.12 Case Study 12

The case of a female infant, whose mother had allegedly consumed ketamine during gestation, is described [35]. The concentrations of ketamine and its metabolite norketamine were 0.141 ng/mg and 0.063 ng/mg of hair, respectively, and their NK/K ratio being 0.45.

11.3.1.13 Case Study 13

Eight authentic hair specimens collected from abusers who had consumed different types of drugs were analyzed [32]. In one case, K was found at trace levels and NK was not detected. Other three samples tested positive both for K and NK at higher concentration (range 6.74–23.6 ng/mg and 0.68–2.90 ng/mg, respectively). If the samples without NK is excluded, the NK/K ratio was in the range 0.03–0.20 (mean 0.09).

11.3.1.14 Case Study 14

Hair samples were collected from five patients admitted within a hospital unit taking care of burned people and treated with ketamine at different dosages [37]. The investigated subjects, who received ketamine for medical reasons, volunteered to provide hair samples and signed an informed consent. All the patients received from 40 up to 430 mg ketamine intravenously in 1 or 2 dosages within 3 months before hair sampling. Ketamine and norketamine were detectable in only two cases out of five. In both positive cases, ketamine was below 0.2 ng/mg and norketamine was lower than 0.1 ng/mg.

11.3.2 Discussion

Ketamine is a weakly basic ($pK_a = 7.5$) substance, which is generally sold as hydrochloride salt and is mainly present in blood as a cation. In its neutral form, ketamine strongly interacts with melanin, facilitating its incorporation into hair. Therefore, its concentration in hair from chronic abusers is expected to be relatively high, compared to other common drugs of abuse.

In our laboratory, we collectively found 15 positive samples for ketamine and norketamine (case studies 1–3). The minimum levels were 0.11 ng/mg and 0.02 ng/mg, respectively. The metabolite was detected in all samples, suggesting that the hair samples were presumably collected from active consumers. In 9 cases out of 15, ketamine levels were above 0.5 ng/mg, while norketamine levels were above 0.05 ng/mg. Their concentration ratio NK/K was higher than 0.05 in 14 cases out of 15.

In the published literature (case studies 4–11 and 13), the results obtained in 9 independent studies are examined, totalizing 97 positive samples for ketamine and norketamine. The minimum detected level for ketamine was 0.022 ng/mg. In 94 cases out of 97, the ketamine hair concentration was higher than 0.5 ng/mg. In one sample NK was not found. Among the other positive specimens, the lowest detected amount for norketamine was 0.02 ng/mg. Besides this case for which the ratio NK/K was still 0.04, the lowest amount of norketamine detected in the remaining 95 samples was 0.1 ng/mg. Overall, the ratio NK/K was higher than 0.05 in 88 specimens out of 97.

In the case of prenatal exposure (described in case study 12), the concentrations of ketamine and its metabolite norketamine were relatively low (0.141 ng/mg and 0.063 ng/mg, respectively), possibly as the result of sporadic exposure. However, this has to be considered as active consume, corresponding to a ratio NK/K of 0.45.

In the case of hospitalized patients (case study 14), ketamine and norketamine could not be detected in three cases out of five, while in the two positive cases, the concentrations were found to be below 0.2 and 0.1 ng/mg, respectively.

11.4 SYNTHETIC CATHINONES

This new group of recreational drugs, often referred as "bath salts," is commonly used both in Europe and North America. "Bath salts" are mainly sold in the form of odorless, white, yellowish or brown powder or fine crystals, but less frequently as tablets or capsules. Most synthetic cathinones originate from China and, to a lesser extent, from India [38]. The active components are cathinone derivatives, example of which is given in Figure 11.1. Synthetic cathinones are β-keto analogues of the natural cathinone (S-(—)-2-amino-1-phenylpropan-1-one), one of the psychoactive phenethylalkylamine alkaloids present in the khat plant *Catha edulis* [38]. Bath salts typically contain at least one of three most prevalent synthetic cathinones: mephedrone,

Common name and abbreviation	Chemical structure	Common name and abbreviation	Chemical structure
Butylone (bk-MBDB)	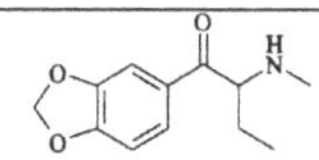	Methedrone (bk-PMMA)	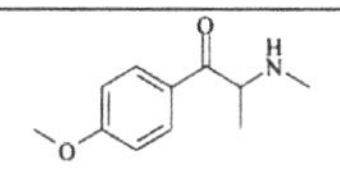
Ethylone (4β-MDEA)	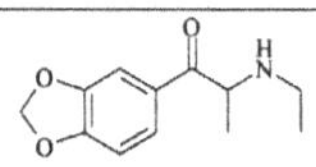	Naphyrone	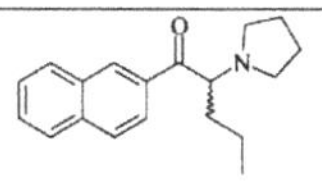
3-Fluoromethcathinone (3-FMC)	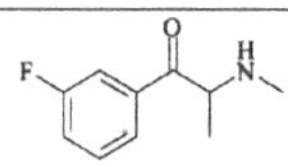	3,4-Methylenedioxypyrovalerone (MDPV)	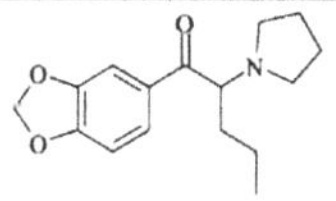
4-Fluoromethcathinone (4-FMC)	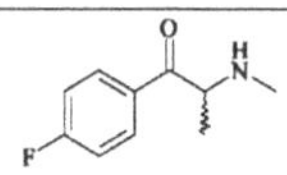	Methylone (bk-MDMA)	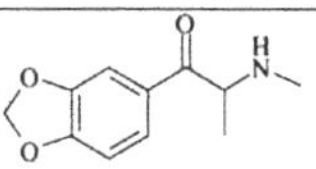
Mephedrone (4-MMC)	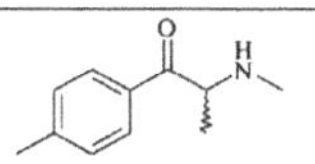	Methcathinone	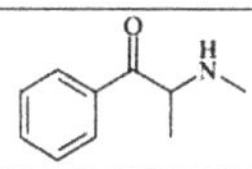

FIGURE 11.1 Names and chemical structures of the most common synthetic cathinones.

methylone, or methylenedioxypyrovalerone. In one of the first studies on mephedrone, this new drug was reported to be one of the most commonly abused psychotropic substances [39]. It was also shown that this drug is more commonly used by males than females. 3-FMC (3-fluoromethcathinone), 4-FMC (4-fluoromethcathinone), buphedrone (α-methylamino-butyrophenone), butylone (β-keto-*N*-methyl-3,4-benzodioxyolybutanamine), methedrone (4-methoxymethcathinone), and naphyrone (naphthylpyrovalerone) are also popular synthetic cathinones [40].

Most common administration routes of synthetic cathinones are insufflation (snorting) and oral ingestion of capsules or tablets, dilution of the substance with water/juice drink, or powder wrapped in cigarette paper and swallowed (so-called "bombing"). Rectal insertion, intravenous, subcutaneous, and intramuscular injections were also reported [38,41]. Finally, there have been occasional reports of synthetic cathinones being inserted in the eye, sometimes referred to as "eyeballing" [38]. Synthetic cathinones consumed in combination with other drugs are intended to intensify desired effects or curtail noxious effects. Routine drug combinations with β-ketones include: cocaine, amphetamines, methamphetamines, cannabis, kratom, GHB, other synthetic cathinones, alcohol, β-blockers, GBL, zopiclone, caffeine, pregabalin, famotidine, omeprazole, domperidone, opiates, and benzodiazepines [40].

Since the late 1990s, pyrrolidinophenone-type designer drugs such as 3,4-methylenedioxypyrrolidinovarelophenone (MDPV), which have a similar structure to cathinone, have appeared on illegal drug markets. New compounds made with small structural modifications such as α-pyrrolidinovalerophenone (α-PVP) and α-pyrrolidinopentiothiophenone (α-PVT) continue to appear on the market [42].

A different group of compounds, which are considered as a new group of designer drugs, is represented by the piperazine-like compounds. These compounds can be structurally divided in two subgroups (benzylpiperazine and phenylpiperazine) [43]. While the best-known piperazine used as an NPS is 1-benzylpiperazine (BZP), other compounds such as 1-(3-chlorophenyl) piperazine (mCPP), 1-(3-trifluoromethylphenyl) piperazine (TFMPP) and, to a lesser extent, 1-benzyl-4-methylpiperazine (MBZP), and 1-(4-fluorophenyl)piperazine (pFPP) have been identified in the market during the last decade [4].

The popularity of these psychoactive drugs has created a demand for sensitive, robust, and reliable analytical methods for their identification and quantification in different matrices, including hair. In a Letter to the Editor, Torrance and Cooper reported the detection of mephedrone in hair samples at 4.2 and 4.7 ng/mg concentration with an ISO/17025 accredited method, but details on the analytical method used and comments or interpretations of results were not included [44]. Several other methods were published afterward, presented in the following paragraphs. The instrumental characteristics of the methods, together with the most significant results from the analysis of real samples (if available), are listed in Table 11.3 and 11.4.

TABLE 11.3 Analytical Procedures for Determination of Synthetic Cathinones in Hair Samples

Reference	Analytes	LLOQ	Hair Sample Preparation	Instrumental Method
[45]	Methylone, MBDB and methcathinone	5.0 pg/mg for methylone and MBDB, 1.0 pg/mg for methcathinone (lowest level of calibration)	Hair washed and extracted with 3 mL of MeOH/5 M HCl solution (20:1) for 1 h under ultrasonication. After dryness, the residue was dissolved in 3 mL of 0.1 M KH_2PO_4 buffer (pH 6.0). The solution was treated with SPE, PFP-derivatized and analyzed.	GC-MS
[46]	Mephedrone	0.2 ng/mg	50 mg washed and incubated overnight at 40°C in 1 mL Sorensen buffer pH 7.0, then after alkalinization with 1 mL of 1N NaOH, 4 mL of ethyl acetate were added. The top organic layer was transferred with 100 μL of a mixture of methanol/hydrochloric acid (99:1, v/v). After dryness, the extract was derivatized with 100 μL of HFBA and 50 μL of ethyl acetate, for 30 min at 60°C.	GC-MS
[47]	Amphetamine, phentermine, methamphamine, cathinone, methcathinone, fenfluramine, desmethylselegiline, 3,4-MDA, 3,4-MDMA, 3,4-MDEA, norketamine, mescaline, 4-bromo-2, 5-dimethoxyphenethylamine	0.01–0.08 ng/mg	20 mg was washed, extracted with 2 mL of 0.25 M methanolic HCl at 50°C for 1 h, then after dryness the trifluoroacetyl-derivatives were formed.	GC-MS
[48]	Mephedrone	Not given	Approximately 100 mg of hair was washed and extracted with methanol at 55°C for 15 h. The organic phase was dried and derivatized with TFAA.	GC-MS

(*Continued*)

TABLE 11.3 (Continued)

Reference	Analytes	LLOQ	Hair Sample Preparation	Instrumental Method
[43]	TFMPP, mCPP, 1-(4-methoxyphenyl) piperazine (MeOPP)	0.05 ng/mg	20 mg was washed and incubated with 1 mL of NaOH 1 M, then neutralized with HCl and 5 mL of KH_2PO_4 was added. After SPE extraction and dryness, 65 μL of MSTFA with 5% of TMS was added.	GC-MS
[49]	Mephedrone and two metabolites	5 pg/mg for mephedrone, 10 pg/mg for the two metabolites	50 mg was washed, digested with Cleland's reagent and enzyme proteinase K, incubated with Tris buffer for 2 h at 37.5°C, extracted with chloroform:ethanol:diethylether 3:1:1. Finally the organic layer was dried and reconstitute with 200 μL of acetonitrile.	LC-MS/MS
[50]	4-fluoroamphetamine, piperazines (BZP, mCPP, and TFMPP), cathinones (4-MMC (mephedrone), methylone, butylone, ethylone, MDPV, methcathinone, and cathinone), methylphenidate and ketamine	LOD: 50 pg/mg for cathinones, 10 pg/mg for the other compounds	20–30 mg of hair were washed and extracted first with MeOH (5 mL, 16 h, ultrasonication) and then with MeOH acidified with 50 μL HCl 33% (3 mL, 3 h, ultrasonication). The extracts were dried and the residue reconstituted with 50 μL MeOH and 500 μL 0.2 mM ammonium formate in water.	LC-MS/MS
[51]	Amphetamine, methamphetamine, MDMA, MDA, methylone, methedrone, mephedrone, MDPV, fluoromethcathinone, fluorometamphetamine, mCPP and TFMPP		30 mg hair was incubated in 2 mL MeOH with 0.1% HCl for 1 h at 60°C. After dryness and reconstitution in 2 mL 2% formic acid, the extracts were submitted to SPE extraction. Eluates were evaporated and reconstituted in 75 μL ammonium formate 2 mM pH 3.	LC-MS/MS

[52]	Methedrone	Lowest calibrator at 2 ng/mg	Segments (7 to 17 mg) were washed and added with 0.5 mL acetonitrile/methanol/20 mM ammonium formate buffer (pH 3) (10:10:80). Samples were incubated in a water bath at 37°C for 18 h.	LC-MS/MS
[42]	α-PBP, MDPBP, MDPV, α-PVP, pyrovalerone	0.5 ng/mg	Hair was washed and extracted with 0.1 mL of 1 M NaOH at 70°C for 20 min. Sorensen's glycine buffer (0.8 mL at 0.3 M, pH 13) was added and the sample loaded onto the column MonoSpin®C18.	LC-MS
			Analytes adsorbed to the column were eluted with 0.1 mL of 50% methanol aqueous solution.	
[53]	MDPV, α-PVP, α-PBP	0.02 ng/10 mm hair	Hair was washed and extracted with 0.1 mL of 1 M NaOH solution, at 70°C for 20 min and the extracted with the Extrelut column. The residue was dissolved in 100 μL of 30% methanol aqueous solution and 5 μL injected into the LC-MS system.	LC-MS

TABLE 11.4 Results for Synthetic Cathinones Obtained from Real Samples

Reference	Real Samples	Number of Positive Samples	Range of Concentrations
[46]	67	13	Mephedrone: 0.2–313.2 ng/mg
[48]	1	1	Mephedrone: 0.25 ng/mg
[49]	154	5 (only 1 quantitative result)	Mephedrone: 21.11 pg/mg
			No metabolites
[50]	325	120	Not given
[51]	7	7 (1 for mCPP)	mCPP: 6,528 pg/mg
[52]	1	1	Methedrone: 29–37 ng/mg
[42]	4	4 (segmental analysis)	7.5–9.4 ng/mg (α-PVP)
			3.1 ng/mg (α-PBP)
[53]	1	1 (segmental analysis)	0.2–1.2 ng/10-mm (α-PVP)
			1–22 ng/10-mm (MDPV)

11.4.1 GC-MS Methods

In a preliminary study in 2007 [45], the incorporation into hair of methylone and other new designer drugs, namely methcathinone and *N*-methyl-1-(3,4-methylenedioxyphenyl)butan-2-amine (MBDB), was performed on an animal model. The authors concluded that new designer drugs such as methylone and MBDB (but not methcathinone) are highly incorporated into the keratin matrix.

Martin et al. [46] developed a specific and accurate method for mephedrone analysis and applied to 67 hair specimens. Thirteen of them were found positive. The mean of the results was 2.9 ng/mg, excluding one outlier sample which resulted at very high concentration (313.2 ng/mg) These levels are comparable to those of amphetamines, namely in the range of nanograms per milligrams, at least in case of repeated abuse.

Thirteen psychotropic phenylalkylamine derivatives, including cathinone and methcathinone, were determined in human hair by Kim et al. [47]. A total of 141 samples were collected from possible drug abusers. Several samples tested positive for amphetamine and methamphetamine, two for phentermine and two for MDMA, only 1 for MDA and 1 for norketamine. Cathinone and methcathinone were not detected in any sample.

In one fatal case in which the concomitant intake of mephedrone, cocaine, and ethanol likely accounted for the death [48], mephedrone was determined in several biological specimens including hair. The analytical finding at 0.25 ng/mg revealed past exposure to mephedrone. However, in this case, the level of mephedrone appears relatively low if compared with existing data [46], leading to the conclusion that the deceased was unlikely a heavy consumer of mephedrone [48].

One of the first examples of detection of three phenylpiperazine in hair was presented by Barroso et al. [43]. The procedure was applied to authentic samples obtained either from autopsies or living subjects. Interestingly, samples belonging to persons undergoing treatment with trazodone were also analyzed, for the detection of mCPP, which in turn can be detected in biological fluids as a metabolite of trazodone.

11.4.2 LC-MS/MS Methods

In 2012, Shah et al. [49] developed a LC-MS/MS method for the quantitative analysis of mephedrone and for two of its metabolites, namely 4-methylephedrine and 4-methylnorephedrine, in hair. The authors screened for mephedrone in 154 healthy volunteers, but only five samples resulted positive. Mephedrone could be successfully quantified in only one sample, while the metabolites were not detected in any of the analyzed samples. The authors also noted that contamination from environmental exposure is not likely because mephedrone is not smoked.

The usefulness of hair analysis for retrospective prevalence studies was shown by Rust et al. [50]. The extracts of all hair samples from 2009 to 2010 that originally tested positive for amphetamines and/or MDMA in the authors' lab ($N = 325$) were reanalyzed concerning NPS. NPS were found in 120 cases (37%). Concerning the piperazine drugs, mCPP could be detected in 34 (10.5%) cases and TFMPP in one case. In 11 cases (3%) 4-MMC (mephedrone) was identified. Concerning smart drugs, methylphenidate was found in 16 cases (5%). In 45 cases (14%) ketamine could be detected. 4-Fluoroamphetamine was identified in 12 cases (4%) and methylone in one case. The antidepressant trazodone was also included into the method in order to discriminate between direct mCPP intake and biotransformation of trazodone.

Recently, Lendoiro et al. [51] developed and validated a method for the determination of amphetamine derivates, synthetic cathinones (methylone, methedrone, mephedrone, MDPV, fluoromethcathinone, and fluorometamphetamine), and piperazines (mCPP and TFMPP) in hair using LC-MS/MS. The method was applied to the analysis of seven hair specimens positive to amphetamine derivatives. All real specimens were positive, including one positive sample for mCPP, with concentration at 6,527.9 pg/mg. Unfortunately, no information was available to elucidate if this subject was under prescription of trazodone. The same authors (personal communication) analyzed nine clinical samples from patients under treatment with trazodone (all of them with DEPRAX® 100). Results for mCPP and trazodone were in the range 16.2–92.0 ng/mg and 367–2,088 ng/mg, respectively.

Some authors have presented the determination of cathinones derivatives or pyrrolidinophenone-type designer drugs in human hair segments to evaluate a history of use. Wikstrom et al. [52] reported two deaths related to the cathinone analogue methedrone (4-methoxymethylcathinone). In one case, hair samples were collected during autopsy and submitted for toxicology analysis. In five short segments of hair, the concentrations of methedrone were evenly distributed (segment one: 37 ng/mg, segment two: 33 ng/mg, segment three: 29 ng/mg, segment four: 29 ng/mg, and segment five: 36 ng/mg) suggesting chronic intake of methedrone over the previous months before death. Namera et al. [42] described a simple quantitative method for identifying cathinone analogues in 10-mm-long hair segments. The method was successfully applied to the determination of α-pyrrolidinophenone-type designer drugs in segmented human hair from drug abusers and confirmed previous drug abuse. In another report, the same group presented a case of fatal poisoning in which a very high concentration of MDPV was found [53]. The authors discuss the quantitative results for cathinone analogues in 10-mm-long hair segments to obtain a history of drug abuse in this case. MDPV and α-PVP were found in all or many of the segments tested. According to authors, it was also demonstrated in this study that MDPV concentrations in hair dramatically decreased after bleaching.

11.5 SYNTHETIC CANNABINOIDS

SCs are the most common drugs among an expanding array of compounds that mimic the effects of traditional illicit psychoactive substances. They are produced in a variety of similar chemical structures, with small changes in the synthetic procedure, in order to escape controls and inclusion on the list of prohibited substances [10]. The SCs appear to be more popular among drug users in United States rather than Europe [54].

Most SCs identified in K2 or Spice products are pharmacologically characterized as agonists to human cannabinoid type I receptors. However, structural differences between SCs and Δ^9-tetrahydrocannabinol (Δ^9-THC) lead these synthetic compounds to possess increased biological activity. This may explain some neurological and cardiovascular complications often reported in human case studies of K2 toxicity [55–58].

Based upon their chemical structure, SCs can be divided into the following main groups (Figure 11.2) [59,60]:

- Classical cannabinoids: Tetrahydrocannabinol, other chemical constituents of cannabis and their structurally related synthetic analogues, for example, AM-411, AM-906, HU-210, O-1184;
- Nonclassical cannabinoids: cyclohexylphenols or 3-arylcyclohexanols, for example, CP-55,244, CP-55,940, CP-47,497 (and C6-9 homologues);

Group of synthetic cannabinoid	Chemical structure
Classical cannabinoids (e.g., THC, HU-210)	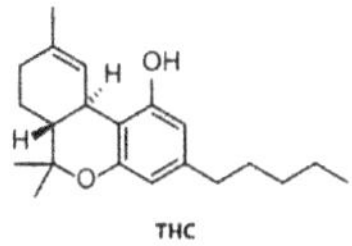 THC 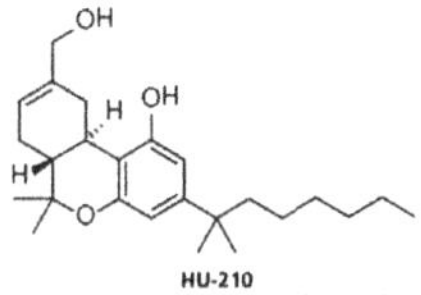HU-210
Nonclassical cannabinoids (e.g., CP-55,940, CP-47,497, and C6-9 homologues)	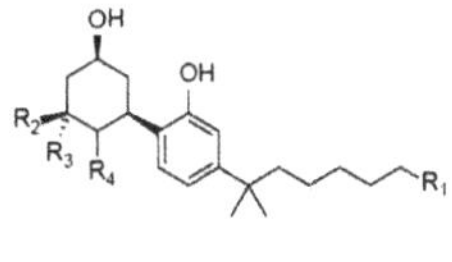 **CP-47,497** ($R_2=R_3=R_4$=H, R_1=methyl) **CP-47,497-C6** ($R_1=R_2=R_3=R_4$=H) **CP-47,497-C8** ($R_2=R_3=R_4$=H, R_1=ethyl) **CP-47,497-C9** ($R_2=R_3=R_4$=H, R_1=propyl) **CP-55,940** ($R_2=R_3$=H, $R_1=CH_3$, R_4=3-hydroxypropyl) **Dimethyl CP-47,497-C8** ($R_2=R_3=CH_3$, R_4=H, R_1=ethyl)
Hybrid cannabinoids (e.g., AM-4030)	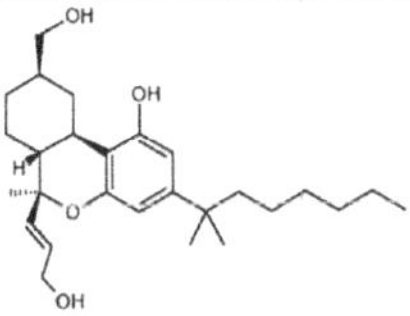
Naphthoylindoles (e.g., JWH-015, JWH-018, JWH-073, JWH-081, JWH-122, JWH-200, JWH-210, JWH-398)	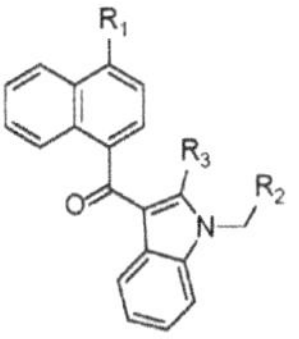 **$R_1=R_3$=H** **AM-1220** (R_2=1-methylpiperidin-2-yl) **AM-2201** (R_2=4-fluorobutyl) **AM-2232** (R_2=butanenitrile) **JWH-018** (R_2=butyl) **JWH-019** (R_2=pentyl) **JWH-020** (R_2=hexyl) **JWH-022** (R_2=3-buten-1-yl) **JWH-072** (R_2=ethyl) **JWH-073** (R_2=propyl) **JWH-200** (R_2=4-morpholinylmethyl) **R_2=butyl, R_3=H** **JWH-081** (R_1=methoxy) **JWH-122** (R_1=methyl) **JWH-210** (R_1=ethyl) **JWH-387** (R_1=Br) **JWH-398** (R_1=Cl) **JWH-412** (R_1=F) **JWH-007** (R_1=H, R_2=butyl, R_3=methyl) **JWH-015** (R_1=H, R_2=ethyl, R_3=methyl) **JWH-073 4-methylnaphthyl** (R_1=methyl, R_2=propyl, R_3=H) **MAM-2201** (R_1=methyl, R_2=4-fluorobutyl, R_3=H)
Phenylacetylindoles (e.g., JWH-250, JWH-251)	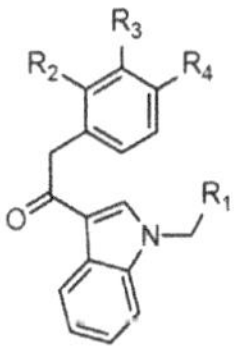 **$R_3=R_4$=H** **Cannabipiperidiethanone** (R_1=1-methylpiperidin-2-yl, R_2=methoxy) **JWH-203** (R_1=butyl, R_2=Cl) **JWH-250** (R_1=butyl, R_2=methoxy) **JWH-251** (R_1=butyl, R_2=methyl) **RCS-8** (R_1=cyclohexylmethyl, R_2=methoxy) **R_1=butyl, R_2=H** **JWH-201** (R_3=H, R_4=methoxy) **JWH-302** (R_3=methoxy, R_4=H)
Benzoylindoles (e.g., pravadoline, AM-694, RSC-4)	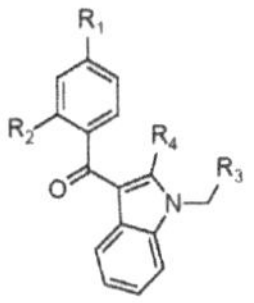 **AM-694** ($R_1=R_4$=H, R_2=I, R_3=4-fluorobutyl) **AM-694 chloro derivative** ($R_1=R_4$=H, R_2=I, R_3=4-chlorobutyl) **AM-2233** ($R_1=R_4$=H, R_2=I, R_3=1-methylpiperidin-2-yl) **RCS-4** (R_1=methoxy, $R_2=R_4$=H, R_3=butyl) **RCS-4-ortho isomer** ($R_1=R_4$=H, R_2=methoxy, R_3=butyl) **RCS-4 butyl homolog** (R_1=methoxy, $R_2=R_4$=H, R_3=propyl) **WIN 48,098** (R_1=methoxy, R_2=H, R_3=4-morpholinylmethyl, R_4=methyl)

FIGURE 11.2 Classification of SCs according to the United Nations Office on Drugs and Crime (UNODC) [59].

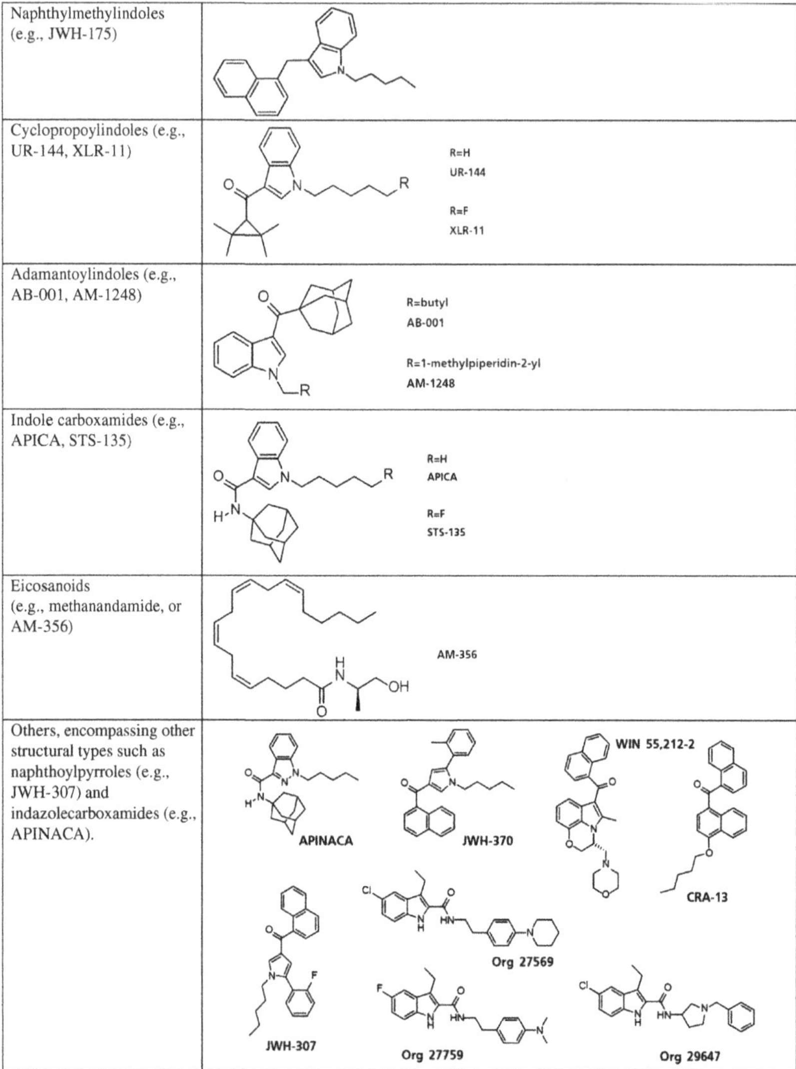

FIGURE 11.2 *(Continued)*

- Hybrid cannabinoids: structural combinations of both classical and non-classical cannabinoids, for example, AM-4030;
- Aminoalkylindoles, which can be further divided into the following groups:
  a. Naphthoylindoles (e.g., JWH-015, JWH-018, JWH-073, JWH-081, JWH-122, JWH-200, JWH-210, JWH-398)
  b. Phenylacetylindoles (e.g., JWH-250, JWH-251)
  c. Benzoylindoles (e.g., pravadoline, AM-694, RSC-4)

d. Naphthylmethylindoles (e.g., JWH-184)
e. Cyclopropoylindoles (e.g., UR-144, XLR-11)
f. Adamantoylindoles (e.g., AB-001, AM-1248)
g. Indole carboxamides (e.g., APICA, STS-135)

- Eicosanoids: endocannabinoids such as anandamide and their synthetic analogues, for example, methanandamide (AM-356);
- Others: Encompassing other structural types such as diarylpyrazoles (e.g., Rimonabant®), naphthoylpyrroles (e.g., JWH-307), naphthylmethylindenes (e.g., JWH-176), and indazole carboxamides (e.g., APINACA).

Many derivatives and analogues in the above classes of compounds could be synthesized by the addition of a halogen, alkyl, alkoxy, or other substituents to one of the aromatic ring systems. Other small changes such as variation of the length and arrangement of the alkyl chain can also be made. The aminoalkylindoles are by far the most common class of SCs found in herbal products as they are easy to synthesize, compared to the other classes of compounds [59].

In order to circumvent the legislation controlling the drugs of abuse, SCs are usually added to "incenses" or "house scents," often with the warning "not for human consumption," However, they need to be smoked to produce strong THC-like effects.

Knowing that commercially available immunoassays do not cross-react with the extremely wide range of existing compounds, laboratories usually develop their own mass spectrometry-based methods. For reasons of specificity and sensitivity, all methods for SC in hair are based on LC-MS/MS.

There is scarce literature on hair analysis for SCs. Few studies tried to estimate the penetration of SC into population through hair analysis (see Table 11.5). However, the range of analytes covered by these methods needs to be expanded, as the number of different SCs identified in herbal mixtures is continuously increasing.

The first multiclass screening for SCs in hair samples was published in 2012 [61]. In this study, a UHPLC-MS/MS procedure was developed and validated in order to determine the possible presence of five SCs (i.e., consistently present in the black market during the period considered: JWH-018, JWH-073, JWH-200, JWH-250, and HU-210) in 179 real hair samples collected from previously proven cannabis consumers. Fourteen (7.82%) samples were found positive to at least one SC demonstrating significant diffusion among drug abusers. Concentrations of SCs ranged from 0.50 to 730 pg/mg (see also Table 11.6).

In a second study [10], a UHPLC-MS/MS method was developed to detect JWH-018, JWH-073, JWH-200, JWH-250, JWH-007, JWH-015, JWH-019, JWH-020, JWH-081, JWH-122, JWH-203, JWH-210, JWH-251, JWH-398, JWH-307, AM-694, AM-1220, AM-2201, RCS-4, RCS-8, WIN-48,098, WIN 55,212-2, and HU-210. Then, 344 samples previously tested in 2011, were reanalyzed; 264 of them had resulted positive for at least one common drug of abuse or tested for ethylglucuronide in the diagnosis of alcohol abuse.

TABLE 11.5 Analytical Procedures for Determination of SCs in Hair Samples

Reference	Analytes	LLOQ	Hair Sample Preparation
[14]	JWH-007, JWH-015, JWH-018, JWH-019, JWH-020, JWH-073, JWH-081, JWH-122, JWH-200, JWH-203, JWH-210, JWH-250, JWH-251, JWH-398, AM-694, AM-2201, methanandamide, RCS-4, RCS-4 orthoisomer, RCS-8, WIN-48,098, and WIN 55,212-2	5.0 pg/mg for methanandamide and JWH-398, 0.5 pg/mg for others compounds	50 mg washed and extracted with 1.5 mL of ethanol for 3 h. Subsequently, 1 mL of the extract dried and reconstituted in 100 μL of liquid chromatography solvents A/B, 50/50 (v/v)
[61]	JWH-018, JWH-073, JWH-200, JWH-250, HU-210	0.07–9.9 pg/mg	50 mg washed and extracted with 3 mL NaOH 1N at 95°C for 10 min, then with 5 mL of *n*-hexane/ethylacetate 90:10 (v/v). Finally, the organic phase is dried and reconstituted with 50 mL of methanol
[10]	JWH-018, JWH-073, JWH-200, JWH-250, JWH-007, JWH-015, JWH-019, JWH-020, JWH-081, JWH-122, JWH-203, JWH-210, JWH-251, JWH-398, JWH-307, AM-694, AM-1220, AM-2201, RCS-4, RCS-8, WIN-48,098, WIN 55,212-2, and HU-210.	0.7–80 pg/mg	10 mg washed and extracted with 3 mL NaOH 1N at 95°C for 10 min, then with 5 mL of *n*-hexane/ethylacetate 90:10 (v/v). Finally, the organic phase is dried and reconstituted with 200 mL of methanol
[62]	JWH-018, JWH-122, JWH-081, JWH-200, JWH-210, JWH-250, JWH-073, AM-694	20 pg/mg for JWH-122 and AM-694, 10 pg/mg for all the others	About 100 mg of hair was incubated overnight at 45°C in 1 mL of 0.5 M NaOH, then extracted for 10 min with 5 mL of a mixture of hexane/ethyl acetate (9:1)
			After collection of the organic phase, the aqueous phase was re-extracted with the same organic mixture. The pooled organic layers were evaporated and then the dried residues were reconstituted in 100 μL of methanol

[63]	JWH-018, JWH-073, and their metabolites	1 pg/10 mg hair	10 mg was washed and extracted by methanol at 38°C for 16 h. The extracts were dried and reconstituted in 100 μL of a 1:1 (v/v) mixture of methanol and mobile phase component A
[64]	AM-694; AM-2201;CP 47,497; HU-210; JWH-007; JWH-015; JWH-018 and its *N*-pentanoic acid metabolite; JWH-019; JWH-020; JWH-073; JWH-081; JWH-122; JWH-200; JWH-203; JWH-210; JWH-250, and WIN 55, 212-2	0.5–5.0 pg/mg	20 mg was washed and incubated in methanol, overnight at 40°C. Finally, the organic phase was evaporated to dryness and reconstituted in mobile phase
[65]	JWH-018, AM-2201, JWH-122, JWH-073, MAM-2201, and their metabolites	1.0 pg/10 mg in human hair	10 mg was washed and cut, extracted with 2 mL of methanol at 38°C for at least 16 h. The extracts were dried and the residues were dissolved with 100 μL of 50% (v/v) mobile phase A in methanol
[66]	AM-2201, AM-2232, JWH-081, JWH-122, JWH-210, JWH-307, MAM-2201, RCS-4, UR-144, XLR-11	0.5 pg/mg	Hair was washed by shaking for 4 min with 4 mL of water, followed by 4 mL of acetone, and 4 mL of petroleum ether, then 50 mg was cut and extracted with 1.5 mL of ethanol. Finally, Finally, 1 mL of the extract was dried and reconstituted in 100 μL of mobile phase A/B (50:50, v/v)

TABLE 11.6 Results for SCs Obtained from Real Samples

Reference	Real Samples	Number of Positive Samples	Gender	Age	Range of Measurable Concentrations
[14]	Obtained from forensic psychiatry inpatients in 2011. All patients admitted chronic consumptions of SC	8 (7 with polyabuse)	Male (8/8)	20–37	5.1–78 pg/mg (JWH-081) 0.5–24 pg/mg (JWH-250) 0.7–21 pg/mg (JWH-073) 5.1–5.7 pg/mg (JWH-018) 0.5–5.2 pg/mg (JWH-210)
[61]	179 real samples (152 head hair, 27 pubic hair) collected in 2010, arising from proved frequent users of THC-containing products	14 (7 with polyabuse)	Male (14/14)	18–48	0.6–70.5 pg/mg (JWH-018) 0.5–413 pg/mg (JWH-073) 1.5–729 pg/mg (JWH-250)
[10]	344 real hair samples collected in 2011 and randomly selected from two groups of subjects, namely driving relicensing and drug abuse/withdrawal control subjects	15 (10 with polyabuse)	12 male, 3 female	18–32	3.1–17.3 pg/mg (JWH-018) 1.6–50.5 pg/mg (JWH-073) 4.8–83.4 pg/mg (JWH-250) 8.0–194 pg/mg (JWH-081) 7.4–2,800 pg/mg (JWH-122) 2.3–5.1 pg/mg (JWH-210) 3.8–4.1 pg/mg (JWH-019) 1.3 pg/mg (AM-1220)
[62]	435 hair samples collected in 2010 for driving relicensing	8 (2 with polyabuse)	Not specified	Not specified	16–1,280 pg/mg (JWH-081) 125 pg/mg (JWH-122) 12 pg/mg (JWH-250) 17–750 pg/mg (JWH-073) 10–11 pg/mg (JWH-018)
[63]	18 (individuals suspected of SC use)	18	7 male, 11 female	22–34	<LOQ-1,700 pg/mg (JWH-018) 2–55 pg/mg (JWH-073)

[64]	232 subjects suspected of narcotic abuse, 131 of them originated from French forensic cases and 101 from foreign legal hair cases	3	Not specified	Not specified	< 0.5–1.0 pg/mg (AM-2201) < 0.5 pg/mg (JWH-201)
[67]	1 fatal case	1	Male	36	Estimated concentrations: 0.05 ng/mg (JWH-122) 13 ng/mg (MAM-2201) 3.0 ng/mg (AM-2201) 0.7 ng/mg (UR-144) 0.05 pg/mg (JWH-018) 0.01 ng/mg (JWH-210)
[65]	9 subjects submitted by the law enforcement agency and suspected of SCs abuse. Some hair samples were equally divided into 3 cm-length segments upon request and subsequently 23 hair samples were analyzed	9	5 male and 4 female	21–30 (male) 20–25 (female)	0.4–38.9 pg/mg (JWH-018) 0.3–37.2 pg/mg (JWH-018 N-5-OH 0.2–1.1 pg/mg (JWH-018 N-COOH) 0.1–2.0 pg/mg (JWH-073) 0.3 pg/mg (JWH-073 N-COOH) 0.1–402 pg/mg (JWH-122) 0.1–3.5 pg/mg (JWH-122 N-5-OH) 1.7–739 pg/mg (AM-2201) 0.2-3.1 pg/mg (AM-2201 N-6-OH indole) 0.4 pg/mg (AM-2201 N-4-OH) 0.2–276 pg/mg (MAM-2201)
[66]	A: 8 participants involved in the analysis of herbal mixturesB: 5 persons living in the same households with participants from group AC: 9 participants from laboratory staff not directly in contact with the drug materials	8 from Group A and 2 from Group C	Not specified	Not specified	0.7–5.5 pg/mg (JWH-081) 0.6–21 pg/mg (JWH-122) 1.0–170 pg/mg (JWH-210) 0.6–43 pg/mg (JWH-307) 0.5–3.2 pg/mg (AM-2201) 0.5–1.7 pg/mg (AM-2232) 0.6–27 pg/mg (MAM-2201) 0.6–1.0 pg/mg (RCS-4) 0.7 pg/mg (UR-144) 0.9–6.5 pg/mg (XLR-11)

Comprehensively, 15 samples were found positive for at least one SC. The majority of positive samples were from young people, particularly males, former or still active cannabis consumers. More detailed results are shown in Table 11.6.

The detection of the parent drugs in hair samples has been also proposed from other authors [14,62,64]. Hutter et al. [14] presented a validated method for the quantitative determination of 22 SCs in human hair based on liquid chromatography–tandem mass spectrometry. The method was successfully applied to authentic hair samples obtained from forensic psychiatry patients, and SCs were detected in hair of eight patients (see Table 11.6).

Gottardo et al. [62] developed a liquid chromatography-quadrupole-time of flight mass spectrometry (LC-QTOF MS) method for the screening of NPS in hair, to investigate NPS-related histories, and support the epidemiological surveys on the penetration of the NPS use in the population. Among 435 hair samples submitted to LC-QTOF analysis for SCs, 8 were found "positive" for the following compounds: JWH-018, JWH-073, JWH-081, JWH-250, and JWH-122, in a broad range of concentrations (0.010–1.28 ng/mg).

A recent study was proposed by Cirimele and colleagues [64], who developed and validated an LC-MS/MS method for the quantitative detection of 18 SCs in human hair. More than 200 authentic specimens were screened by the validated method. Three of them revealed the presence of two different SCs, AM-2201 and JWH-210, with concentrations ranging from below their quantification limits to 1.0 pg/mg.

A single fatal case involving SC was reported by Schaefer et al. [67]. The hair sample resulted positive for six different compounds, whose estimated concentrations are reported in Table 11.6.

Quite remarkably, the majority of the presented studies reported several cases of polyabuse of SC. As no systematic correspondence can be found between herbal blend name and content of SCs, the consumers are rarely aware of the actual composition of the products. This is not surprising because herbal blends are not standardized but rather consist in semiclandestine preparations obtained from a mixture of herbal leaves of different origin, on which the cannabinoids are sprayed. Thus, according to the availability of cannabinoids, the active ingredients vary from lot to lot, also within the same trade name [62]. It was also shown that concentrations vary among different packages of the same brand [68] and that some blends may contain two or more active compounds [12,69–71]. Different amounts or combinations of these substances are probably used in different "Spice" products to generate cannabis-like effects. It is possible that substances from these or other chemical groups with a cannabinoid agonist or other pharmacological activity could be added to any herbal mixture [9].

11.5.1 Detection of Metabolites

In the previous paragraph, several methods describing the detection of SC in human hair have been presented. All studies have included only the parent drugs in the list of targeted analytes, but this does not exclude the possibility of external contamination, nor does it provide conclusive evidence of active drug consumption. As a matter of fact, and similarly to what has happened in the last two decades for THC, only the presence of metabolites and possibly the evaluation of concentration ratios between parent drugs and metabolites, can prove the active use of SC and exclude external contamination from side-stream smoke or handling material.

The first study to investigate the presence of metabolites in hair was presented by Kim et al. [63]. The authors established and validated an analytical method for simultaneous detection of JWH-018 and JWH-073, and their most abundant monohydroxylated and carboxylated metabolites. The incorporation of metabolites of SCs into hair was also investigated, together with the effect of pigmentation on the deposition of these compounds in hair, by means of an animal model. For the latter purpose, JWH-073 was chosen as representative for SCs. Finally, the developed method was applied to 18 hair samples from individuals suspected of use of SCs. Among the positive results, only the *N*-(5-hydroxypentyl) metabolite of JWH-018 (JWH-018 N-5-OH M) was found, suggesting its prevalence in hair. The concentrations varied widely, and so did the ratios between parent drug and metabolite. Even in those hair samples containing relatively high concentrations of JWH-018 (above 50 pg/mg), JWH-018 N-5-OH M was not detected. The highest concentrations of JWH-018 N-5-OH M was 85 pg/mg; in this case, the JWH-018 concentration was 151 pg/mg. Overall, in samples positive both to JWH-018 and JWH-018 N-5-OH M, parent drug-to-metabolite ratios were highly variable, ranging from 1.1 to 62.8. Noteworthy, JWH-018 N-5-OH is also the product of metabolism of AM-2201 [72]. Therefore, some high concentrations in hair could also be generated by the co-ingestion of other SC. Therefore, results about JWH-018 N-5-OH levels in hair might be inconclusive without a comprehensive screening of the most popular SC, including AM-2201.

Very recently, the same group published a new study in which they expended their previous method to AM-2201, JWH-122, MAM-2201, and their monohydroxylated metabolites in hair [65]. The method was also applied to investigate the distribution of five naphthoylindole-based SCs and their metabolites in authentic human hair samples from forensic cases and the deposition of AM-2201 and its metabolites in pigmented and nonpigmented rat hair. In real samples, JWH-018, JWH-018 N-5-OH M, JWH-018 N-COOH M, JWH-073, JWH-073 N-COOH M, AM-2201, AM-2201 N-4-OH M, AM-2201 N-6-OH indole M, JWH-122, JWH-122 N-5-OH M, and MAM-2201 were simultaneously or individually detected. The concentration range of parent drugs (e.g., AM-2201) was much wider than that of metabolites (e.g., JWH-018 N-5-OH M). Parent SCs and their monohydroxylated metabolites were identified in the hair samples of all nine cases, to confirm that the simultaneous determination of both parent drugs and metabolites in hair is helpful for the interpretation of results. Indeed, this allows to exclude the possibility of a passive contamination and provide information on the ingested parent SC [65].

The issue of possible external contamination has been recently raised [66,73]. These studies aimed to evaluate the extent of external contamination caused by handling of SC containing drug material under realistic conditions in a forensic laboratory. Hair of laboratory workers involved in the analysis of 670 herbal mixture samples (covering 31 brands and 12 different SCs) within a 2-week period was analyzed for SCs with a validated LC-MS/MS method. In addition, hair samples of laboratory staff not directly in contact with the drug material and close relatives of exposed subjects were analyzed to check for cross-contamination. All samples of persons who were in direct contact with drug material were tested positive for at least one of the SCs. Concentrations ranged from trace amounts up to a maximum of 170 pg/mg (JWH-210) and roughly reflected duration of exposure. Unexpectedly, subjects without direct contact to drug material also showed measurable hair concentrations. In one case, despite a JWH-210 concentration of less than 0.5 pg/mg in the hair sample of participant who was involved in the work, up to 11 pg/mg was detected in the hair sample of his girlfriend, who lived in the same household, but had no contact with the drug materials. One possible explanation from the authors for these results is the direct transfer through contaminated fingers; for example, from a head massage or by sleeping on pillows accidentally contaminated by the hands of the partner [66,73]. Overall, concentrations caused by contamination are in the typical range found in known users of these drugs, although the majority of them is below 50 pg/mg. Therefore, only the detection of metabolites in hair (or the simultaneous analysis of body fluids) can strongly suggest an actual consumption.

Recently, the first results from a comprehensive screening of metabolites on a large group of subjects were presented [74]. Initially, 15 samples which were previously found positive to SC were reprocessed [10]. The new results are presented in Table 11.7. In 10 cases (#1, 3–5, 7, 8, 12–15 in Table 11.7), low concentrations for the parent drug (below 50 pg/mg) were measured and no metabolites were found. For these cases, either sporadic exposure or external contamination could be suggested. In two cases (#6 and 10), where very high concentrations for JWH-122 were found, metabolites were present at very low concentration. For these two cases, it can be suggested that there was frequent exposure to SC, proven by the high levels, and active use, because of the presence of metabolites. Remarkably, JWH-122 metabolite can be also produced by the metabolism of MAM-2201 [75].

TABLE 11.7 Results from Real Samples, Including Metabolites

#	JWH-018 (pg/mg)	JWH-073 (pg/mg)	JWH-250 (pg/mg)	JWH-081 (pg/mg)	JWH-122 (pg/mg)	JWH-210 (pg/mg)	JWH-019 (pg/mg)	AM-1220 (pg/mg)	Metabolites (pg/mg)	Washing Solutions
1	–	1.6	–	–	–	–	–	–	No	n/a
2	17.3	7.6	83.4	12.3	–	–	–	–	No	n/a
3	–	1.9	26.9	–	–	–	–	–	No	n/a
4	–	1.8	–	–	–	–	–	–	No	n/a
5	–	5.2	5.8	–	11.7	–	–	–	No	n/a
6	10.4	2.0	6.0	–	2,800	2.3	–	1.3	2.5	n/a
7	–	1.8	–	–	–	–	–	–	No	n/a
8	–	–	–	8.0	–	–	–	–	No	n/a
9	–	50.5	6.4	194	713	–	–	–	No	n/a
10	–	1.6	–	–	760	–	–	–	0.7	n/a
11	3.1	1.6	–	81.4	–	5.1	–	–	No	n/a
12	–	9.0	4.8	–	40.9	–	–	–	No	n/a
13	–	–	–	–	7.4	–	–	–	No	n/a
14	–	–	–	–	11.2	–	3.8	–	No	n/a
15	–	–	–	47.8	15.8	–	4.1	–	No	n/a

n/a: not applicable

The remaining three cases (#2, 9, 11) were positive to different compounds at relatively high concentrations (above 50 pg/mg), but negative for metabolites. For these cases, frequent exposure to SC can be suggested, even though no metabolites were found.

Afterward, 153 hair samples taken in 2012 and 2013 from young habitual THC consumers were tested. Results for hair and washing solution are shown in Table 11.8. Sample #4 was positive to several SC, of which two (JWH-073 and JWH-122) were present at high concentrations, but no metabolites were detected. The washing solutions were also negative. Therefore, frequent exposure can be suggested, but no conclusion can be definitely drawn about the active use. In the other four cases, very low concentrations (much below 50 pg/mg) were found and no metabolites were detected. For these cases, the washing solutions were also analyzed and all were found to be negative. Therefore, for these cases, sporadic exposure can be speculated and no conclusion can be drawn about possible external contamination.

Lastly, 47 hair samples taken in 2014 from frequent THC consumers were processed (Table 11.9). Merely three positive samples were found, but only #3 (which was also positive to ketamine) is of interest. Some compounds of the JWH-series were detected at low concentrations, while the only high concentration was obtained for AM-2201, proving frequent exposure to this SC. The metabolites of AM-2201 were also present, confirming active use. Finally, the positive result for the washing solutions might likely indicate a recent exposure.

11.6 CONCLUSIONS

The elusive and changeable profile of the synthetic drugs progressively introduced on the black market makes any tentative study of the diffusion of these new drugs within our communities quite uncertain and incomplete. The use of hair analysis to investigate their diffusion among selected populations of drug abusers represents a considerably practical tool to obtain significant information with limited investment.

Considering the wide range of compounds forming the class of NPS, it would be difficult to speculate about the binding capacity of each molecule to the keratin matrix. Since their pharmacological potency *in vitro* is extremely high, it is likely that also *in vivo* these compounds are active at relatively low doses, reducing the detectable levels in hair. Some laboratories have set the limit of detection as the minimum criterion to establish the use of NPS. These positive samples, at very low levels, should be interpreted with caution. For example, the SCs are predominantly smoked, and in consequence it is not inconceivable that passive inhalation of the smoke, external contamination from sidestream smoke, or handling material can result in positive testing.

TABLE 11.8 Results from Real Samples taken in 2012 and 2013, Including Metabolites

#	JWH-018 (pg/mg)	JWH-073 (pg/mg)	JWH-250 (pg/mg)	JWH-081 (pg/mg)	JWH-122 (pg/mg)	JWH-210 (pg/mg)	AM-694 (pg/mg)	Metabolites (pg/mg)	Washing Solutions
1	–	–	4.92	–	–	–	–	No	Negative
2	2.27	<LOQ	–	–	–	–	–	No	Negative
3	–	–	–	<LOQ	2.79	<LOQ	–	No	Negative
4	2.55	287	32.8	22.4	61.6	–	0.78	No	Negative
5	2.15	1.89	<LOQ	3.16	–	–	–	No	Negative

TABLE 11.9 Results from Real Samples taken in 2014, Including Metabolites

#	JWH-018 (pg/mg)	JWH-073 (pg/mg)	JWH-250 (pg/mg)	JWH-081 (pg/mg)	JWH-122 (pg/mg)	AM-2201 (pg/mg)	AM-694 (pg/mg)	Metabolites	Washing Solutions
1	–	–	–	–	1.9	–	–	Negative	Negative
2	–	–	–	–	4.9	–	–	Negative	Negative
3	9.9	0.9	2.8	–	2.4	715	1.6	JWH-018/ AM-2201 JWH-073/ AM-2201 JWH-122	Positive (AM-2201)

To date, the international scientific community is still in need of guidelines to establish:

- If metabolites should be measured in order to discriminate between active intake or passive exposure;
- If cutoffs should be used in order to discriminate between chronic consumption and occasional use (or even single-scouting intake).

From the very preliminary results available in published literature, it can be noted that, similarly to metabolites of THC, the metabolites of SC are also present at very low levels in real hair samples. This seems to be the case even when the parent drug is detected at high concentration. Scarce biotransformation of the parent drug, poor incorporation of metabolites into the keratin matrix, and external contamination all present a plausible explanation to this imbalanced parent drug-to-metabolite ratio. In the next scenario, it is foreseeable that several new research studies will address the complex interpretation of real samples' results. In this effort, a sound understanding of possible presence of metabolites in hair is indispensable.

In conclusion, laboratories must develop and validate their own very sensitive methods for detecting NPS in biological fluids, especially hair. As long as this new class of substances is not routinely screened in the context of roadside control, workplace testing or driving relicensing, an increasing risk exists that drug consumers will be induced to substitute the traditional drugs of abuse with these new synthetic substances. Therefore, the progressive introduction of efficient screening and confirmation tests for the detection of acute and chronic abuse of the new designer drugs appears to be crucial within the whole drug prevention policy.

ACKNOWLEDGMENTS

I would like to thank Julie Busch and Enrico Gerace for revising the manuscript.

REFERENCES

[1] Davidson C. New psychoactive substances. Prog Neuropsychopharmacol Biol Psychiatry 2012;39:219–20.

[2] King L, Kicman T. A brief history of "new psychoactive substances". Drug Test Anal 2011;3:401–3.

[3] Favretto D, Pascali JP, Tagliaro F. New challenges and innovation in forensic toxicology: focus on the "new psychoactive substances". J Chromatogr A 2013;1287:84–95.

[4] The challenge of new psychoactive substances (2013). Available from: <www.unodc.org/documents/scientific/NPS_2013_SMART.pdf>.

[5] World drug report. Available from: <http://www.unodc.org/wdr2013/>; 2013.

[6] Wohlfarth A, Weinmann W. Bioanalysis of new designer drugs. Bioanal 2010;2:965–79.

[7] Dresen S, Kneisel S, Weinmann W, Zimmermann R, Auwärter V. Development and validation of a liquid chromatography-tandem mass spectrometry method for the quantitation of synthetic cannabinoids of the aminoalkylindole type and methanandamide in serum and its application to forensic samples. J Mass Spectrom 2011;46:163–71.

[8] Archer RP, Treble R, Williams K. Reference materials for new psychoactive substances. Drug Test Anal 2011;3:505–14.

[9] Vardakou I, Pistos C, Spiliopoulou C. Spice drugs as a new trend: mode of action, identification and legislation. Toxicol Lett 2010;197:157–62.

[10] Salomone A, Luciano C, Di Corcia D, Gerace E, Vincenti M. Hair analysis as a tool to evaluate the prevalence of synthetic cannabinoids in different populations of drug consumers. Drug Test Anal 2014;6:126–34.

[11] Kerwin J. Doors of deception: the diaspora of designer drugs. Drug Test Anal 2011;3: 527–31.

[12] Papanti D, Schifano F, Botteon G, Bertossi F, Mannix J, Vidoni D, et al. "Spiceophrenia": a systematic overview of "spice"-related psychopathological issues and a case report. Hum Psychopharmacol 2013;28:379–89.

[13] Crews B. Synthetic cannabinoids: the challenges of testing for designer drugs. Available from: <http://www.aacc.org/publications/cln/2013/february/Pages/Cannabinoids.aspx#>; 2013.

[14] Hutter M, Kneisel S, Auwärter V, Neukamm MA. Determination of 22 synthetic cannabinoids in human hair by liquid chromatography-tandem mass spectrometry. J Chromatogr B Analyt Technol Biomed Life Sci 2012;903:95–101.

[15] Wu Y-H, Lin K-L, Chen S-C, Chang Y-Z. Integration of GC/EI-MS and GC/NCI-MS for simultaneous quantitative determination of opiates, amphetamines, MDMA, ketamine, and metabolites in human hair. J Chromatogr B Analyt Technol Biomed Life Sci 2008;870: 192–202.

[16] Wu Y, Lin K, Chen S, Chang Y. Simultaneous quantitative determination of amphetamines, ketamine, opiates and metabolites in human hair by gas chromatography/mass spectrometry. Rapid Commun Mass Spectrom 2008;22:887–97.

[17] Harun N, Anderson RA, Cormack PA. Analysis of ketamine and norketamine in hair samples using molecularly imprinted solid-phase extraction (MISPE) and liquid chromatography-tandem mass spectrometry (LC-MS/MS). Anal Bioanal Chem 2010;396:2449–59.

[18] Leong HS, Tan NL, Lui CP, Lee TK. Evaluation of ketamine abuse using hair analysis: concentration trends in a Singapore population. J Anal Toxicol 2010;29:314–18.

[19] Miyaguchi H, Inoue H. Determination of amphetamine-type stimulants, cocaine and ketamine in human hair by liquid chromatography/linear ion trap-Orbitrap hybrid mass spectrometry. Analyst 2011;136:3503–11.

[20] Tabernero MJ, Felli ML, Bermejo AM, Chiarotti M. Determination of ketamine and amphetamines in hair by LC/MS/MS. Anal Bioanal Chem 2009;395:2547–57.

[21] Lin Y-H, Lee M-R, Lee R-J, Ko W-K, Wu S M. Hair analysis for methamphetamine, ketamine, morphine and codeine by cation-selective exhaustive injection and sweeping micellar electrokinetic chromatography. J Chromatogr A 2007;1145:234–40.

[22] Favretto D, Vogliardi S, Stocchero G, Nalesso A, Tucci M, Terranova C, et al. Determination of ketamine and norketamine in hair by micropulverized extraction and liquid chromatography-high resolution mass spectrometry. Forensic Sci Int 2013;226:88–93.

[23] Gentili S, Cornetta M, Macchia T. Rapid screening procedure based on headspace solid-phase microextraction and gas chromatography–mass spectrometry for the detection of many recreational drugs in hair. J Chromatogr B 2004;801:289–96.

[24] Zhu KY, Leung KW, Ting AKL, Wong ZCF, Fu Q, Ng WYY, et al. The establishment of a highly sensitive method in detecting ketamine and norketamine simultaneously in human hairs by HPLC-Chip-MS/MS. Forensic Sci Int 2011;208:53–8.

[25] Parkin MC, Longmoore AM, Turfus SC, Braithwaite RA, Cowan DA, Elliott S, et al. Detection of ketamine and its metabolites in human hair using an integrated nanoflow liquid chromatography column and electrospray emitter fritted with a single porous 10 μm bead. J Chromatogr A 2013;1277:1–6.

[26] Park M, Kim J, Park Y, In S, Kim E, Park Y. Quantitative determination of 11-nor-9-carboxy-tetrahydrocannabinol in hair by column switching LC-ESI-MS(3). J Chromatogr B Analyt Technol Biomed Life Sci 2014;947–948C:179–85.

[27] Kim JY. RCM Letter to the Editor, 2006;20:3159–62.

[28] Lendoiro E, Quintela O, de Castro A, Cruz A, López-Rivadulla M, Concheiro M. Target screening and confirmation of 35 licit and illicit drugs and metabolites in hair by LC-MSMS. Forensic Sci Int 2012;217:207–15.

[29] Merola G, Gentili S, Tagliaro F, Macchia T. Determination of different recreational drugs in hair by HS-SPME and GC/MS. Anal Bioanal Chem 2010;397:2987–95.

[30] Nielsen MKK, Johansen SS, Dalsgaard PW, Linnet K. Simultaneous screening and quantification of 52 common pharmaceuticals and drugs of abuse in hair using UPLC-TOF-MS. Forensic Sci Int 2010;196:85–92.

[31] Shen M, Xiang P, Shi Y, Pu H, Yan H, Shen B. Mass imaging of ketamine in a single scalp hair by MALDI-FTMS. Anal Bioanal Chem 2014;406:4611–16.

[32] Chang Y-J, Chao M-R, Chen S-C, Chen C-H, Chang Y-Z. A high-throughput method based on microwave-assisted extraction and liquid chromatography-tandem mass spectrometry for simultaneous analysis of amphetamines, ketamine, opiates, and their metabolites in hair. Anal Bioanal Chem 2014;406:2445–55.

[33] Inagaki S, Makino H, Fukushima T, Min JZ, Toyo'oka T. Rapid detection of ketamine and norketamine in rat hair using micropulverized extraction and ultra-performance liquid chromatography-electrospray ionization mass spectrometry. Biomed Chromatogr 2009;23: 1245–50.

[34] Xiang P, Shen M, Zhuo X. Hair analysis for ketamine and its metabolites. Forensic Sci Int 2006;162:131–4.

[35] Su P-H, Chang Y-Z, Chen J-Y. Infant with in utero ketamine exposure: quantitative measurement of residual dosage in hair. Pediatr Neonatol 2010;51:279–84.

[36] Zhu KY, Leung KW, Ting AKL, Wong ZCF, Fu Q, Ng WYY, et al. The establishment of a highly sensitive method in detecting ketamine and norketamine simultaneously in human hairs by HPLC-Chip-MS/MS. Forensic Sci Int 2011;208:53–8.

[37] Salomone A, Gerace E, Diana P, Romeo M, Malvaso V, Di Corcia D, et al. Cut-off proposal for the detection of ketamine in hair. Forensic Sci Int 2015. Available from: http://dx.doi.org/10.1016/j.forsciint.2014.12.030.

[38] Zawilska JB, Wojcieszak J. Designer cathinones—an emerging class of novel recreational drugs. Forensic Sci Int 2013;231:42–53.

[39] Vardakou I, Pistos C, Spiliopoulou C. Drugs for youth via Internet and the example of mephedrone. Toxicol Lett 2011;201:191–5.

[40] Katz DP, Bhattacharya D, Bhattacharya S, Deruiter J, Clark CR, Suppiramaniam V, et al. Synthetic cathinones: "A khat and mouse game". Toxicol Lett 2014;229:349–56.

[41] Miotto K, Striebel J, Cho AK, Wang C. Clinical and pharmacological aspects of bath salt use: a review of the literature and case reports. Drug Alcohol Depend 2013;132:1–12.

[42] Namera A, Konuma K, Saito T, Ota S, Oikawa H, Miyazaki S, et al. Simple segmental hair analysis for α-pyrrolidinophenone-type designer drugs by MonoSpin extraction for evaluation of abuse history. J Chromatogr B Analyt Technol Biomed Life Sci 2013; 942–943:15–20.

[43] Barroso M, Costa S, Dias M, Vieira DN, Queiroz JA, López-Rivadulla M. Analysis of phenylpiperazine-like stimulants in human hair as trimethylsilyl derivatives by gas chromatography-mass spectrometry. J Chromatogr A 2010;1217:6274–80.

[44] Torrance H, Cooper G. The detection of mephedrone (4-methylmethcathinone) in 4 fatalities in Scotland. Forensic Sci Int 2010;202:e62–3.

[45] Kikura-Hanajiri R, Kawamura M, Saisho K, Kodama Y, Goda Y. The disposition into hair of new designer drugs; methylone, MBDB and methcathinone. J Chromatogr B Analyt Technol Biomed Life Sci 2007;855:121–6.

[46] Martin M, Muller JF, Turner K, Duez M, Cirimele V. Evidence of mephedrone chronic abuse through hair analysis using GC/MS. Forensic Sci Int 2012;218:44–8.

[47] Kim JY, Jung KS, Kim MK, Lee JI, In MK. Simultaneous determination of psychotropic phenylalkylamine derivatives in human hair by gas chromatography/mass spectrometry. Rapid Commun Mass Spectrom 2007;21:1705–20.

[48] Gerace E, Petrarulo M, Bison F, Salomone A, Vincenti M. Toxicological findings in a fatal multidrug intoxication involving mephedrone. Forensic Sci Int 2014;243C:68–73.

[49] Shah SB, Deshmukh NIK, Barker J, Petróczi A, Cross P, Archer R, et al. Quantitative analysis of mephedrone using liquid chromatography tandem mass spectroscopy: application to human hair. J Pharm Biomed Anal 2012;61:64–9.

[50] Rust KY, Baumgartner MR, Dally AM, Kraemer T. Prevalence of new psychoactive substances: a retrospective study in hair. Drug Test Anal 2012;4:402–8.

[51] Lendoiro E, Jiménez-Morigosa C, Cruz A, López-Rivadulla M, de Castro A. O20: hair analysis of amphetamine-type stimulant drugs (ATS), including synthetic cathinones and piperazines, by LC-MSMS. Toxicol Anal Clin 2014;26:S13.

[52] Wikström M, Thelander G, Nyström I, Kronstrand R. Two fatal intoxications with the new designer drug methedrone (4-Methoxymethcathinone) autopsy cases. J Anal Toxicol 2010;34:594–8.

[53] Namera A, Urabe S, Saito T, Torikoshi-Hatano A, Shiraishi H, Arima Y, et al. A fatal case of 3,4-methylenedioxypyrovalerone poisoning: coexistence of α-pyrrolidinobutiophenone and α-pyrrolidinovalerophenone in blood and/or hair. Forensic Toxicol 2013;31: 338–43.

[54] King L. Legal controls on cannabimimetics: an international dilemma? Drug Test Anal 2013; 6:80–7.

[55] Seely K, Patton AL, Moran CL, Womack ML, Prather PL, Fantegrossi WE, et al. Forensic investigation of K2, Spice, and "bath salt" commercial preparations: a three-year study of new designer drug products containing synthetic cannabinoid, stimulant, and hallucinogenic compounds. Forensic Sci Int 2013;233:416–22.

[56] Every-Palmer S. Synthetic cannabinoid JWH-018 and psychosis: an explorative study. Drug Alcohol Depend 2011;117:152–7.

[57] Gunderson EW, Haughey HM, Ait-Daoud N, Joshi AS, Hart CL. "Spice" and "K2" herbal highs: a case series and systematic review of the clinical effects and biopsychosocial implications of synthetic cannabinoid use in humans. Am J Addict 2012;21:320–6.

[58] Lapoint J, James LP, Moran CL, Nelson LS, Hoffman RS, Moran JH. Severe toxicity following synthetic cannabinoid ingestion. Clin Toxicol (Phila) 2011;49:760–4.

[59] Recommended methods for the Identification and Analysis of Synthetic Cannabinoid Receptor Agonists in Seized Materials. Available from: <www.unodc.org/documents/scientific/STNAR48_Synthetic_Cannabinoids_ENG.pdf>; 2013.

[60] De BN, Deventer K, Stove V, Van Eenoo P. Synthetic cannabinoids: general considerations. Proc Belgian R Aademies Med 2013;2:209–25.

[61] Salomone A, Gerace E, D'Urso F, Di Corcia D, Vincenti M. Simultaneous analysis of several synthetic cannabinoids, THC, CBD and CBN, in hair by ultra-high performance liquid chromatography tandem mass spectrometry. Method validation and application to real samples. J Mass Spectrom 2012;47:604–10.

[62] Gottardo R, Sorio D, Musile G, Trapani E, Seri C, Serpelloni G, et al. Screening for synthetic cannabinoids in hair by using LC-QTOF MS: a new and powerful approach to study the penetration of these new psychoactive substances in the population. Med Sci Law 2014;54:22–7.

[63] Kim J, In S, Park Y, Park M, Kim E, Lee S. Deposition of JWH-018, JWH-073 and their metabolites in hair and effect of hair pigmentation. Anal Bioanal Chem 2013;405:9769–78.

[64] Cirimele V, Klinger N, Etter M, Duez M, Humbert L, Gaulier J-M, et al. O21: testing for 18 synthetic cannabinoids in hair using HPLC-MS/MS: method development and validation, its application to authentic samples and preliminary results. Toxicol Anal Clin 2014;26:S13.

[65] Kim J, Park Y, Park M, Kim E, Yang W, Baeck S, et al. Simultaneous determination of five naphthoylindole-based synthetic cannabinoids and metabolites and their deposition in human and rat hair. J Pharm Biomed Anal 2014. Available from: http://dx.doi.org/10.1016/j.jpba.2014.09.013.

[66] Moosmann B. Hair analysis of synthetic cannabinoids: does the handling 3 of herbal mixtures affect the analyst's hair concentration? Forensic Toxicol 2014, doi:10.1007/s11419-014-0244-7.

[67] Schaefer N, Peters B, Bregel D, Kneisel S, Schmidt PH, Ewald AH. A fatal case involving several synthetic cannabinoids, Toxichem Krimtech 2013;80:248–51.

[68] Seely KA, Lapoint J, Moran JH, Fattore L. Spice drugs are more than harmless herbal blends: a review of the pharmacology and toxicology of synthetic cannabinoids. Prog Neuropsychopharmacol Biol Psychiatry 2012;39:234–43.

[69] Zuba D, Byrska B, Maciow M. Comparison of "herbal highs" composition. Anal Bioanal Chem 2011;400:119–26.

[70] Shanks KG, Behonick GS, Dahn T, Terrell A. Identification of novel third-generation synthetic cannabinoids in products by ultra-performance liquid chromatography and time-of-flight mass spectrometry. J Anal Toxicol 2013;37:517–25.

[71] Zamengo L, Frison G, Bettin C, Sciarrone R. Understanding the risks associated with the use of new psychoactive substances (NPS): high variability of active ingredients concentration, mislabelled preparations, multiple psychoactive substances in single products. Toxicol Lett 2014;229:220–8.

[72] Hutter M, Moosmann B, Kneisel S, Auwärter V. Characteristics of the designer drug and synthetic cannabinoid receptor agonist AM-2201 regarding its chemistry and metabolism. J Mass Spectrom 2013;48:885–94.

[73] Auwärter V, Hutter M, Neukamm MA, Moosmann B. O23: hair analysis for synthetic cannabinoids: how does handling of herbal mixtures during forensic analysis affect the analyst's hair concentrations? Toxicol Anal Clin 2014;26:S14.

[74] Salomone A, Gerace E, Luciano C, Di Corcia D, Vincenti M. O22: quantification of 22 synthetic cannabinoids and 10 metabolites in human hair. Toxicol Anal Clin 2014;26:S14.

[75] Jang M, Shin I, Yang W, Chang H, Yoo HH, Lee J, et al. Determination of major metabolites of MAM-2201 and JWH-122 in in vitro and in vivo studies to distinguish their intake. Forensic Sci Int 2014;244C:85–91.

Chapter 12

New Challenges and Perspectives in Hair Analysis

Marco Vincenti[1,2] and Pascal Kintz[3,4]
[1]*Dipartimento di Chimica, Università di Torino, Turin, Italy,* [2]*Centro Regionale Antidoping e di Tossicologia "A. Bertinaria", Orbassano (TO), Italy,* [3]*X-Pertise Consulting, Oberhausbergen, France,* [4]*Institute of Legal Medicine, University of Strasbourg, Strasbourg, France*

12.1 INTRODUCTION

The relatively short history of hair analysis has gradually evolved from an early stage, when most research efforts were addressed to the discovery of the new opportunities offered by this unique memory-keeping biological matrix, to the recent consolidation stage, in which the search for innovative procedures and applications for hair analysis is balanced by a thoughtful reconsideration of unexplored variables that may influence the final analytical results. This two-pronged character of scientific research dedicated to hair analysis will most likely persist in the future, because the continuous innovation of instrumental technology offers unprecedented investigation opportunities but, on the other hand, the scientific community is requested to provide solid interpretation foundation to their analytical outcomes.

These two components of scientific research on hair analysis, respectively addressed to innovative and conservative issues, emblematically correspond to the counterparts of any lawsuit—prosecution and defense—and, under such circumstances, provide good reasoning elements for each one. Both sides of the coin offer challenging objectives to be pursued, which will represent the focus of most research struggles in the forthcoming years. These two perspective aspects are examined in the subsequent paragraphs.

12.2 CONDITIONING FACTORS AND SOURCES OF VARIABILITY

The two extreme steps of hair analysis are respectively (i) the intake of a certain substance by a subject and (ii) the quantitative determination of the same

TABLE 12.1 Influencing Factors and Processes

- Mean of intake/administration (ingestion, injection, inhalation, absorption)
- Metabolism of the substance (kinetics and metabolites' distribution)
- Distribution of the marker into the body fluids (blood, sweat, sebum—plus urine, etc.)
- Mechanisms of marker transfer from blood, sweat, sebum to hair and relative importance
- Mechanism of incorporation of the marker into the keratin structure
- Interaction of the marker with melanin by hydrogen bonding (hair color)
- Longitudinal and radial diffusion of the marker
- Washing-out phenomena (porosity)
- External contamination
- Physical transformation processes (heat, light)
- Chemical transformation processes (oxidants, dyes, strong alkaline agents)
- Type of hair (head, pubic, axillary, chest, legs)
- Growing rate of hair and phases
- Site of sampling (for head hair)
- Length of hair sampled and investigated
- Decontamination procedure
- Hair fragmentation
- Extraction of the marker
- Purification of the extract
- Analytical method
- Quantification method

substance and/or its metabolites in the hair of this subject, collected after a delay of days/weeks/months from the intake. The practical success of hair testing in a variety of circumstances, together with its demonstrated trustworthiness in providing crucial information, does not conceal the fact that the final analytical result is conditioned by a large number of biological, chemical, and physical processes, which in turn depend on several causes, ranging from individual and environmental to methodological. Each process generates a source of variability that must be taken into account when the analytical data have to be interpreted [1,2].

A possible sequence of steps and processes, occurring along the way that the designated substance (and/or its metabolites) may be subjected to, before it is detected at the final mass spectrometric instrumentation, is reported in Table 12.1. A classification of the factors that produce these processes, and the consequent variability of the expected results, is reported in Table 12.2. Quite obviously, the variability associated with each of these factors varies from negligible to relevant, depending on the specific substances under investigation, and the population of subjects considered. Each factor could ideally be treated as a single element of contribution to the expanded uncertainty of hair analysis or, more practically, their clustered effect can be evaluated by holistic approaches.

In the past, the effect of various experimental aspects of hair analysis has been studied in detail [3,4], and the procedures developed to decontaminate,

TABLE 12.2 Classification of Influencing Factors

- Individual:
  - Genetic determiners and polymorphisms (and their expression)
  - Personal factors (gender, age, body mass index)
  - Physiopathological factors
  - Behavioral factors:
    - Mode and frequency of substance intake
    - Hygienic habits
    - Cosmetic habits
    - Diet habits
    - Intake of interfering medicines
    - Clothing habitually used (fabric, heavy/light, special clothes, i.e., hat, foulard, scarf)
    - Other frequent sources of (self)-contamination (i.e., pillow, armchair with headrest)
- Environmental (domestic, work-related, leisure-related)
- Climatic (meteorological, geographical, seasonal)
- Methodological (sampling, sample treatment, analysis)

fragment, extract, purify, and analyze hair samples are extensively reviewed in Chapter 2 of the present book. Also the physiology of hair growth and the mechanisms by which xenobiotic substances are transferred from body fluids to the hair surface and bulb, and then incorporated into the keratin structure, have been extensively investigated throughout the years. The remaining factors listed in Table 12.1 still represent subjects of controversial debate [5–7], or either the discussion about their effect just started within the scientific community. These themes characterize a significant part of the recent research on hair analysis and will most likely constitute upcoming challenges for the future. Chapter 3 of this book is devoted to the problem of external contamination, where the reader can find extensive review of the pertinent scientific literature.

A group of interrelated processes potentially undermines the permanence inside the hair structure of the incorporated compounds. Various chemical and physical phenomena may partially modify or remove the original substances from the hair, ultimately leading to biased results. These phenomena may produce the degradation of the analytes' molecular structure due to the effects of heat, light, and strong oxidants, acids, and bases. Besides direct decomposition processes, the same agents may also cause the degradation of the keratin structure, leading to increased porosity of the hair surface, and facilitated release of the incorporated substances. The extent of these phenomena also depends on the chemical stability of the investigated compounds and the strength of their interactions with hair constituents, in particular melanin. Recent experimental studies have investigated both single factors and aggregated phenomena.

12.2.1 Physical and Chemical Agents

The effect of heating on ethyl glucuronide (EtG) concentration, consequent to the repeated application of a hair straightener under mild conditions, was investigated by *in vitro* experiments [8]. Although significant changes were observed with respect to nontreated hair locks, opposite variations were recorded for each half of the investigated population, with a prominent dependence on hair color. Even for such a relatively simple experiment, at least two coexisting phenomena have to be recalled to explain the results, namely thermal degradation (for EtG drop) and more efficient extraction from the keratin matrix (for EtG increase) [8]. On the other hand, EtG proved to be extremely stable within the keratin matrix over extended periods of time, as was clearly demonstrated by its determination on mummy hair samples, collected several hundred years after death [9].

Another physical effect that recently has drawn the attention was the potential action of the solar light [10], possibly playing a role in the degradation of hair-incorporated substances, especially during the summer season. Real positive hair samples were exposed to UV-B radiation [11] and the change of drug content (methadone, morphine, cocaine, and their metabolites) was recorded, revealing particularly extensive decomposition for methadone, and higher stability for drugs incorporated in thick dark hair. Further experimental work, using true solar light and *in vivo* conditions, appears to be needed in the future, before a definite answer may be given about the potential transformation of hair content, induced by exposition to solar light.

While physical agents do not appear to produce substantial modification of the hair cuticle [12], considerable alteration of hair porosity [13] is induced by the chemical products utilized in several cosmetic treatments, including perms, bleaching, and dyeing [14–16]. Under these circumstances, it is virtually impossible to distinguish the release of the incorporated substances through the pores and cavities produced at the hair surface [12,15], from their possible loss by chemical reaction with cosmetic products. The earlier findings [14–16] on the effects of cosmetic treatments have been confirmed by the most recent literature, even if the risk of observing a large number of false-negative results from cosmetically treated hair has been recently contradicted, on the basis of statistical criteria, founded on a large population of treated and nontreated authentic hair samples [17]. Particularly for alcohol abuse testing by hair EtG determination, the percentage of positive samples turned out to be the same for treated and nontreated hair, demonstrating the reliability of the EtG biomarker also under stressed conditions. Once more, this study highlights the differences occurring between the investigations conducted under real *in vivo* conditions [17] and those performed by *in vitro* experiments [18], which occasionally lead to opposite conclusions.

Recent studies carried out to investigate the effects of cosmetic treatments on hair content stability have chiefly concerned amphetamines

[13,19,20] and cocaine [12,13,21]. In general, the variability of hair features in humans (thickness, color, curly/straight), together with the vast assortment of cosmetic products, have prevented any systematic study of their effects, but rather spot conditions have been tested. For amphetamines, modest but statistically significant loss was recorded upon the single application of a bleaching agent (for 40 min) [19], or different dyes [20]. Slightly lower amphetamines release was also produced by repeated (10 times) washing with liquid soap, followed by UV irradiation (for 2 h) [20]. The effect of hydrogen peroxide on hair containing cocaine, from both active user and external contamination, was recently investigated by matrix-assisted laser desorption/ionization-time-of-flight (MALDI-TOF) imaging on single hairs [21]. It was deduced that extensive oxidation and removal of cocaine and its metabolites/oxidation products occurred in both cases, making the detectability of cocaine abuse from bleached hair problematic; the overall effect was attributed to a combination of the cleavage of melanin binding and direct oxidation processes [21]. However, a possibility was considered that cocaine removal could be partially promoted by treating previously cut hair, unlike in real cases.

Another important point, evidenced by several studies, is that contamination from external sources is enhanced in highly porous hair, such as that previously subjected to repeated cosmetic treatments [12,15]. This increases the chance of producing false-positive results, whenever an unsuited decontamination policy is executed [13,22], taking into account that incorrect hair washing can itself produce significant hair damage [23].

A well-structured strategy has been recently devised [13] to measure the degree of hair porosity, to deduce the differences occurring between "normal" cosmetic treatments from "extreme" treatments—possibly carried out by drug users to circumvent positive hair testing—to distinguish external contamination from actual intake for both normal and damaged hair, and to establish a correct washing/decontamination protocol for each of these cases. As previously cited [17], mild cosmetic treatments applied according to the vendor instructions appear to be compatible with hair testing for drug and alcohol abuse [13]. Likewise, the sporadic use of special cleansing shampoos, promoted as capable of "cleaning" the hair from drugs and alcohol biomarkers, was demonstrated not to affect EtG hair concentration [24], nor to remove a significant amount of any drug, including Δ^9-tetrahydrocannabinol (THC), cocaine, heroin, 6-monoacetylmorphine (6-MAM), morphine, codeine, methadone, and several amphetamines [25].

12.2.2 Interaction Between Hair Constituents and Incorporated Substances

A fundamental aspect of hair analysis that has repercussions on a variety of observed phenomena is the nature of the bonds that the incorporated substances form with hair constituents, particularly melanin. These chemical interactions have been investigated in the past, mainly using *in vitro* experiments [2,26,27], animal models [28–31], and statistical inference [2,32,33], but the increasing interest captured by hair analysis and its ever-expanding applications will propel more systematic work on the basics of human hair capturing capability in the future.

Human hair contains essentially two types of melanin oligomers: eumelanin, comprising 5,6-dihydroxyindole and 5,6-dihydroxyindole-2-carboxylic acid units, and pheomelanin, incorporating benzothiazine and benzothiazole units. Both have acidic properties, but the higher density of carboxylic groups confers to eumelanin a considerably stronger acidic character. Eumelanin is more abundant in black and brown hair, whereas pheomelanin predominates in blond and red hair. Detailed description of melanin structures and their binding properties may be found in a specific text [34]. On the basis of theoretical considerations and some experimental evidence, basic substances are expected to strongly bind to eumelanin via hydrogen bonds and charge–transfer interactions, allegedly resulting in biased outcomes that depend on the hair color of the tested subject. On the other hand, acidic and neutral substances are supposed to produce only weak interaction with both types of melanin, yielding similar incorporation from hair of any color. For these substances, the structural hindrance provided by the compact keratin net of hair plays a substantial role in their entrapment and prevents their easy removal.

Among drugs of abuse, a direct comparison between basic amphetamine and cocaine, and their neutral or zwitterion counterparts, namely *N*-acetylamphetamine and benzoylecgonine, was conducted by measuring their *in vitro* affinity toward various types of melanin [26]. It was confirmed that basic drugs are strongly bound by eumelanin-rich substrates by means of noncovalent multiple interactions with ionic character, whereas pure pheomelanin does not bind to the same substances to any extent. Likewise, neutral *N*-acetylamphetamine and benzoylecgonine showed little or no interaction with any type of melanin, as they cannot establish highly polarized interaction with the substrates [26]. Some evidence of covalent binding between amphetamines and melanin was observed in experiments of melanin synthesis in the presence of amphetamine, namely under *in vitro* conditions very distant from reality [27].

Animal models represent a practical way to bypass the ethical problem of administering illegal drugs to humans for experimental trials, but the extension to mankind of conclusions drawn from animal models is frequently questioned. An interesting recent study involved the administration of codeine and morphine to rats carrying white and dark gray spots of hair on their body. Although the two substances could be detected in both types of hair, the concentrations measured in dark gray hair were 8 to 40 times higher than in white hair [30]. An analogous approach was utilized to verify the possible influence of hair pigmentation on EtG [28] and fatty acid ethyl esters (FAEE) [29] concentrations, after ethanol administration to 11 and 6 rats, respectively. Opposite conclusions with respect to opiates were drawn, since highly correlated and almost coincident results were obtained from pigmented and white hair, demonstrating that melanin does not bind EtG [28] nor FAEE [29]. Noteworthy, EtG is highly hydrophilic, whereas FAEE are highly hydrophobic, but both biomarkers have neither basic properties nor positive charge, resulting in scarce interaction with melanin. Likewise, no influence of pigmentation was observed in the hair distribution of the new cannabimimetic agent JWH-073 (and its metabolites), following its administration to five rats [31].

Some statistical studies were undertaken around the year 2000, where the positivity rate among different ethnic groups was compared from hair and urine testing for several drugs of abuse [2,32,33]. Mieczkowski and coworkers [32] recorded almost identical results from urine and hair testing, while modest differences of positivity rates from the expected values could be entirely attributed to preferences for specific drugs among ethnic groups. No hair color bias, nor selective binding of drugs to hair of a certain color, nor metabolic difference among ethnic groups was evidenced [32,33]. While it is admitted that color may play a role in the accumulation of drugs in hair, large within-group variations occurring in other determining factors nonetheless obscure any clear evidence of color bias [33]. Kidwell and coworkers argued that ethnic differences in drug hair-binding capacity do exist, but they are not due to the hair color (which is typically black for both Asian and African people), but rather to different hair permeability, hair care and personal hygiene habits, and also the route of drug administration or passive exposure [2]. They also argued that the "positivity rate" is not a valuable means of detecting differences, because it is largely dependent on the cutoff threshold used for rating a positive outcome. Indeed, while the high sensitivity of modern analytical instrumentation has overcome the problem of achieving sufficiently low detection thresholds, it has emphasized the need to discriminate between active intake of drugs from passive exposure. The problem of the correct choice of cutoff values was also posed in an *in vivo* study that compared the hair concentration of codeine for 44 subjects after repeated administration of the drug [35]. This study evidenced large differences in codeine hair-binding for groups with different hair color, and high correlation with melanin hair content, possibly resulting in biased positive versus negative judgment, unless the results are corrected for melanin content [35,36].

The latter studies demonstrated the advantages of *in vivo* experiments conducted on humans, with respect to model and statistical investigations, to provide direct answer and practical solutions to the interpretation queries that forensic toxicology consistently poses. Moreover, *in vivo* studies provide easier control of variability sources and avoid any cruelty to animals, an issue of increasing concern. The application of safe experimental setting, ethical committee control, and clear informed consent policy has made *in vivo* trials on humans gradually accepted, encouraging their use in the recent past and, most likely, in forthcoming years.

A controlled administration of codeine and cocaine to 10 volunteers demonstrated that the concentrations in hair of codeine, cocaine, and their metabolites is dose-related and linearly correlated with total melanin content in the hair [37]. In contrast, routine analysis of 8,687 hair samples, found positive to both cocaine and benzoylecgonine, showed no difference between brown and black hair samples for cocaine, and little difference for benzoylecgonine, with no practical consequences on the positivity rate [38].

The strategy to compare results obtained from hair of different color but arising from the same subject is appropriately applied to eliminate all the sources of interindividual variability from the experimental data. Following this strategy, samples are collected from subjects with gray hair, then the white and pigmented hairs are separated, and processed independently [39–41] Significantly higher concentrations were observed in pigmented hair, with respect to white hair, for cocaine, amphetamines [41], and several prescription drugs with basic properties, such as amitriptyline [40], chlorpromazine, and clozapine [39]. Highly dissimilar (1/100) hair incorporation of zolpidem, after single administration, was hypothetically attributed to ethnic differences—in terms of melanin content—between Asian and European volunteers; [42,43] indeed, largely different zolpidem concentrations were observed also between white and black hairs of a single subject [43]. In the same way, Appenzeller and coworkers collected samples from 21 deceased persons with grizzled hair, with which they convincingly demonstrated that the concentration of EtG in hair does not depend on its melanin content [44]. The actual incorporation of all these substances in white hair, not containing melanin, even though at lower concentration, together with the incorporation of substances that do not interact with melanin, such as EtG, prove that keratin and other hair proteins play an important role in the binding of drugs in hair [39].

The recent introduction of a huge number of new synthetic drugs into the illegal market has already opened the problem of their updated detection in hair samples [45–48] but leaves the investigation field of their interaction with hair components still unexplored. Large differences exist in the structure and chemical properties between mostly neutral cannabimimetic agents and basic phenethylamines and cathinones, but also among the substances of the same class, which are likely to be reflected into their binding to melanin and keratin. The study of these interactions by both theoretical and experimental approaches represents one of the emerging challenges for the upcoming toxicology research.

12.2.3 Distribution of the Xenobiotic Substances Within the Hair

Incorporation of drugs and other xenobiotic substances inside the hair structure occurs from several sources (blood, sweat, and sebum) with different mechanisms, each contributing to the final hair composition, depending on the specific substance considered. The second variable that determines the distribution of substances along the hair length is its growth rate and the succession of anagen, catagen, and telogen phases. The third important parameter to be considered is the frequency of intake: single, occasional, variable in dose, or continuous. All these aspects are extensively reviewed in Chapter 1 of the present book; although the hair inclusion mechanisms are still partially unknown, their fundamental principles have been studied with sufficient detail to constitute a solid foundation for drawing quite sophisticated interpretations.

However, other confounding factors are occasionally present under specific circumstances, which may modify the expected distribution of the substances under investigation within the hair. For example, while blood can release its constituents only inside the bulb, sweat and sebum might be dispersed along a larger portion of the hair shaft, from which diffuse uptake of xenobiotics could arise. Other body fluids, specifically urine, may contaminate hair, determining wide alteration of its composition, as is frequently observed for pubic hair. Moreover, longitudinal and radial diffusion phenomena may enlarge the spot where the incorporated substances are initially present [49].

By keeping most of the cited factors as known and controlled as possible, deductions about a missing parameter can be inferred: for example, the time at which a certain drug was taken, the compliance of a patient to follow a certain therapy, the period during which a worker was exposed to a certain industrial pollutant. Most of these queries are tentatively answered by means of segmental analysis, namely the reconstruction of the chronological sequence of the investigated substances' uptake from the analysis of small segments of a hair lock grown homogeneously. Due to the numbers of controlling factors that determine the hair content distribution, interpretation of segmental analysis should be conducted with caution, and much more experimental evidence needs to be collected in the future to improve its reliability further.

A key application of segmental hair analysis is represented by the forensic investigation of drug-facilitated crimes (DFCs), particularly in sexual assaults [50]. An extensive review of the published literature in this field is presented in Chapter 9 of this book. Taking into account only the major issues of these investigations, it is important to note that hair analysis is particularly problematic because (i) the incapacitating drug has allegedly been taken only once, leading to extremely low hair concentration and high analytical sensitivity requirements, (ii) the drug to be detected is potentially comprised within a list of tens of psychoactive substances and is generally not known in advance by the analyst [51–54], and (iii) the circumstantial evidence is frequently disjointed, due to the amnesic effects of the drug. Thus, increasingly sensitive and multitargeted analytical methods have recently been proposed to deal with these challenges [55–58]. While general recommendations for segmental analysis suggest the use of 10–30 mm segments [59,60] in order to cope with peak broadening phenomena (diffusion, differential hair growth, etc.) [49,61], more condensed segmentations are frequently useful in DFC investigations in order not to dilute the limited amount of trapped drug into an excess of drug-free matrix [59,62]. Indeed, under apparently comparable conditions, much lower concentrations were found in real forensic cases [56,63] than in single controlled administrations [42,43].

Another important application of segmental analysis with progressively increasing practice is related to therapeutic drug monitoring [64,65]. The pioneering work of Sato and coworkers showed that, for each patient, changes in the concentration of haloperidol and chlorpromazine concentration in hair segments corresponded to variations of the administered dose [66–68]. This observation disclosed the opportunity of monitoring dosage changes by following the drug concentration pattern in the sequence of hair segments, and to reveal episodes of noncompliance behavior. In rehabilitation programs, segmental analysis also reveals compliance to abstinence and the degree of noncompliance conduct. It is worth noting that a strong correspondence between drug dosage and hair concentration exists for the same patient, even for the substances that exhibited virtually no correlation at interindividual level, due to genetic, personal, and behavioral factors (including hair melanin content) that modify on a large scale the absolute degree of hair incorporation.

During later research, therapeutic drug monitoring was applied to carbamazepine [69–71], selegiline [41], clozapine [72], buprenorphine [73–75], and methadone [76,77]. Buprenorphine received considerable interest recently because it is increasingly used to replace methadone in opiate addiction treatments. In its Suboxone® formulation, buprenorphine is mixed with naloxone, a strong opioid antagonist, to discourage drug diversion and abuse. Therefore, self-administration of Suboxone® is recurrently committed to the patients with a reliable history of compliance. For these patients, frequent urine tests to check compliance appear a pointless burden, while bimonthly hair analysis represents a viable and cheap alternative. In the stabilization phase of buprenorphine maintenance therapy, the drug dosage is adjusted for each patient and frequently varied, making hair analysis highly valuable as a useful method to monitor these dosage changes. In a recent study, the summed buprenorphine and norbuprenorphine (its main metabolite) hair concentrations appeared to be highly sensitive to dosage changes and strictly correlated with them, along an intraindividual time scale [74]. The case of buprenorphine illustrates the reasons why hair analysis is likely to earn the chance to substitute urine and blood testing in many long-term drug monitoring programs in the future: more valuable information will be gained at less expense, as long as the preliminary treatment steps for hair samples receive some degree of automation, to reduce production costs and increase sample throughput. These themes will undoubtedly receive growing attention in the forthcoming years.

As in the case of administered prescriptions, segmental hair analysis is also being increasingly utilized to control compliance to illegal drug abstinence or occasional intake of drugs in the mid-term [77]. Recently published papers have focused their interest on heroin [78], cocaine [79], cannabis [80,81], amphetamines [82], and new designer drugs [83], showing a diverse pattern of crucial evidence that can be acquired from segmental hair analysis. Twenty-eight cases of lethal heroin overdose were investigated with the aim of disclosing whether previous abstinence from opiates played a role in the fatalities, as a result of reduced tolerance to the drug [78]. Segmental hair analysis allowed investigators to distinguish continuous consumers from previous abstinent subjects, and ascertain the recurrent intake of additional drugs. The two groups of deceased heroin users were found to have homogeneous morphine concentrations in postmortem blood, proving that abstinence may not represent a critical factor in the fatalities, whereas the concurrent intake of other drugs possibly represents a more significant cause for the deaths [78]. The timeline of cocaine disappearance from growing hair was investigated within the implementation of rehabilitation programs involving drug abstinence. It was found that both cocaine and benzoylecgonine could be detected in the hair of former cocaine abusers for several months after abstinence started, suggesting caution in the assessment of alleged abstinence violation [79]. The absolute and relative hair concentrations of THC and 11-nor-9-carboxy-Δ^9-tetrahydrocannabinol (THC-COOH) were studied in both self-reporting and controlled administration of cannabis products, revealing limited dependence on consumption frequency [80], scattered absolute concentration values, and generalized decrease of THC-COOH from proximal to distal segments [81]. More precise correlation between consumption pattern and head hair concentration was found for methamphetamine [82], and the first reports that connect segmental hair analysis to the consumption history of cathinone- and pyrrolidinophenone-type new designer drugs in real cases of chronic intoxication have already been published [83,84]. Quite obviously, many more studies on segmental hair analysis for cannabimimetics, cathinones, phenethylamines, and other designer drugs are anticipated for the future, as long as these new classes of substances find substantial diffusion among the population of drug consumers.

12.2.4 Individual Factors

Besides hair color and melanin content, many other personal factors, listed in Table 12.2, are expected to contribute to the variability of hair composition. Up to now, these factors have almost always been considered using a holistic approach, knowingly or unconsciously. This is to say that when the experimental results could not be predicted nor rationalized, then the reason was generally attributed to several aspects of individual variability, whose particular contributions cannot be ignored and whose effects should be considered as a whole. In the preceding pages, several instances have been discussed, where the expected correlation between the dose of a certain substance (or exposure level to it) and hair concentration was not found or was extremely weak.

For practical applications, it is generally sufficient to know how reliable the conclusions that can be drawn from the experimental data are, and not to overestimate the predictability of the dependent variable, when most of the contributing factors are unknown. However, some of these "individual factors" have been extensively studied in the past, because they had been suspected of imparting substantial bias to the data. This has been the case for natural hair color and cosmetic treatments, whose effect has occasionally been overestimated, as discussed previously.

Whenever a decision has to be made on the basis of a cutoff value, these issues acquire crucial relevance. Quantitative determinations with legal impact are in fact the focal topic of most studies aiming to highlight the role of individual variability on the hair concentration of the targeted substance, as is the case for EtG and FAEE. This explains why recent research efforts are trying to single out the specific contribution of some other individual factors (ideally, each individual factor) to the overall data variability, in the attempt to build a more "parametric" modeling of uncertainty quantification. This endeavor also represents a tremendous open challenge for future research.

Among individual factors, genetic polymorphisms and various forms of genetic expression constitute the most difficult elements to deal with, because the source of information is generally missing, except for circumstances when some metabolic dysfunctions are evident. On the other hand, the first studies about the dependence of hair content on drug-metabolizing enzyme phenotyping have only started [85,86]. At least within ethnic groups, individuals are assumed to have relatively similar metabolic pathways toward toxicology relevant substances, but this similarity may not hold for specific drugs. Genotyping and genotyping characterization might be increasingly used in the future to uncover a potential source of interindividual variability in hair analysis.

It has been sometimes questioned if the gender and/or the age of the controlled subjects may influence their metabolic response to the targeted substances. For example, the possible effect of gender differences on EtG and FAEE hair concentration has been recently examined by several scientists [87–89]. Gareri and coworkers studied a population of 199 female and 73 male subjects, particularly to verify if the hair EtG and FAEE cutoff values commonly adopted for a masculine-prevalent population could be appropriate also for a feminine cohort, taking into account that the application of cosmetic products and treatments are suspected to decrease EtG hair concentration and increase FAEE [87]. They concluded that colored hair actually exhibited partly biased results for EtG, and that slightly lower average EtG values were anyhow determined for females with respect to males, whereas comparable FAEE results were observed for the two populations. Consequently, just for female subjects, they suggested (i) decreasing the hair EtG cutoff value to 20 pg/mg and (ii) combining EtG and FAEE determinations [87]. An almost opposite suggestion was made by Crunelle and coworkers, who investigated a cohort of 36 alcohol-dependent patients: they found the same strong correlation between alcohol consumption and hair EtG for males and females, and no gender effect [88]. They concluded that an identical evaluation scale and cutoff values can be used for both genders. Absence of gender differences for hair EtG response to alcohol intake was also confirmed in a retrospective study on a population of over 20,000 subjects (18,920 males and 1,373 females), prevalently examined for hair EtG within the driving license regranting procedure [89]. Average alcohol consumption is however slightly lower for women in both the addict [88] and driving under the influence (DUI) offender [89] populations, leading to a lower incidence of positive samples among women [89]. The same study also considered the influence of the age of the controlled subjects on the hair EtG positivity rate: the statistically significant higher incidence observed for the oldest subjects was nevertheless attributed to behavioral and habit differences (older people are more reluctant to refrain from alcohol than young, even when they are being controlled), rather than to metabolic changes that intervene with aging [89]. A comparative study of hair EtG values after controlled alcohol ingestion on a statistically significant cohort of subjects of different ages would be necessary to clarify the possible existence of metabolic changes.

Also a variety of physiopathological factors may also be taken into account in evaluating the results of hair analysis, especially when a decision is taken on the quantitation of a target biomarker. For example, Høiseth and coworkers recently studied how a decreased kidney function may alter the physiological level of hair EtG concentration [90]. Although very limited cases of positive EtG concentrations (EtG>30 pg/mg) in certain social low-alcohol drinkers were reported in the past, this study provides clear evidence that patients with serious renal disease may have hair EtG values far above the cutoff, even if they observe a low-alcohol diet. This evidence is possibly attributed to delayed excretion of EtG, resulting in increased hair incorporation [90].

While the hair concentration of cortisol [91,92], testosterone, and various metals [93,94] have been recently compared for obese subjects with respect to a reference population, analogous investigation has not yet been conducted for abused substances. Dependence of hair concentration on body mass index (BMI) is not expected for hydrophilic substances, such as EtG [89] and amphetamines, but highly hydrophobic substances are likely to accumulate into the fat tissues of the body to be released with delay, ultimately resulting in a modified chronological profile of hair concentration. This is surely the case for THC, but possibly also for FAEE, cocaine, and morphine. However, the BMI effect on hair analysis has still to be investigated.

A seasonal factor has recently been disclosed for hair EtG, since the average EtG values measured in hair grown in the warm season are sensibly lower than those measured in the hair grown in the cold season [89]. The hypothetical reasons for this difference are various, and range from a lower consumption of alcohol in the warm months, or a more abundant perspiration leading to EtG dilution, up to a more recurrent removal of sweat, due to frequent showers and sea bathing occurring during summer. Quite evidently, these reasons include effective lower exposure to the substance (i.e., no bias) together with climatic and behavioral factors, possibly associated with a bias. Thus, the seasonal effect has to be investigated further.

Other conceivable sources of individual variability should be considered under specific circumstances. These include the already-cited cosmetic and hygiene habits, but also the mode of substance consumption (i.e., smoking, inhalation, injection, and so on) and its frequency (e.g., regular drinking versus binge drinking), together with the concomitant intake of physiological or metabolic interfering drugs or special diet regimes. Lastly, the clothes habitually worn might induce anomalous hair perspiration, especially hats, foulards, and scarves, and become a source of self-contamination, because the entire hair length is maintained in close contact with the head, skin, and its sebaceous and sweat emissions. The same might occur with pillows and armchairs with headrests, especially on the posterior vertex region of the head, which is commonly sampled [59].

Increasing interest is also devoted to the problem of collecting representative hair samples. For head hair, this problem has been quite extensively examined in the past, leading to the recommendation to collect hair from the posterior vertex region of the head, where the least variations of hair growth are observed [59]. However, recent concerns have been raised as to whether a single lock of hair is representative of the real hair concentration of the targeted substances [95], and how much variation in growth rate and sampling procedures might influence their distribution along the measured hair length [96]. Indeed, Dussy and coworkers recorded coefficients of variation up to 28% and 62%, respectively for EtG and caffeine, from 10 hair locks collected from various sites of the skull of the same individuals [95]. Thus, it is always advisable to collect more than one hair lock and, in critical situations (e.g., concentration close to the cutoff value), to repeat the analysis on further hair locks. LeBeau and coworkers evaluated the variability associated with growth rate of human head hair, as well as the ability to uniformly collect hair next to the scalp, the latter being affected by several sources of errors [96]. From both contributions, they deduced that a quite large range of uncertainty for chronological attribution exists, and that the first segment close to the cut actually corresponds to hair formed 1.3 ± 0.2 to 2.2 ± 0.4 months earlier [96].

Whenever it is impossible to collect a hair lock from the head, either because of complete baldness or because head hair has been subjected to strong cosmetic treatments, alternative sources of hair sampling should be found, possibly providing experimental results as close as possible to those that would have been obtained had head hair been available. However, nonhomogeneous hair growth and impossibility to align hair shafts prevent any chronological assignment from nonhead hair. For most drugs of abuse, accurate quantitation from nonhead hair is generally not requested nor achievable and the analysis merely assesses the presence or absence of the drug, even if roughly comparable results were recently reported for amphetamines having been determined from head, pubic, and axillary hair collected from the same subjects [97]. More critical is the case of hair EtG, whose quantitative result has to be compared with cutoff values for excessive drinking [98]. Several recent studies [89,99–101] confirmed that pubic and axillary hair cannot substitute head hair for EtG determination, because they respectively over- and underestimate the correct value, whereas chest hair [89,99,100], leg and arm hair [100] may adequately represent the correct head hair EtG content. Together with these matrices, beards can also be used to verify teetotalism [101]. It has also been recently assessed that head locks of hair which are longer than the prescribed 3 cm can be profitably utilized to measure EtG, because no significant washing-out effect occurs on more distal segments from regular hygiene practices [102]. Further research to investigate the chemical and physical properties of the keratin matrix, together with its three-dimensional architecture [103], is foreseen.

12.3 INNOVATIVE TECHNOLOGIES AND INSTRUMENTAL ADVANCEMENTS

Crucial issues of hair analysis in the past years were represented by the number of different analytical procedures needed to accomplish an exhaustive toxicological screening, and the minimal amount of hair needed to execute these procedures on separate aliquots, with sufficient sensitivity. The continuous innovation and rapid improvement of analytical instrumentation has radically modified this scenario in recent years, allowing increasingly comprehensive procedures to be run on progressively smaller hair aliquots. In particular, the unceasing development of new chromatographic and mass spectrometric technologies has considerably improved the separation of complex mixture components, their mass spectra resolution, and overall instrumental sensitivity, insomuch as to make the limit of quantification for most target substances substantially lower than their effective concentration in hair, at least for the majority of real applications [104,105]. Among innovations, the emergence of ultra-high-pressure liquid chromatography (UHPLC), together with TOF and Orbitrap mass analyzers, and a wide range of new devices for efficient ion generation, accumulation, and transmission, epitomize the technological milestones toward the progressive accomplishments of hair analysis. On the other hand, the resulting high sensitivity has made even more mandatory the use of extreme care in both sample handling, to avoid contamination, and data interpretation, to avoid false-positive judgments, as discussed in the preceding chapters.

12.3.1 Broad-Spectrum Toxicological Analysis

Among the changes introduced by the new instrumental technologies, the issue of major impact, in terms of broad applicability, is possibly represented by the achievement of general toxicological analysis, with the aim of making it totally untargeted. Significant steps toward the completion of large multianalyte and multiclass screenings by single analysis have already been taken up by combining the high chromatographic resolution of UHPLC with the fast electronics and high sensitivity of modern triple quadrupole mass spectrometers. While in the years 2007–2009, several methods capable of screening 15–20 substances within a single liquid chromatography (LC) run were proposed, already in 2012 the concomitant detection and quantification of an extended panel of 35 licit and illicit drugs and metabolites was developed on 50 mg of hair, achieving limit of quantitation (LOQ) concentrations in the range of 0.5–100 pg/mg, and was applied to 17 real forensic cases [55]. More recently, another LC-MS/MS method was developed to determine as many as 87 psychoactive drugs and metabolites in 20 mg of hair, reaching 0.3–45 pg/mg LOQ values, which was applied to real postmortem specimens [106]. Similarly, a UHPLC-MS/MS protocol was used to determine 96 psychoactive drugs on a 10 mg hair specimen, yielding 2–50 pg/mg LOQ values for most analytes of interest [107]. Taking into account that for the latter methods limit of detection (LOD) concentrations were found in the low pg/mg range for most psychoactive drugs, or even below 1 pg/mg for some specific substances, application of these methods in several forensic inquiries, including DFCs, appears to be at hand.

Fit-for-purpose multiclass and comprehensive UHPLC-MS/MS methods, with a relatively restricted panel of analytes, were developed to substitute multiple GC-MS procedures, with the final scope to improve sample throughput and decrease costs. These are crucial issues of increasing relevance in hair analysis, particularly for the application in workplace drug testing. In such a context, the main difficulty is to create a single analytical method adequate to determine simultaneously polar and basic (i.e., amphetamines), acidic (i.e., benzoylecgonine), and nonpolar (i.e., THC) drugs. Two recent papers undertook this task with similar outcomes: respectively 13 [108] and 16 [109] drugs of abuse and metabolites, including THC, were separated in about 5 min, and detected after MS/MS analysis down to LOQ concentrations of 20–80 pg/mg, using triple quadrupole instruments. A further published method for hair analysis determined as many as 33 drugs of abuse and metabolites (without THC) in less than 9 min, with LOD values ranging from 6 to 63 pg/mg [110].

The increasing performances of high-resolution mass spectrometers, in terms of sensitivity and acquisition speed, have made these instruments potentially capable of executing a truly general toxicological analysis on hair samples, allowing factual untargeted screening and retrospective reexamination of acquired data, following the appearance of upcoming investigation elements. As a matter of fact, in most DFCs, acute intoxications, and postmortem investigations, the intoxicating substance cannot be anticipated, or even the actual occurrence of an intoxicating agent is doubtful. Thus, negative results from a targeted screening leave the chance open that an untargeted substance is present in the investigated sample.

Untargeted screening of hair samples performed by hybrid quadrupole time-of-flight (QTOF)-MS/MS devices, or similar instrumental arrangements, requires some forms of data-dependent [111] or data-independent [112,113] acquisition software, and the availability of large tandem mass spectra libraries [111,114]. For example, MS and MS/MS modes of acquisition are alternated within short cycle periods, where the preliminary low-resolution MS spectrum is used to identify the most abundant precursor ions at a certain retention time and the subsequent high-resolution MS/MS spectra of the automatically selected precursors (recorded upon collisional activation) are compared with dedicated libraries of collision-induced dissociation spectra of toxicologically relevant substances. Other strategies of data collection and library inquiry have been implemented on single- and double-stage mass spectrometers, and software of increasing sophistication will be made commercially available in the near future to accomplish these tasks. However, the benefits of comprehensive screening are still somewhat compromised by a decreased sensitivity compared to targeted screening protocols [114]. This represents a critical drawback when a limited amount of samples are available, as in hair analysis. For this reason, broad-spectrum targeted screening methods have also been developed for high-resolution mass spectrometers, with which the collection of accurate mass signals is exploited to provide unequivocal identification of the targeted drugs.

Progressively higher performances were obtained as long as improved instrumental technologies were made commercially available. For example, 52 licit and illicit drugs were targeted in 10 mg hair samples using a single TOF instrument that allowed to reach LOD concentrations in the range of

10–100 pg/mg and LOQs around 50–200 pg/mg [115]. In the same period, other groups used Orbitrap mass spectrometers to complete toxicological screenings: in one study, 28 substances were monitored in 2.5 mg hair aliquots with LOQ values ranging from 100 to 500 pg/mg; [116] in another study, an hybrid linear ion trap—Orbitrap tandem mass spectrometer—was used to determine a restricted panel of stimulant drugs on extremely small aliquots (0.2 mg) of hair samples, also yielding LOQ values in the range of 100–500 ng/mg [117]. More recently, LOQ concentrations in the tens of pg/mg range were obtained for a wide panel of new psychoactive substances, using 100 mg of hair and an analytical method based on an hybrid QTOF mass spectrometer [47]. The comparison of these figures-of-merit with those obtained from triple quadrupole instruments, previously reported, apparently indicates that high-resolution mass analyzers, even in the hybrid configuration, still provide slightly lower sensitivity than triple quadrupole instruments for targeted screening investigations on hair samples. However, the sensitivity gap between the two classes of instruments observed in the scientific literature is being progressively filled with the introduction of improved technologies for ion- and energy-focusing, ion accumulation, and synchronization of ion storage and mass detection processes, and reflects their delayed widespread dissemination into the toxicology laboratories.

12.3.2 Highly Demanding Investigations

Besides highly general screening, also highly specific and/or demanding determinations characterize the incipient frontier of hair analysis. The investigation following DFCs represents a good example of a highly demanding objective for hair analysis, as previously mentioned. Further examples of highly demanding investigations are (i) the detection of certain drug metabolites at particularly low concentration, necessary to exclude external contamination; (ii) the quantitative discrimination of enantiomers, when their pharmacological activity is different or only one stereoisomer produces psychoactive effects; and (iii) the determination of licit and illicit drugs on extremely small hair aliquots, or even on a single hair.

Among the drug metabolites whose determination in hair samples is particularly problematic, the most extensively studied is certainly THC-COOH, whose hair concentration is frequently below 1 pg/mg, and cutoff values to prove cannabis abuse are set at 50–100 fg/mg. While the traditional methods to determine THC-COOH in hair are based on gas chromatography–electron capture negative ionization–tandem mass spectrometry (GC-ECNI-MS/MS), after chemical derivatization, new alternative approaches have been recently proposed to achieve extremely low detection limits. These include GC × GC-MS [118], surface-activated chemical ionization combined with electrospray ionization and LC-MS [119], negative-ion electrospray ionization LC-MS/MS of the unmodified THC-COOH [120] or its methyl ester, obtained after selective derivatization that leaves the hydroxyl group unchanged and still available for negative ionization [121]. The latter approach is particularly skillful, because it entirely removes the interference of fatty acids, and opens the way to further reduction of the THC-COOH detection limit, as long as the chromatographic and mass spectrometric instrumentation keeps improving. The discrimination between drug intake and external contamination by means of metabolites detection also applies to cocaine, but the sensitivity requirements for the determination of cocaine metabolites are less severe, and consequently the struggle to set up innovative analytical methods is comparatively less forceful [122].

Chiral drug analysis is another topic of increasing investigation because of its relevance in clinical and forensic toxicology. Up to now, a large number of these studies has been conducted on blood and urine specimen [123], but very few on the keratin matrix, despite its unique property of providing a time-integrated perspective of drug intake. Thus, more frequent recourse to chiral separation in hair analysis is expected in the forthcoming years. Stereoisomer discrimination has been completed for amphetamines present in hair samples by both GC-MS [124,125] and LC-MS [126] methods. GC-MS analysis of amphetamines enantiomers was conducted after derivatization with (S)-heptafluorobutyrylprolyl chloride, which generated couples of diastereoisomers, that could be separated with a nonchiral stationary phase and ionized in the electron-capture negative ion mode [124]. Similarly, a new chiral derivatization agent of improved efficiency was synthesized [namely (2S,4R)-*N*-heptafluorobutyryl-4-heptafluorobutoyloxyprolyl chloride] and utilized under comparable conditions [125]. For the five amphetamines prevalent on the illegal market, LOQ values in the range of 7–150 pg/mg were obtained with the former derivatization agent, but LODs were reduced to 2–9 pg/mg when the new and optimized derivatization agent was applied to the hair extracts [125]. Derivatization of the amine group of amphetamines was applied also before LC separation to produce a 5-fold sensitivity increase, but a nonchiral reagent was used (i.e., trifluoroacetic anhydride); thus, the resulting derivative enantiomers had still to be separated on a chiral stationary phase [126].

The enantioselective metabolism of levomethorphan (a narcotic drug) and its discrimination from the enantiomeric dextromethorphan (an antitussive medicine) was investigated on various biological samples including hair, collected from rats [127]. The LC-MS/MS analytical procedure involved the use of a chiral LC column and achieved complete separation of enantiomers for the parent substances and their *O*-demethyl and *N*,*O*-didemethyl metabolites in 12 min, showing that the drug metabolism does not induce its racemization. This represents a crucial aspect for the prospective application to human hair, in view of toxicological applications [127]. Also the enantioselective metabolism of methadone, commonly administered as a racemic mixture, was demonstrated in a study founded on the setup of a dedicated LC-MS/MS method, where three couples of enantiomers (methadone and its two main metabolites) were extracted from hair samples and chromatographically separated on a chiral column [76]. This complex task was proficiently accomplished by using factorial analysis experimental design and artificial neural networks to optimize the chromatographic separation from response surfaces analysis [76]. In fact, the widespread use of multivariate chemometric tools represents another challenging perspective for hair analysis, both to deduce optimal experimental conditions and planning from a reduced set of preliminary experiments, as in the study previously cited, and to interpret the resultant data on the basis of sound statistical principles. For example, principal component analysis can be extensively employed for data mining [128], while statistical discriminant analysis can be exploited to make predictions and estimate likelihood ratios to distinguish real drug consumption from external contamination of hair [129]. The Bayesian concept of likelihood ratio was also proposed to support the interpretation of hair testing for alcohol abuse with solid probabilistic foundation [130].

12.3.3 Minute Hair Availability and Single Hair Analysis

Another demanding request for future hair analysis is the opportunity to determine drugs on extremely minute amounts of hair sample, down to the limit of a single hair. Currently, single hair analysis is no longer a simple wishful dream, but is nowadays a concrete objective, persistently pursued with increasing success. However, the need to operate on tiny amounts of keratin matrix is frequently combined with a requisite of low detection limits, in terms of concentration, making single hair analysis not yet adequate to fulfill both requirements. The most promising approach to decrease the amount of sampled hair and the drug detection limits at the same time utilized a microfluidic chip-based nano-HPLC system, coupled to tandem mass spectrometry [131,132]. In the most recent publication, the nano-LC-MS/MS technique was applied to the determination of 14 illicit drugs and metabolites extracted from only 2 mg of hair, and obtained low detection limits for all analytes that ranged from 0.10 to 0.75 pg/mg [131]. A similar procedure, involving a chromatographic run time of 15 min with an eluent flow rate of 4 μL/min, was applied to the determination of ketamine and norketamine in 10 mg of hair. Again, LODs of 0.5 and 1.0 pg/mg were obtained, with very little solvent consumption [132].

Single hair analysis was recently conducted to detect both organic and inorganic analytes, mostly for toxicological purposes. The content of heavy metals (Cd and Pb) in a single hair was investigated by coupling of a tungsten coil electrothermal vaporizer with an argon–hydrogen flame for atomic fluorescence spectrometry. In particular, an outstanding absolute detection limit of 50 fg was observed for Cd [133]. In another study, a wide panel of essential and toxic metals were determined in single hair by laser ablation—inductively coupled plasma—mass spectrometry, reaching unequal LODs in the range of 1–900 pg/mg [134]. The advantage of laser ablation lies in its micrometric spatial resolution that allows investigation of the metal longitudinal distribution along the hair length.

The detection of drugs on single hairs has been handled both by (i) direct spatially resolved exposure to impinging particles such as laser photons or electrospray droplets and (ii) preliminary minute segmentation of the hair followed by application of dedicated small-volume extraction and treatment methods for each hair segment. In both approaches, the absolute amount of drug to be detected is extremely small and, consequently, sensitivity limitations exist that fundamentally restrict single hair investigations to targeted analysis. On the other hand, the analysis of a single hair furnishes a more detailed chronological profile of drug intake than a hair lock, because several confounding factors are not present, such as alternating hair growth phases and incorrect alignment of hair bundles during sampling. The absence of these smoothing effects on peak concentrations partly compensate for the reduced sensitivity due to minimal sampling.

Thieme and Sachs introduced single hair segmental analysis in 2007 to investigate a case of repeated clozapine poisoning [135]. Fine segmentation (1.0–2.5 mm) of individual hairs provided an extremely detailed chronology of clozapine administration, with a resolution within a few days and a detection limit of about 1 pg/mg. This study also revealed that synchronization of sharp clozapine blood and hair peaks occurred, thus showing the marginal role of drug incorporation from sweat and sebum and excluding the incidence of any longitudinal diffusion of the drug along the hair shaft. Circumstantial evidence also allowed calculation of the hair growth rate for the investigated subject [135]. The same scientists recently expanded the application of single hair segmental analysis to the screening of further psychoactive substances, potentially used to perpetrate DFCs [57]. The proposed investigation strategy for DFC cases involved the preliminary screening of 183 substances on a hair lock of adequate weight, followed by fine segmental analysis on single hairs, targeted to the drugs whose presence was ascertained. Segments of 0.25–0.50 mm, and weight of 10–50 μg, proved optimal to balance sensitivity, quantitative accuracy, and time resolution of this chronological profiling. Quite obviously, all the analytical steps had to be optimized to work on a microscopic scale [57].

Since 2011, MALDI has been proposed to obtain a punctual analysis of single hair, resulting in a mass-spectrometric image of drug distribution along the hair shaft [136–138]. Even before, MALDI was applied to pulverized hair (1.0–2.5 mg) to execute rapid and high-throughput screening of hair samples for cocaine and metabolites [139,140]. The first method developed to prepare single hairs for MALDI-MS imaging involved attaching the hair to a glass slide using a conductive carbon adhesive tape, then manually cutting the hair shaft lengthwise using a razor and a microscope and applying α-cyano-4-hydroxycinnamic acid as the MALDI matrix. After completing this procedure, the arranged hair was exposed to the laser beam, and the desorbed ions were detected either with a TOF or a Fourier transform ion cyclotron resonance (FTICR) mass analyzer. The method was applied to create highly resolved drug concentration profiles of the hair taken from amphetamine abusers [136,137]. The complex manual step of microscopic hair slicing was avoided in the MS imaging study conducted on the hair collected from cocaine abusers, considerably shortening the sample preparation time [138]. Cocaine and its main metabolites were nevertheless determined with a LOD of about 5 ng/mg (corresponding to 20 fg of cocaine per laser shot), adequate for chronological drug profiling of the hair of habitual drug consumers, not for the detection of a single and episodic intake. Both axial and radial diffusion of the analytes during the acquisition time proved to be modest, allowing a temporal resolution of hair growth of a few days [138]. In a subsequent study, the use of a MALDI-MS/MS instrument with high resolution capability in the second stage of mass analysis allowed the registration of the chronological profile for both the parent drug and two metabolites in the hair of an episodic cocaine consumer [141]. However, it was observed that, while two hairs furnished similar profiles, two others from the same subject turned out totally negative to the presence of cocaine. This observation alerted researchers to the fact that the results from single hair analysis have to be carefully evaluated in the light of different growth phases, which might produce false-negative results if the intake occurs during the quiescent telogen and catagen periods [141,142]. Image profile was determined also for ketamine in three hairs out of four collected from a single individual. A MALDI source interfaced to 9.4 Tesla FTICR mass spectrometer was used to this purpose. Gentle scraping using a scalpel somewhat damaged the hair surface, so as to allow the incorporated substances to be more easily desorbed outside the keratin structure [142].

Unlike clozapine [135], single hair analysis of samples collected 30 days after unique intake of tilidine revealed quite homogeneous distribution of the drug along the hair length and large concentration differences among five hairs collected from the same subject. These two observations suggested that these hairs had been randomly coated on their external surface by the sweat containing tilidine shortly after its administration; then, the tilidine coated at the surface was incorporated inside the keratin structure, simulating chronic administration [143]. The dissimilar hair distribution obtained from single intake of clozapine and tilidine (sharp concentration peak versus homogeneous distribution) is likely to reflect the different predominant way by which these drugs are incorporated, respectively blood or sweat. Consequently, the practical utility of highly resolved segmental hair analysis becomes questionable when the prevalent way leading to drug incorporation is sweat, suggesting the need (i) to use extreme care in the interpretation of the inherent data and (ii) to conduct further research on the mechanisms of drug transfer from biological fluids to hair, for each class of substances.

Further direct ionization methods prior to MS analysis, including low temperature plasma ionization (LTP) [144], desorption electrospray ionization (DESI) [145], and direct analysis in real time (DART) [146], had been proposed to perform fast and cheap drug screening on hair locks and hair extracts. However, the first attempts to work directly on the keratin material failed, and some form of extraction proved to be necessary before applying ambient ionization to the extracts [144,145]. The more recent study showed the successful application of DART to the detection of THC directly on hair locks collected from chronic drug abusers [146]. Although the instrumentation used was not ideal to achieve the highest sensitivity toward a single target compound, this study demonstrated the potential of the DART-MS technique for the fast screening of hair samples. Easily achievable improvements are predictable using dedicated mass spectrometers and further refinement of the experimental conditions.

These are only few of the technological advancements expected in the forthcoming years that will influence the advancements of hair analysis. A weak point of almost any current analytical method is the extensive need of manual operations during the initial hair sample preparation. This weakness limits the widespread application of hair analysis and has a major impact on the final cost of any analytical protocol. Therefore, decisive technological improvements are waited in the automation of hair sample processing, in order to reduce the manual intervention and its associated cost and variability, to assure more constant operating conditions, and to increase sample throughput. Under such conditions, the role of hair analysis in the assessment of human exposure to a variety of drugs and/or toxic substances will rapidly grow in any field of clinical and toxicological domains, making hair a customary biological matrix to investigate, as is presently true for urine and blood.

REFERENCES

[1] Wennig R. Potential problems with the interpretation of hair analysis results. Forensic Sci Int 2000;107:5–12.

[2] Kidwell DA, Lee EH, DeLauder SF. Evidence for bias in hair testing and procedures to correct bias. Forensic Sci Int 2000;107:39–61.

[3] Musshoff F, Madea B. Analytical pitfalls in hair testing. Anal Bioanal Chem 2007;388:1475–94.

[4] Musshoff F, Madea B. New trends in hair analysis and scientific demands on validation and technical notes. Forensic Sci Int 2007;165:204–15.

[5] Tagliaro F, Bortolotti F, Viel G, Ferrara SD. Caveats against an improper use of hair testing to support the diagnosis of chronic excessive alcohol consumption, following the "Consensus" of the Society of Hair Testing 2009 [Forensic Science International 196 (2010) 2]. Forensic Sci Int 2011;207:69–71.

[6] LeBeau MA, Montgomery MA. Considerations on the utility of hair analysis for cocaine. J Anal Toxicol 2009;33:343–4.

[7] Pragst F, Sachs H, Kintz P. Hair analysis for cocaine continues to be a valuable tool in forensic and clinical toxicology. J Anal Toxicol 2010;34:354–6.

[8] Ettlinger J, Kirchen L, Yegles M. Influence of thermal hair straightening on ethyl glucuronide content in hair. Drug Test Anal 2014;6(Suppl. 1):74–7.

[9] Musshoff F, Brockmann C, Madea B, Rosendahl W, Piombino-Mascali D. Ethyl glucuronide findings in hair samples from the mummies of the Capuchin Catacombs of Palermo. Forensic Sci Int 2013;232:213–17.

[10] Skopp G, Pötsch L, Mauden M. Stability of cannabinoids in hair samples exposed to sunlight. Clin Chem 2000;46:1846–8.

[11] Favretto D, Tucci M, Monaldi A, Ferrara SD, Miolo G. A study on photodegradation of methadone, EDDP, and other drugs of abuse in hair exposed to controlled UVB radiation. Drug Test Anal 2014;6(Suppl. 1):78–84.

[12] Gerace E, Veronesi A, Martra G, Salomone A, Vincenti M. Study of the incorporation of cocaine in damaged hair samples. Proceeding 18th Scientific Meeting of the Society of Hair Testing (SoHT), Geneva, Switzerland, August 28–30, 2013.

[13] Hill V, Loni E, Cairns T, Sommer J, Schaffer M. Identification and analysis of damaged or porous hair. Drug Test Anal 2014;6(Suppl. 1):42–54.

[14] Cirimele V, Kintz P, Mangin P. Drug concentrations in human hair after bleaching. J Anal Toxicol 1995;19:331–2.

[15] Jurado C, Kintz P, Menéndez M, Repetto M. Influence of the cosmetic treatment of hair on drug testing. Int J Legal Med 1997;110:159–63.

[16] Skopp G, Pötsch L, Moeller MR. On cosmetically treated hair—aspects and pitfalls of interpretation. Forensic Sci Int 1997;84:43–52.

[17] Agius R. Utility of coloured hair for the detection of drugs and alcohol. Drug Test Anal 2014;6(Suppl. 1):110–19.

[18] Morini L, Zucchella A, Polettini A, Politi L, Groppi A. Effect of bleaching on ethyl glucuronide in hair: an in vitro experiment. Forensic Sci Int 2010;198:23–7.

[19] Martins LF, Yegles M, Thieme D, Wennig R. Influence of bleaching on the enantiomeric disposition of amphetamine-type stimulants in hair. Forensic Sci Int 2008;176:38–41.

[20] Baeck S, Han E, Chung H, Pyo M. Effects of repeated hair washing and a single hair dyeing on concentrations of methamphetamine and amphetamine in human hairs. Forensic Sci Int 2011;206:77–80.

[21] Cuypers E, Flinders B, Bosman IJ, Lusthof KJ, Van Asten AC, Tytgat J, et al. Hydrogen peroxide reactions on cocaine in hair using imaging mass spectrometry. Forensic Sci Int 2014;242C:103–10.

[22] Hill V, Cairns T, Schaffer M. Hair analysis for cocaine: factors in laboratory contamination studies and their relevance to proficiency sample preparation and hair testing practices. Forensic Sci Int 2008;176:23–33.

[23] Stout PR, Ropero-Miller JD, Baylor MR, Mitchell JM. Morphological changes in human head hair subjected to various drug testing decontamination strategies. Forensic Sci Int 2007;172:164–70.

[24] Binz TM, Baumgartner MR, Kraemer T. The influence of cleansing shampoos on ethyl glucuronide concentration in hair analyzed with an optimized and validated LC-MS/MS method. Forensic Sci Int 2014;244:20–4.

[25] Röhrich J, Zörntlein S, Pötsch L, Skopp G, Becker J. Effect of the shampoo ultra clean on drug concentrations in human hair. Int J Legal Med 2000;113:102–6.

[26] Borges CR, Roberts JC, Wilkins DG, Rollins DE. Cocaine, benzoylecgonine, amphetamine, and *N*-acetylamphetamine binding to melanin subtypes. J Anal Toxicol 2003;27:125–34.

[27] Claffey DJ, Ruth JA. Amphetamine adducts of melanin intermediates demonstrated by matrix-assisted laser desorption/ionization time-of-flight mass spectrometry. Chem Res Toxicol 2001;14:1339–44.

[28] Kharbouche H, Steiner N, Morelato M, Staub C, Boutrel B, Mangin P, et al. Influence of ethanol dose and pigmentation on the incorporation of ethyl glucuronide into rat hair. Alcohol 2010;44:507–14.

[29] Kulaga V, Velazquez-Armenta Y, Aleksa K, Vergee Z, Koren G. The effect of hair pigment on the incorporation of fatty acid ethyl esters (FAEE). Alcohol Alcohol 2009;44:287–92.

[30] Lee S, Han E, Kim E, Choi H, Chung H, Oh SM, et al. Simultaneous quantification of opiates and effect of pigmentation on its deposition in hair. Arch Pharm Res 2010;33:1805–11.

[31] Kim J, In S, Park Y, Park M, Kim E, Lee S. Deposition of JWH-018, JWH-073 and their metabolites in hair and effect of hair pigmentation. Anal Bioanal Chem 2013;405:9769–78.

[32] Kelly RC, Mieczkowski T, Sweeney SA, Bourland JA. Hair analysis for drugs of abuse. Hair color and race differentials or systematic differences in drug preferences? Forensic Sci Int 2000;107:63–86.

[33] Mieczkowski T, Newel R. Statistical examination of hair color as a potential biasing factor in hair analysis. Forensic Sci Int 2000;107:13–38.

[34] Nordlund JJ, Boissy RE, Hearing VJ, King RA, Oetting WS, Ortonne J-P. The pigmentary system: physiology and pathophysiology. 2nd ed. Malden, MA: Blackwell Publishing; 2006. p. 1328.

[35] Rollins DE, Wilkins DG, Krueger GG, Augsburger MP, Mizuno A, O'Neal C, et al. The effect of hair color on the incorporation of codeine into human hair. J Anal Toxicol 2003;27:545–51.

[36] Kronstrand R, Förstberg-Peterson S, Kågedal B, Ahlner J, Larson G. Codeine concentration in hair after oral administration is dependent on melanin content. Clin Chem 1999;45:1485–94.

[37] Scheidweiler KB, Cone EJ, Moolchan ET, Huestis MA. Dose-related distribution of codeine, cocaine, and metabolites into human hair following controlled oral codeine and subcutaneous cocaine administration. J Pharmacol Exp Ther 2005;313:909–15.

[38] Mieczkowski T, Kruger M. Interpreting the color effect of melanin on cocaine and benzoylecgonine assays for hair analysis: brown and black samples compared. J Forensic Leg Med 2007;14:7–15.

[39] Kronstrand R, Roman M, Hedman M, Ahlner J, Dizdar N. Dose–hair concentration relationship and pigmentation effects in patients on low-dose clozapine. Forensic Sci Med Pathol 2007;3:107–14.

[40] Rothe M, Pragst F, Thor S, Hunger J. Effect of pigmentation on the drug deposition in hair of grey-haired subjects. Forensic Sci Int 1997;84:53–60.

[41] Kronstrand R, Ahlner J, Dizdar N, Larson G. Quantitative analysis of desmethylselegiline, methamphetamine and amphetamine in hair and plasma from Parkinson patients on long-term selegiline medication. J Anal Toxicol 2003;27:135–41.

[42] Cui X, Xiang P, Zhang J, Shi Y, Shen B, Shen M. Segmental hair analysis after a single dose of zolpidem: comparison with a previous study. J Anal Toxicol 2013;37:369–75.

[43] Villain M, Chèze M, Tracqui A, Ludes B, Kintz P. Windows of detection of zolpidem in urine and hair: application to two drug facilitated sexual assaults. Forensic Sci Int 2004;143:157–61.

[44] Appenzeller BMR, Schuman M, Yegles M, Wennig R. Ethyl glucuronide concentration in hair is not influenced by pigmentation. Alcohol Alcohol 2007;42:326–7.

[45] Hutter M, Kneisel S, Auwärter V, Neukamm MA. Determination of 22 synthetic cannabinoids in human hair by liquid chromatography-tandem mass spectrometry. J Chromatogr B Analyt Technol Biomed Life Sci 2012;903:95–101.

[46] Salomone A, Gerace E, D'Urso F, Di Corcia D, Vincenti M. Simultaneous analysis of several synthetic cannabinoids, THC, CBD and CBN, in hair by ultra-high performance liquid chromatography tandem mass spectrometry. Method validation and application to real samples. J Mass Spectrom 2012;47:604–10.

[47] Gottardo R, Sorio D, Musile G, Trapani E, Seri C, Serpelloni G, et al. Screening for synthetic cannabinoids in hair by using LC-QTOF MS: a new and powerful approach to study the penetration of these new psychoactive substances in the population. Med Sci Law 2014;54:22–7.

[48] Salomone A, Luciano C, Di Corcia D, Gerace E, Vincenti M. Hair analysis as a tool to evaluate the prevalence of synthetic cannabinoids in different populations of drug consumers. Drug Test Anal 2014;6:126–34.

[49] Kintz P. Issues about axial diffusion during segmental hair analysis. Ther Drug Monit 2013;35:408–10.

[50] Scott KS. The use of hair as a toxicological tool in DFC casework. Sci Justice 2009;49:250–3.

[51] Scott-Ham M, Burton FC. Toxicological findings in cases of alleged drug-facilitated sexual assault in the United Kingdom over a 3-year period. J Clin Forensic Med 2005;12:175–86.

[52] Hall JA, Moore CBT. Drug facilitated sexual assault—a review. J Forensic Leg Med 2008;15:291–7.

[53] Madea B, Musshoff F. Knock-out drugs: their prevalence, modes of action, and means of detection. Dtsch Arztebl Int 2009;106:341–7.

[54] Parkin MC, Brailsford AD. Retrospective drug detection in cases of drug-facilitated sexual assault: challenges and perspectives for the forensic toxicologist. Bioanalysis 2009;1:1001–13.

[55] Lendoiro E, Quintela O, de Castro A, Cruz A, López-Rivadulla M, Concheiro M. Target screening and confirmation of 35 licit and illicit drugs and metabolites in hair by LC-MS/MS. Forensic Sci Int 2012;217:207–15.

[56] Salomone A, Gerace E, Di Corcia D, Martra G, Petrarulo M, Vincenti M. Hair analysis of drugs involved in drug-facilitated sexual assault and detection of zolpidem in a suspected case. Int J Legal Med 2012;126:451–9.

[57] Thieme D, Baumer C, Sachs H, Teske J. Screening and long-term retrospection for psychoactive drugs in presumptive drug-facilitated crimes using segmented single hairs. Drug Test Anal 2013;5:736–40.

[58] Maublanc J, Dulaurent S, Morichon J, Lachâtre G, Gaulier J-M. Identification and quantification of 35 psychotropic drugs and metabolites in hair by LC-MS/MS: application in forensic toxicology. Int J Legal Med 2015;129:259–68.

[59] Cooper GAA, Kronstrand R, Kintz P. Society of hair testing guidelines for drug testing in hair. Forensic Sci Int 2012;218:20–4.

[60] Kintz P. Bioanalytical procedures for detection of chemical agents in hair in the case of drug-facilitated crimes. Anal Bioanal Chem 2007;388:1467–74.

[61] Xiang P, Sun Q, Shen B, Chen P, Liu W, Shen M. Segmental hair analysis using liquid chromatography-tandem mass spectrometry after a single dose of benzodiazepines. Forensic Sci Int 2011;204:19–26.

[62] Jakobsson G, Kronstrand R. Segmental analysis of amphetamines in hair using a sensitive UHPLC-MS/MS method. Drug Test Anal 2014;6(Suppl. 1):22–9.

[63] Kintz P, Villain M, Dumestre-Toulet V, Ludes B. Drug-facilitated sexual assault and analytical toxicology: the role of LC-MS/MS a case involving zolpidem. J Clin Forensic Med 2005;12:36–41.

[64] Beumer JH, Bosman IJ, Maes RA. Hair as a biological specimen for therapeutic drug monitoring. Int J Clin Pract 2001;55:353–7.

[65] García-Algar Ó, Pichini S, de la Torre R. Clinical applications of hair analysis. In: Kintz P, editor. Analytical and practical aspects of drug testing in hair. Boca Raton, FL: CRC Press; 2006. p. 201–22.

[66] Uematsu T, Sato R, Suzuki K, Yamaguchi S, Nakashima M. Human scalp hair as evidence of individual dosage history of haloperidol: method and retrospective study. Eur J Clin Pharmacol 1989;37:239–44.

[67] Sato R, Uematsu T, Yamaguchi S, Nakashima M. Human scalp hair as evidence of individual dosage history of haloperidol: prospective study. Ther Drug Monit 1989;11:686–91.

[68] Sato H, Uematsu T, Yamada K, Nakashima M. Chlorpromazine in human scalp hair as an index of dosage history: comparison with simultaneously measured haloperidol. Eur J Clin Pharmacol 1993;44:439–44.

[69] Kintz P, Marescaux C, Mangin P. Testing human hair for carbamazepine in epileptic patients: is hair investigation suitable for drug monitoring? Hum Exp Toxicol 1995;14:812–15.

[70] Williams J, Patsalos PN, Wilson JF. Hair analysis as a potential index of therapeutic compliance in the treatment of epilepsy. Forensic Sci Int 1997;84:113–22.

[71] Williams J. The assessment of therapeutic compliance based upon the analysis of drug concentration in hair. In: Mieczkowski T, editor. Drug testing technology. Boca Raton, FL: CRC Press; 1999. p. 1–32.

[72] Shen M, Xiang P, Wu H, Shen B, Huang Z. Detection of antidepressant and antipsychotic drugs in human hair. Forensic Sci Int 2002;126:153–61.

[73] Skopp G, Kniest A, Haisser J, Mann K, Hermann D. Buprenorphine and norbuprenorphine findings in hair during constant maintenance dosage. Int J Legal Med 2011;125:277–81.

[74] Pirro V, Fusari I, Di Corcia D, Gerace E, De Vivo E, Salomone A, et al. Hair analysis for long-term monitoring of buprenorphine intake in opiates withdrawal. Ther Drug Monit 2014,36.796–807.

[75] Belivanis S, Tzatzarakis MN, Vakonaki E, Kovatsi L, Mantsi M, Alegakis A, et al. Buprenorphine and nor-buprenorphine levels in head hair samples from former heroin users under Suboxone® treatment. Drug Test Anal 2014;6(Suppl. 1):93–100.

[76] Kelly T, Doble P, Dawson M. Chiral analysis of methadone and its major metabolites (EDDP and EMDP) by liquid chromatography-mass spectrometry. J Chromatogr B Analyt Technol Biomed Life Sci 2005;814:315–23.

[77] Seldèn T, Berglund L, Druid H, Håkansson A, Kronstrand R. O9: segmental hair analysis from patients in opiate substitution treatment is useful to investigate drug use history, abstinence and compliance with treatment. Toxicol Anal Clin 2014;26:S8–9.

[78] Druid H, Strandberg JJ, Alkass K, Nyström I, Kugelberg FC, Kronstrand R. Evaluation of the role of abstinence in heroin overdose deaths using segmental hair analysis. Forensic Sci Int 2007;168:223–6.

[79] Garcia-Bournissen F, Moller M, Nesterenko M, Karaskov T, Koren G. Pharmacokinetics of disappearance of cocaine from hair after discontinuation of drug use. Forensic Sci Int 2009;189:24–7.

[80] Huestis MA, Gustafson RA, Moolchan ET, Barnes A, Bourland JA, Sweeney SA, et al. Cannabinoid concentrations in hair from documented cannabis users. Forensic Sci Int 2007;169:129–36.

[81] Han E, Chung H, Song JM. Segmental hair analysis for 11-nor-Δ9-tetrahydrocannabinol-9-carboxylic acid and the patterns of cannabis use. J Anal Toxicol 2012;36:195–200.

[82] Han E, Yang H, Seol I, Park Y, Lee B, Song JM. Segmental hair analysis and estimation of methamphetamine use pattern. Int J Legal Med 2013;127:405–11.

[83] Namera A, Konuma K, Saito T, Ota S, Oikawa H, Miyazaki S, et al. Simple segmental hair analysis for α-pyrrolidinophenone-type designer drugs by MonoSpin extraction for evaluation of abuse history. J Chromatogr B Analyt Technol Biomed Life Sci 2013;942-943:15–20.

[84] Martin M, Muller JF, Turner K, Duez M, Cirimele V. Evidence of mephedrone chronic abuse through hair analysis using GC/MS. Forensic Sci Int 2012;218:44–8.

[85] De Kesel PMM, Lambert WE, Stove CP. O8: metabolite-to-parent drug concentration ratios in hair to study metabolism? The case of CYP1A2 phenotyping. Toxicol Anal Clin 2014;26:S8.

[86] Fisichella M, Steuer AE, Kraemer T, Baumgartner MR. O18: chiral analysis of methadone and its main metabolite EDDP in hair: incorporation depending on hair colour and metabolizer status. Toxicol Anal Clin 2014;26:S12.

[87] Gareri J, Rao C, Koren G. Examination of sex differences in fatty acid ethyl ester and ethyl glucuronide hair analysis. Drug Test Anal 2014;6(Suppl. 1):30–6.

[88] Crunelle CL, Cappelle D, Covaci A, van Nuijs AL, Maudens KE, Sabbe B, et al. Hair ethyl glucuronide as a biomarker of alcohol consumption in alcohol-dependent patients: role of gender differences. Drug Alcohol Depend 2014;141:163–6.

[89] Salomone A, Pirro V, Lombardo T, Di Corcia D, Pellegrino S, Vincenti M. Interpretation of group-level factors from a large population dataset in the determination of ethyl glucuronide in hair. Drug Test Anal 2015; Available from: http://dx.doi.org/10.1002/dta.1697.

[90] Høiseth G, Morini L, Ganss R, Nordal K, Mørland J. Higher levels of hair ethyl glucuronide in patients with decreased kidney function. Alcohol Clin Exp Res 2013;37(Suppl. 1): E14–16.

[91] Chan J, Sauvé B, Tokmakejian S, Koren G, Van Uum S. Measurement of cortisol and testosterone in hair of obese and non-obese human subjects. Exp Clin Endocrinol Diabetes 2014;122:356–62.

[92] Veldhorst MA, Noppe G, Jongejan MH, Kok CB, Mekic S, Koper JW, et al. Increased scalp hair cortisol concentrations in obese children. J Clin Endocrinol Metab 2014;99:285–90.

[93] González-Reimers E, Martín-González C, Galindo-Martín L, Aleman-Valls R, González-Pérez JM, Jorge-Ripper C, et al. Hair copper in normal individuals: relationship with body mass and dietary habits. Trace Elem Electrolytes 2014;31:67–72.

[94] Gonzalez-Reimers E, Martín-González C, Galindo-Martín L, Aleman-Valls MR, Velasco-Vázquez J, Arnay-de-la-Rosa M, et al. Lead, cadmium and zinc in hair samples: relationship with dietary habits and urban environment. Biol Trace Elem Res 2014;157:205–10.

[95] Dussy F, Carson N, Hangartner S, Briellmann T. Is one hair lock really representative? Drug Test Anal 2014;6(Suppl. 1):5–8.

[96] LeBeau MA, Montgomery MA, Brewer JD. The role of variations in growth rate and sample collection on interpreting results of segmental analyses of hair. Forensic Sci Int 2011;210:110–16.

[97] Han E, Yang W, Lee J, Park Y, Kim E, Lim M, et al. Correlation of methamphetamine results and concentrations between head, axillary, and pubic hair. Forensic Sci Int 2005;147:21–4.

[98] Kintz P. Consensus of the Society of Hair Testing on hair testing for chronic excessive alcohol consumption 2011. Forensic Sci Int 2012;218:2.

[99] Pirro V, Di Corcia D, Pellegrino S, Vincenti M, Sciutteri B, Salomone A. A study of distribution of ethyl glucuronide in different keratin matrices. Forensic Sci Int 2011;210:271–7.

[100] Pianta A, Liniger B, Baumgartner MR. Ethyl glucuronide in scalp and non-head hair: an intra-individual comparison. Alcohol Alcohol 2013;48:295–302.

[101] Baumgartner MR, Binz TM, Kraemer T. O15: determination of EtG in body hair samples for monitoring of teetotalism: a suitable alternative to scalp hair locks? Toxicol Anal Clin 2014;26:S11.

[102] Agius R, Ferreira LM, Yegles M. Can ethyl glucuronide in hair be determined only in 3 cm hair strands?. Forensic Sci Int 2012;218:3–9.

[103] Harland DP, Walls RJ, Vernon JA, Dyer JM, Woods JL, Bell F. Three-dimensional architecture of macrofibrils in the human scalp hair cortex. J Struct Biol 2014;185:397–404.

[104] Vincenti M, Salomone A, Gerace E, Pirro V. Application of mass spectrometry to hair analysis for forensic toxicological investigations. Mass Spectrom Rev 2013;32:312–32.

[105] Vincenti M, Salomone A, Gerace E, Pirro V. Role of LC-MS/MS in hair testing for the determination of common drugs of abuse and other psychoactive drugs. Bioanalysis 2013;5:1919–38.

[106] Fisichella M, Morini L, Sempio C, Groppi A. Validation of a multi-analyte LC-MS/MS method for screening and quantification of 87 psychoactive drugs and their metabolites in hair. Anal Bioanal Chem 2014;406:3497–506.

[107] Montesano C, Johansen SS, Nielsen MK. Validation of a method for the targeted analysis of 96 drugs in hair by UPLC-MS/MS. J Pharm Biomed Anal 2014;88:295–306.

[108] Di Corcia D, D'Urso F, Gerace E, Salomone A, Vincenti M. Simultaneous determination in hair of multiclass drugs of abuse (including THC) by ultra-high performance liquid chromatography-tandem mass spectrometry. J Chromatogr B Analyt Technol Biomed Life Sci 2012;899:154–9.

[109] Koster RA, Alffenaar J-W, Greijdanus B, VanDernagel JE, Uges DR. Fast and highly selective LC-MS/MS screening for THC and 16 other abused drugs and metabolites in human hair to monitor patients for drug abuse. Ther Drug Monit 2014;36:234–43.

[110] Fernández MDMR, Di Fazio V, Wille SMR, Kummer N, Samyn N. A quantitative, selective and fast ultra-high performance liquid chromatography tandem mass spectrometry method for the simultaneous analysis of 33 basic drugs in hair (amphetamines, cocaine, opiates, opioids and metabolites). J Chromatogr B Analyt Technol Biomed Life Sci 2014;965:7–18.

[111] Broecker S, Herre S, Wüst B, Zweigenbaum J, Pragst F. Development and practical application of a library of CID accurate mass spectra of more than 2,500 toxic compounds for systematic toxicological analysis by LC-QTOF-MS with data-dependent acquisition. Anal Bioanal Chem 2011;400:101–17.

[112] Gillet LC, Navarro P, Tate S, Röst H, Selevsek N, Reiter L, et al. Targeted data extraction of the MS/MS spectra generated by data-independent acquisition: a new concept for consistent and accurate proteome analysis. Mol Cell Proteomics 2012;11: O111.016717.

[113] Röst HL, Rosenberger G, Navarro P, Gillet L, Miladinović SM, Schubert OT, et al. OpenSWATH enables automated, targeted analysis of data-independent acquisition MS data. Nat Biotechnol 2014;32:219–23.

[114] Broecker S, Herre S, Pragst F. General unknown screening in hair by liquid chromatography-hybrid quadrupole time-of-flight mass spectrometry (LC-QTOF-MS). Forensic Sci Int 2012;218:68–81.

[115] Nielsen MKK, Johansen SS, Dalsgaard PW, Linnet K. Simultaneous screening and quantification of 52 common pharmaceuticals and drugs of abuse in hair using UPLC-TOF-MS. Forensic Sci Int 2010;196:85–92.

[116] Favretto D, Vogliardi S, Stocchero G, Nalesso A, Tucci M, Ferrara SD. High performance liquid chromatography-high resolution mass spectrometry and micropulverized extraction for the quantification of amphetamines, cocaine, opioids, benzodiazepines, antidepressants and hallucinogens in 2.5 mg hair samples. J Chromatogr A 2011;1218:6583–95.

[117] Miyaguchi H, Inoue H. Determination of amphetamine-type stimulants, cocaine and ketamine in human hair by liquid chromatography/linear ion trap-Orbitrap hybrid mass spectrometry. Analyst 2011;136:3503–11.

[118] Moore C, Rana S, Coulter C, Feyerherm F, Prest H. Application of two-dimensional gas chromatography with electron capture chemical ionization mass spectrometry to the detection of 11-nor-delta9-tetrahydrocannabinol-9-carboxylic acid (THC-COOH) in hair. J Anal Toxicol 2006;30:171–7.

[119] Conti M, Tazzari V, Bertona M, Brambilla M, Brambilla P. Surface-activated chemical ionization combined with electrospray ionization and mass spectrometry for the analysis of cannabinoids in biological samples. Part I: analysis of 11-nor-9-carboxytetrahydrocannabinol. Rapid Commun Mass Spectrom 2011;25:1552–8.

[120] Mercolini L, Mandrioli R, Protti M, Conti M, Serpelloni G, Raggi MA. Monitoring of chronic cannabis abuse: an LC-MS/MS method for hair analysis. J Pharm Biomed Anal 2013;76:119–25.

[121] Thieme D, Sachs H, Uhl M. Proof of cannabis administration by sensitive detection of 11-nor-delta(9)-tetrahydrocannabinol-9-carboxylic acid in hair using selective methylation and application of liquid chromatography- tandem and multistage mass spectrometry. Drug Test Anal 2014;6:112–18.

[122] Harrison R, Fu S. A review of methodology for testing hair for cocaine. J Forensic Investig 2014;2:1–8.

[123] Schwaninger AE, Meyer MR, Maurer HH. Chiral drug analysis using mass spectrometric detection relevant to research and practice in clinical and forensic toxicology. J Chromatogr A 2012;1269:122–35.

[124] Martins L, Yegles M, Chung H, Wennig R. Simultaneous enantioselective determination of amphetamine and congeners in hair specimens by negative chemical ionization gas chromatography-mass spectrometry. J Chromatogr B Analyt Technol Biomed Life Sci 2005;825:57–62.

[125] Martins LF, Yegles M, Chung H, Wennig R. Sensitive, rapid and validated gas chromatography/negative ion chemical ionization-mass spectrometry assay including derivatisation with a novel chiral agent for the enantioselective quantification of amphetamine-type stimulants in hair. J Chromatogr B Analyt Technol Biomed Life Sci 2006;842:98–105.

[126] Nishida K, Itoh S, Inoue N, Kudo K, Ikeda N. High-performance liquid chromatographic-mass spectrometric determination of methamphetamine and amphetamine enantiomers, desmethylselegiline and selegiline, in hair samples of long-term methamphetamine abusers or selegiline users. J Anal Toxicol 2006;30:232–7.

[127] Kikura-Hanajiri R, Kawamura M, Miyajima A, Sunouchi M, Goda Y. Chiral analyses of dextromethorphan/levomethorphan and their metabolites in rat and human samples using LC-MS/MS. Anal Bioanal Chem 2011;400:165–74.

[128] Pirro V, Valente V, Oliveri P, De Bernardis A, Salomone A, Vincenti M. Chemometric evaluation of nine alcohol biomarkers in a large population of clinically-classified subjects: pre-eminence of ethyl glucuronide concentration in hair for confirmatory classification. Anal Bioanal Chem 2011;401:2153–64.

[129] Hoelzle C, Scheufler F, Uhl M, Sachs H, Thieme D. Application of discriminant analysis to differentiate between incorporation of cocaine and its congeners into hair and contamination. Forensic Sci Int 2008;176:13–18.

[130] Bossers DLCAM, Paul DR. Application of Bayesian theory to the reporting of results in alcohol hair testing. Forensic Sci Int 2014;242:e56–8.

[131] Zhu KY, Leung KW, Ting AKL, Wong ZCF, Ng WYY, Choi RCY, et al. Microfluidic chip based nano liquid chromatography coupled to tandem mass spectrometry for the determination of abused drugs and metabolites in human hair. Anal Bioanal Chem 2012;402:2805–15.

[132] Zhu KY, Leung KW, Ting AKL, Wong ZCF, Fu Q, Ng WYY, et al. The establishment of a highly sensitive method in detecting ketamine and norketamine simultaneously in human hairs by HPLC-Chip-MS/MS. Forensic Sci Int 2011;208:53–8.

[133] Chen Y, Li M, Fu L, Hou X, Jiang X. Simultaneous determination of trace cadmium and lead in single human hair by tungsten electrothermal vaporization-flame atomic fluorescence spectrometry. Microchem J 2014;114:182–6.

[134] Dressler VL, Pozebon D, Mesko MF, Matusch A, Kumtabtim U, Wu B, et al. Biomonitoring of essential and toxic metals in single hair using on-line solution-based calibration in laser ablation inductively coupled plasma mass spectrometry. Talanta 2010;82:1770–7.

[135] Thieme D, Sachs H. Examination of a long-term clozapine administration by high resolution segmental hair analysis. Forensic Sci Int 2007;166:110–14.

[136] Miki A, Katagi M, Kamata T, Zaitsu K, Tatsuno M, Nakanishi T, et al. MALDI-TOF and MALDI-FTICR imaging mass spectrometry of methamphetamine incorporated into hair. J Mass Spectrom 2011;46:411–16.

[137] Miki A, Katagi M, Shima N, Kamata H, Tatsuno M, Nakanishi T, et al. Imaging of methamphetamine incorporated into hair by MALDI-TOF mass spectrometry. Forensic Toxicol 2011;29:111–16.

[138] Porta T, Grivet C, Kraemer T, Varesio E, Hopfgartner G. Single hair cocaine consumption monitoring by mass spectrometric imaging. Anal Chem 2011;83:4266–72.

[139] Vogliardi S, Favretto D, Frison G, Ferrara SD, Seraglia R, Traldi P. A fast screening MALDI method for the detection of cocaine and its metabolites in hair. J Mass Spectrom 2009;44:18–24.

[140] Vogliardi S, Favretto D, Frison G, Maietti S, Viel G, Seraglia R, et al. Validation of a fast screening method for the detection of cocaine in hair by MALDI-MS. Anal Bioanal Chem 2010;396:2435–40.

[141] Musshoff F, Arrey T, Strupat K. Determination of cocaine, cocaine metabolites and cannabinoids in single hairs by MALDI Fourier transform mass spectrometry—preliminary results. Drug Test Anal 2013;5:361–5.

[142] Shen M, Xiang P, Shi Y, Pu H, Yan H, Shen B. Mass imaging of ketamine in a single scalp hair by MALDI-FTMS. Anal Bioanal Chem 2014;406:4611–16.

[143] Poetzsch M, Baumgartner MR, Steuer AE, Kraemer T. Segmental hair analysis for differentiation of tilidine intake from external contamination using LC-ESI-MS/MS and MALDI-MS/MS imaging. Drug Test Anal 2015;7:143–9.

[144] Jackson AU, Garcia-Reyes JF, Harper JD, Wiley JS, Molina-Díaz A, Ouyang Z, et al. Analysis of drugs of abuse in biofluids by low temperature plasma (LTP) ionization mass spectrometry. Analyst 2010;135:927–33.

[145] Nielen MW, Nijrolder AW, Hooijerink H, Stolker AA. Feasibility of desorption electrospray ionization mass spectrometry for rapid screening of anabolic steroid esters in hair. Anal Chim Acta 2011;700:63–9.

[146] Duvivier WF, van Beek TA, Pennings EJ, Nielen MW. Rapid analysis of Δ-^{9}-tetrahydrocannabinol in hair using direct analysis in real time ambient ionization Orbitrap mass spectrometry. Rapid Commun Mass Spectrom 2014;28:682–90.

Index

Note: Page numbers followed by "*f*" and "*t*" refer to figures and tables, respectively.

B

C

D

Printed in the United States
By Bookmasters